INTRODUCTION TO
STRUCTURAL DYNAMICS

McGRAW-HILL BOOK COMPANY
NEW YORK
SAN FRANCISCO
TORONTO
LONDON

JOHN M. BIGGS

Professor of Civil Engineering
Massachusetts Institute of Technology

INTRODUCTION TO

STRUCTURAL

DYNAMICS

INTRODUCTION TO STRUCTURAL DYNAMICS

Copyright © 1964 by McGraw-Hill, Inc. All Rights Reserved. Printed in the United States of America. This book, or parts thereof, may not be reproduced in any form without permission of the publishers.
Library of Congress Catalog Card Number 64-21068

05255

Preface

This book has been written to provide a self-contained text on structural dynamics for use in courses offered to seniors or first-year graduate students in civil engineering. The material is based on the author's notes for such a course given at MIT during the past few years. It is presumed that the student has completed the normal undergraduate work in applied mechanics, mathematics, and structural engineering.

The emphasis in this text is on the practical analysis and design of real structures rather than on the mathematical techniques of dynamic analysis. Throughout the book examples are given to illustrate application of the theory to actual structural problems. Much of the material has been drawn from the author's experience as a consulting engineer. With this emphasis the book should be useful to practicing engineers, as well as to students whose objective is to become structural designers.

In the author's opinion, structural dynamics is too often taught as a course in advanced mathematics for engineers. For some students this approach makes the subject unnecessarily difficult. Other students find the mathematical manipulation so intriguing that they fail to develop the physical understanding essential for good design. In this text the author has avoided mathematical complexities, which, although they may be useful in advanced research, are unnecessary for most design purposes.

Chapters 1 and 2, in some respects, are a review of the dynamics normally taught in applied mechanics. In addition, however, two other purposes are served: (1) the basic theory is related to actual structures; and (2) numerical analysis, which is not normally covered in applied mechanics, is introduced. The order of presentation, i.e., numerical analysis before closed solution, is somewhat unorthodox. The author believes, as a result of his teaching experience, that this order is preferable, because numerical analysis executed by hand develops a physical "feel" for dynamic behavior much more rapidly than does the solution of differential equations.

Chapters 3 and 4 are, perhaps, the heart of the book, since they contain the theory of analysis for multidegree systems. The author has chosen not to use matrix notation, which is currently so popular, because in his opinion it is pedagogically unwise to do so at this introductory level. For those teachers who prefer matrix formulation, the Appendix may be helpful. Chapter 4 contains considerable material on beams of various types, because this is believed to be particularly important to structural engineers.

Chapter 5 is devoted to approximate methods of design, which are developed on the basis of the theory presented in earlier chapters. Because many dynamic problems in civil engineering involve uncertain loading conditions, these methods are often more appropriate than the more precise but time-consuming procedures.

Chapters 6, 7, and 8 contain applications of the theory to some important types of structural problems. These treatments are incomplete, but they are believed to be sufficiently thorough to provide a sound introduction to the subjects.

The author wishes to acknowledge with gratitude the assistance of his wife, Margaret C. Biggs, who not only typed the manuscript, but provided encouragement throughout the writing of this book. Daniel Beltran-Maldonado was extremely helpful in preparing the figures and proofreading the manuscript.

The author is particularly indebted to his teacher and colleague Prof. Charles H. Norris for instruction and inspiration over an extended period of time.

John M. Biggs

Contents

Preface v
Introduction ix
List of Symbols xi

Chapter 1 Numerical Analysis of Simple Systems 1

    1.1 Introduction 1
    1.2 One-degree Elastic Systems 3
    1.3 Two-degree Elastic Systems 11
    1.4 One-degree Elastic System with Damping 17
    1.5 One-degree Elasto-plastic Systems 20
    1.6 Alternative Methods of Numerical Analysis 26

Chapter 2 Rigorous Analysis of One-degree Systems 34

    2.1 Introduction 34
    2.2 Undamped Systems 35
    2.3 Various Forcing Functions (Undamped Systems) 40
    2.4 Damped Systems 51
    2.5 Response to a Pulsating Force 58
    2.6 Support Motions 65
    2.7 Elasto-plastic Systems 69
    2.8 Charted Solutions for Maximum Response of One-degree Undamped Elasto-plastic Systems 76

Chapter 3 Lumped-mass Multidegree Systems 85

    3.1 Introduction 85
    3.2 Direct Determination of Natural Frequencies 89
    3.3 Characteristic Shapes 93
    3.4 Stodola-Vianello Procedure for Natural Frequencies and Characteristic Shapes 97
    3.5 Modified Rayleigh Method for Natural Frequencies 105
    3.6 Lagrange's Equation 111
    3.7 Modal Analysis of Multidegree Systems 116
    3.8 Multistory Rigid Frames Subjected to Lateral Loads 125
    3.9 Elasto-plastic Analysis of Multidegree Systems 138
    3.10 Damping in Multidegree Systems 140

Chapter 4 Structures with Distributed Mass and Load 150

    4.1 Introduction 150
    4.2 Single-span Beams—Normal Modes of Vibration 151

viii   Contents

            4.3 Forced Vibration of Beams   158
            4.4 Beams with Variable Cross Section and Mass   170
            4.5 Continuous Beams   174
            4.6 Beam-girder Systems   183
            4.7 Plates or Slabs Subjected to Normal Loads   188
            4.8 Elasto-plastic Analysis of Beams   192

Chapter 5 Approximate Design Methods   199

            5.1 Introduction   199
            5.2 Idealized System   202
            5.3 Transformation Factors   205
            5.4 Dynamic Reactions   217
            5.5 Response Calculations   219
            5.6 Design Examples   224
            5.7 Approximate Design of Multidegree Systems   233

Chapter 6 Earthquake Analysis and Design   245

            6.1 Introduction   245
            6.2 Response of Multidegree Systems to Support Motion   246
            6.3 Multistory-building Analysis   250
            6.4 Response Spectra   257
            6.5 Earthquake Ground Motions   263
            6.6 Earthquake Spectrum Analysis of Multidegree Systems   265
            6.7 Practical Design for Earthquake   269

Chapter 7 Blast-resistant Design   276

            7.1 Introduction   276
            7.2 Loading Effects of Nuclear Explosions   277
            7.3 Aboveground Rectangular Structures   282
            7.4 Aboveground Arches and Domes   297
            7.5 Belowground Structures   308
            7.6 Ground Motions   309

Chapter 8 Beams Subjected to Moving Loads   315

            8.1 Introduction   315
            8.2 Constant Force with Constant Velocity   315
            8.3 Pulsating Force with Constant Velocity   318
            8.4 Beam Traversed by a Rolling Mass   321
            8.5 Beam Vibration Due to Passage of Sprung Masses   322
            8.6 Bridge Vibration Due to Moving Vehicles   323

Appendix Matrix Formulation of Modal Analysis   329

References   333
Index   336

Introduction

The subject of this text is the analysis and design of structures subjected to dynamic loads, i.e., loads which vary with time. Although the majority of civil-engineering structures can properly be designed as though the loads were static, there are some important exceptions, and it is obviously imperative that the designer be able to distinguish between static and dynamic loads.

In fact, no structural loads (with the possible exception of dead load) are really static, since they must be applied to the structure in some manner, and this involves a time variation of force. It is obvious, however, that if the magnitude of force varies slowly enough, it will have no dynamic effect and can be treated as static. "Slowly enough" is not definite, and apparently the question of whether or not a load is dynamic is a relative matter. It turns out that the natural period of the structure is the significant parameter, and if the load varies slowly relative to this period, it may be considered to be static. The natural period, loosely defined, is the time required for the structure to go through one cycle of free vibration, i.e., vibration after the force causing the motion has been removed or has ceased to vary.

The interest in structural design for dynamic loads has been increasing steadily over the years. This is in part due to advancing technology, which has made possible more accurate design. It is also due to the fact that more daring structures (larger, lighter, etc.) are being attempted, and these are more susceptive to dynamic effects because they are generally more flexible and have longer natural periods. Examples of situations in which dynamic loading must be considered include (1) structures subjected to alternating forces caused by oscillating machinery, (2) structures which support moving loads such as bridges, (3) structures subjected to suddenly applied forces such as blast pressure or wind gust, and (4) cases where the supports of the structure move, e.g., a building during an earthquake.

The basic principles of structural analysis are of course not invalidated by the fact that the load is dynamic. The same relationships between deflection and stress apply under both dynamic and static conditions. Dynamic analysis consists primarily of the determination of the time variation of deflection, from which stresses can be directly computed. Since the natural period depends upon the mass and stiffness of the structure, these two quantities are of perhaps greater importance in dynamic analysis.

Introduction

In this text considerable attention is given to the inelastic behavior of structures, i.e., behavior beyond the elastic limit. This is particularly important in dynamic design because it is often impractical, or at least uneconomical, to design the structure so as to remain completely elastic. The energy absorption which results from the plastic deformation of the material permits a much lighter structure than would be required if all energy had to be absorbed by elastic strain.

Chapters 1 and 2, which deal with simple dynamic systems, contain the more elementary theory of structural dynamics. In Chapters 3 and 4 this theory is extended to more complex structural systems. Chapter 5 is a presentation of approximate design procedures, which are often more suitable for practical purposes than direct application of the theory. Finally, Chapters 6, 7, and 8 contain applications of the material in preceding chapters to practical problems of importance.

List of Symbols

A_n	modal amplitude
A_{nst}	static modal deflection
$\mathcal{A}$	constant
a	characteristic amplitude, dimension
$\mathcal{B}$	constant
b	width
$C, \mathcal{C}$	constants
C_d	drag coefficient
c	damping coefficient
c_{cr}	critical damping coefficient
c_s	seismic velocity
D_{ij}	flexibility coefficient
D_c	total concrete thickness
$\mathcal{D}$	constant
DLF	dynamic load factor
d	effective depth of concrete section
E	modulus of elasticity
F	force
f	natural frequency, cps
$f(t)$	nondimensional time function
G	constant
g	acceleration of gravity
H	constant
h	height
I	moment of inertia
$I(t)$	inertia force
i	impulse
$\mathcal{K}$	kinetic energy
K_L, K_M, K_R, K_{LM}	transformation factors
k	spring constant
L, l	span
M	mass
$\mathcal{M}$	bending moment
$\mathcal{M}_P$	ultimate bending strength
m	mass per unit length
N	number of modes
P	axial stress
p	pressure, distributed load

List of Symbols

p_d	dynamic pressure
p_r	reflected pressure
p_s	side-on overpressure
p_{so}	initial side-on overpressure
q	displacement
R	resistance
R_m	maximum resistance
$\mathcal{R}$	weapon range
r	radius
S	shear
S_c	clearing distance
T	natural period
t	time
Δt	time interval
t_d	load duration
t_r	load rise time
U	shock-front velocity
U_s	velocity of sound
$\mathcal{U}$	strain energy
u	relative displacement
$\dot{u}$	relative velocity
V	shear or reaction
v	velocity, displacement
W	weight
$\mathcal{W}$	work
w	weight per unit length
x	dimension
Y	weapon yield
y	displacement or deflection
$\dot{y} = dy/dt$	velocity
$\ddot{y} = d^2y/dt^2$	acceleration
y_{el}	elastic-limit deflection
y_{st}	static deflection
y_s	support motion
z	displacement, coordinate
α	constant, phase angle
β	constant, damping coefficient ($c/2M$)
Γ	participation factor
Δ	spring distortion
ϵ, η	nondimensional displacements
θ	angle of rotation
Λ	half-arch central angle
μ	ductility ratio

ν	Poisson's ratio
ξ	time variable
ρ	density
ρ_s	steel ratio in concrete
σ	stress intensity
σ_{st}	static stress
σ_{dy}	dynamic yield strength
σ'_{dc}	dynamic concrete compressive strength
τ	time variable
$\Phi(x), \phi(x)$	characteristic shape
ϕ	characteristic coordinate
ψ	participation factor
Ω	forcing frequency, rad/sec
ω	natural circular frequency

1
Numerical Analysis of Simple Systems

1.1 Introduction

The determination of the dynamic response of simple systems using numerical procedures is discussed in this chapter. The more traditional rigorous methods are introduced in Chap. 2. This order of presentation is followed because numerical analysis, rather than rigorous solution, is believed to be the most general and yet straightforward approach possible and the best for introductory purposes. Only basic principles of physics and the most elementary mathematics are used. Thus the reader should be able to concentrate on the physical phenomena involved rather than on the mathematical techniques employed. It is hoped that this emphasis will help develop a physical "feel," or intuition, for dynamic response, which is necessary for successful analysis of more complicated dynamic problems. The reader is urged to keep this objective in mind during his study of the following sections.

Numerical analysis, that is, solution of the differential equations of motion by arithmetic procedures, is a much more general attack on the problem than rigorous, or closed, solution, because the latter is possible only when the loading and the resistance functions can be expressed in relatively simple mathematical terms. For the type of problems in which we are interested, this is a severe restriction, and thus the rigorous approach is obviously of limited usefulness.

The availability of electronic computers has accelerated the adoption

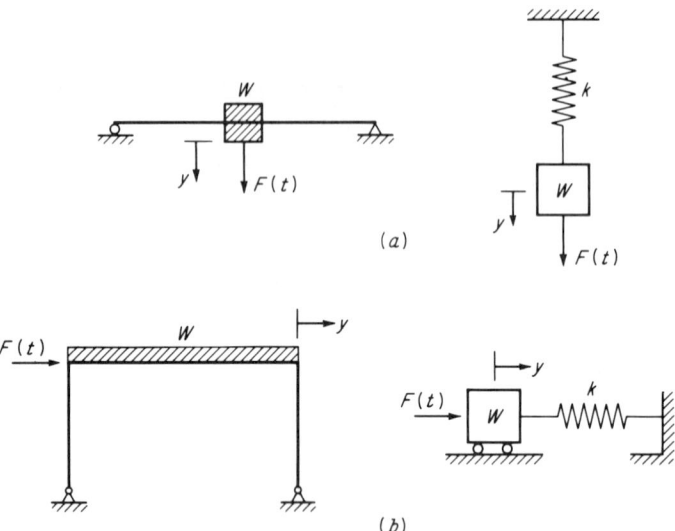

FIGURE 1.1 Structures idealized as spring-mass systems.

of numerical methods by structural engineers. Solutions to many complex dynamic problems which were impossible in earlier times can now be obtained with relative ease. Although computer programming is not discussed in this text and the procedures are illustrated by hand calculations, it is expected that, in practice, many of these computations would be done electronically.

The following sections deal with systems consisting of combinations of springs and masses. It should be emphasized that these are not merely academic exercises, but rather that the system idealized in this manner is a convenient representation of an actual structure. For example, in Fig. 1.1a, the weight W, supported by a beam and subjected to a dynamic load, may often be represented by the simple mass-spring system shown. The same is true of the rigid-frame structure shown in Fig. 1.1b, where the mass is distributed along the girder and only horizontal motions are considered. In order for the idealized system to perform in the same way as the actual structure, it is only necessary to make a proper selection of the system parameters. For example, the *spring constant* k can be determined from the properties of the beam or frame since it is merely the ratio of force to deflection. In the cases shown, the weight, or mass, of the idealized system is the same as that of the actual structure since the weight of the structural members is assumed to be negligible. In other cases this may not be true and a factor must be applied to obtain the equivalent mass for the idealized

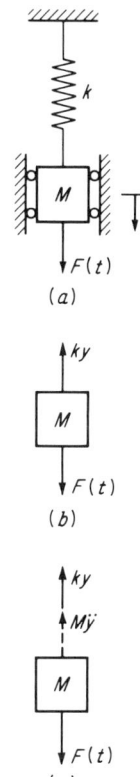

FIGURE 1.2 One-degree system—dynamic equilibrium.

system. Such a factor would be necessary if the mass were distributed over the structure—a situation which will be discussed in later chapters. The load-time relationship, or *load function*, is usually the same for the two systems, although the magnitude of the load may differ. The ideal spring-mass system is selected such that the deflection of the mass is the same as at some point of significance on the structure, for example, the midspan of the beam. The important point is that an idealized system which behaves timewise in exactly the same fashion as the actual structure can be constructed and then analyzed with relative ease.

1.2 One-degree Elastic Systems

A one-degree system is defined as one in which only one type of motion is possible, or in other words, the position of the system at any instant can be defined in terms of a single coordinate. Such a system is shown in Fig. 1.2a, where the mass can move in a vertical direction only and all the mass in the system deflects by the same amount (the spring is assumed massless). As an example of dynamic analysis, let us determine the motion of this mass resulting from the application of a time-varying force.

a. Formulation of the Problem

The first step is to isolate the mass as shown in Fig. 1.2b. To this mass we apply the external forces, in this case the applied force $F(t)$ and the spring force ky. It is assumed here that the spring is linear, i.e., that the force in the spring is always equal to the spring constant times the displacement. Note that the weight, or gravity force, does not appear in the figure. This implies that the displacement y is measured from the neutral position, in other words, the static position which the mass would take if only the force of gravity were acting.

Having isolated the mass, we may write the equation of motion simply by applying the elementary formula $F = Ma$. F is, of course, the net, or algebraic, sum of the forces acting on the mass, and the positive direction of force is the same as that for displacement or acceleration. Thus the equation of motion for this system is*

$$F(t) - ky = M\ddot{y} \qquad (1.1)$$

This differential equation may be solved to determine the variation of displacement with time once the loading function, the initial conditions, and the other parameters are known.

An alternative and very convenient way of writing the equation of motion is by the use of D'Alembert's principle of *dynamic equilibrium*. This method is illustrated in Fig. 1.2c, where an additional imaginary force is applied to the mass. This is the *inertia force*, and is equal to the product of the mass and the acceleration. Note that it must always be applied in the direction of negative acceleration, or opposite to positive displacement. Having added this force, we may treat the situation shown in Fig. 1.2c exactly as a problem in static equilibrium. The equilibrium equation is

$$F(t) - ky - M\ddot{y} = 0 \qquad (1.2)$$

It is seen that this approach results in exactly the same equation as that previously obtained. In general, the second approach given is more convenient, especially when distributed masses are involved.

b. Numerical Integration

Before considering a specific example we shall discuss the process of numerical integration in general terms. This is a procedure by which

---

* Throughout this text $\dot{y}$ and $\ddot{y}$ will be used to designate the first and second derivatives of displacement with respect to time, or in other words, the velocity and acceleration.

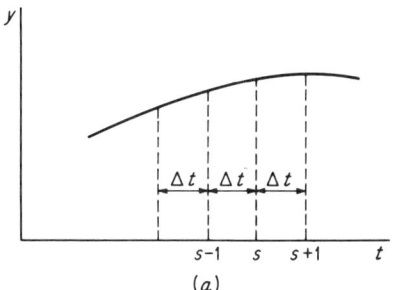
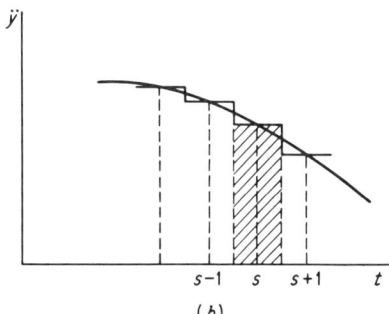

FIGURE 1.3 Numerical integration—lumped-impulse procedure.

the differential equation of motion is solved step by step, starting at zero time, when the displacement and velocity are presumably known. The time scale is divided into discrete intervals, and one progresses by successively extrapolating the displacement from one time station to the next. There are many such methods available, but in keeping with the policy stated for this chapter, only one of the more simple versions will be presented here. This might be called the *constant-velocity*, or *lumped-impulse*, *procedure*.

Suppose an analysis for the determination of the displacement-time variation for a dynamic system was in progress as indicated in Fig. 1.3. Suppose, further, that the displacements $y^{(s)}$ at time station s and $y^{(s-1)}$ at the preceding time station $s - 1$ had been previously determined. The acceleration $\ddot{y}^{(s)}$ at time station s can then be determined using the equation of motion. The problem is to determine the next displacement, $y^{(s+1)}$, by extrapolation. This could be done by the following self-evident formula:

$$y^{(s+1)} = y^{(s)} + \dot{y}_{\text{av}} \Delta t \tag{1.3}$$

where $\dot{y}_{\text{av}}$ is the average velocity between time stations s and $s + 1$, and Δt is the time interval between stations. The average velocity may be expressed by the following approximate formula:

$$\dot{y}_{\text{av}} = \frac{y^{(s)} - y^{(s-1)}}{\Delta t} + \ddot{y}^{(s)} \Delta t \tag{1.4}$$

where the first term is the average velocity in the time interval between $s-1$ and s, and the second term is the increase in the velocity between the two time intervals, assuming that $\ddot{y}^{(s)}$ is an average acceleration throughout that time period. The assumption stated is equivalent to approximating the acceleration curve by a series of straight lines as shown in Fig. 1.3b and, in addition, replacing the area under these lines by a series of pulses concentrated at the time stations. Thus the shaded area represents an impulse applied at station s which equals the change in average velocity between the two adjacent time intervals. Substituting Eq. (1.4) into Eq. (1.3), the following *recurrence formula* is obtained:

$$y^{(s+1)} = 2y^{(s)} - y^{(s-1)} + \ddot{y}^{(s)} (\Delta t)^2 \qquad (1.5)$$

With this equation one is able to extrapolate to find the displacement at the next time station. Note, from Eq. (1.1), that $\ddot{y}^{(s)}$ may be determined since it depends only upon the displacement $y^{(s)}$, which was previously obtained, and $F(t)$, which is known.

The recurrence formula given by Eq. (1.5) is obviously approximate, but it gives sufficiently accurate results provided that the time interval Δt is taken small in relation to the variations in acceleration. In fact, as $\Delta t \to 0$, the solution becomes exact, although the number of computations obviously increases as Δt is reduced, thereby increasing the number of time intervals involved. In general, it has been found that results sufficiently accurate for practical purposes can be obtained if the time interval is taken no larger than one-tenth of the natural period of the system. This point will be discussed in more detail in later sections. Many other recurrence formulas are available for use. Some of these formulas permit larger time intervals but require more elaborate computations in each step. When problems of stability or convergence are encountered, more accurate recurrence formulas may be necessary. These are beyond the scope of the present discussion, but are covered in Sec. 1.6. The author has found the formula presented, Eq. (1.5), to be adequate for most problems in structural dynamics.

Using the recurrence formula, the analyst simply begins at time equals zero and proceeds step by step to determine the displacements at the time stations selected. It is necessary, however, to use a special procedure in the first time interval because, at $t = 0$, no value of $y^{(s-1)}$ is available. Two different procedures may be used. First, the acceleration may be assumed to vary linearly up to the first time station, in which case the displacement at that time is given by the following:*

$$y^{(1)} = \frac{1}{6}(2\ddot{y}^{(0)} + \ddot{y}^{(1)})(\Delta t)^2 \qquad (1.6)$$

---

* Equations (1.6) and (1.7) may be derived by evaluating $y = \int\int \ddot{y}\, dt\, dt$.

Second, the acceleration may be assumed constant during the first time interval and equal to the initial value. For the latter assumption the following equation applies:

$$y^{(1)} = \tfrac{1}{2}\ddot{y}^{(0)} (\Delta t)^2 \tag{1.7}$$

In either case, having established $y^{(1)}$, the analyst then proceeds in the normal way, using Eq. (1.5). Note that Eq. (1.6) must be solved by trial and error since $\ddot{y}^{(1)}$ depends upon $y^{(1)}$. However, Eq. (1.6) must be used if there is zero force (and hence zero acceleration) at zero time, for in no other way can $y^{(1)}$ be determined. If the acceleration at $t = 0$ is not zero, Eq. (1.7), which does not require iteration, may be used without appreciable error, provided that the force does not change greatly in the first interval. This method of numerical analysis is illustrated in the example which follows.

The numerical procedure outlined above will at first seem tedious. However, in the very common cases where no closed solutions are possible, there is no alternative. If one is dealing with a one-degree system, the computations can, with a little practice, be done very quickly. When the system is more complicated, computers are normally used and the length of the calculations is not a serious problem.

c. Example

To illustrate the procedures discussed above, consider the spring-mass system shown in Fig. 1.4a, which is subjected to the load-time function shown in Fig. 1.4b. It is desired to determine the variation of displacement with time, starting with the system at rest at $t = 0$. Substituting into Eq. (1.1) and being careful to keep the units consistent, the equation of motion is written as follows:

$$F(t) - 2000y = \frac{64.4}{32.2} \ddot{y}$$

or
$$\ddot{y} = \tfrac{1}{2}F(t) - 1000y \tag{1.8}$$

Thus knowing the load and the displacement at any time enables one to compute acceleration at that time.

The next step is to select a time interval for the numerical integration. As mentioned above, this should not be greater than one-tenth of the natural period of the system. The *natural period T* of a one-degree system is given by

$$T = 2\pi \sqrt{\frac{W}{kg}} \tag{1.9}$$

which, in this example, is 0.198 sec. The *natural frequency f* is the inverse of the natural period, or 5.04 cps. The *natural circular frequency*

Introduction to Structural Dynamics

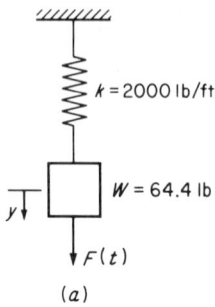

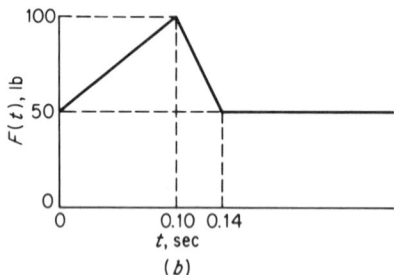

FIGURE 1.4 Example. (a) Spring-mass system; (b) load-time function.

ω is $2\pi f$, or 31.6 rad/sec. Natural periods and frequencies are discussed more fully in later chapters.

One-tenth of the natural period is approximately 0.02 sec, and this value will be used in the computation. However, a second criterion must be considered when selecting the time interval: the interval should be small enough to represent properly the variation of load with time. In this example it will be noted that the time stations ($\Delta t = 0.02$) occur at the sudden breaks in the load function and, furthermore, that the interval is small enough so that points at this spacing accurately represent the function (Fig. 1.4b). Therefore the time interval selected is satisfactory.

The numerical integration for the example problem is shown in Table 1.1. Note that, in the first time interval, Eq. (1.7) is used and, in the following intervals, Eq. (1.5) is employed. Since Eq. (1.8) gives $\ddot{y}(t = 0) = 25$, Eq. (1.7) yields $y(t = 0.02) = \frac{1}{2}(25)(0.02)^2 = 0.0050$. Using the latter value in Eq. (1.8) provides $\ddot{y}(t = 0.02) = 25$, and therefore $y(t = 0.04) = 2(0.0050) - 0 + 25(0.02)^2 = 0.0200$ by Eq. (1.5). The calculations then continue in identical manner. The result of the complete calculation is plotted in Fig. 1.5, where displacement versus time is shown. The ordinate also represents the time variation of the spring force if the displacements are multiplied by the spring constant k. To assist in the interpretation of the result, the hypothetical displace-

Table 1.1 Numerical Integration; Undamped Elastic One-degree System (Fig. 1.4)

t sec	$\frac{1}{2}F(t)$ lb	$1000y$ lb	$\ddot{y}$ Eq. (1.8) ft/sec²	$\ddot{y}(\Delta t)^2$ ft	y Eq. (1.5) ft
0.0	25	0	25.0	0.0100	0
0.02	30	5.0	25.0	0.0100	0.0050*
0.04	35	20.0	15.0	0.0060	0.0200
0.06	40	41.0	−1.0	−0.0004	0.0410
0.08	45	61.6	−16.6	−0.0066	0.0616
0.10	50	75.6	−25.6	−0.0102	0.0756
0.12	37.5	79.4	−41.9	−0.0168	0.0794
0.14	25	66.4	−41.4	−0.0166	0.0664
0.16	25	36.8	−11.8	−0.0047	0.0368
0.18	25	2.5	22.5	0.0090	0.0025
0.20	25	−22.8	47.8	0.0191	−0.0228
0.22	25	−29.0	54.0	0.0216	−0.0290
0.24	25	−13.6	38.6	0.0154	−0.0136
0.26	25	17.2	7.8	0.0031	0.0172
0.28	25	51.1	−26.1	−0.0104	0.0511
0.30	25	74.6	−49.6	−0.0198	0.0746
0.32	25	78.3	−53.3	−0.0123	0.0783
0.34	....	.....	.....	....	0.0607

*Equation (1.7).

ment corresponding to the static application of the load at any instant is also plotted. The maximum displacement which occurs at 0.12 sec corresponds to a spring force of 159 lb, which is 1.59 times the maximum external load. Subsequent peaks are somewhat smaller and correspond to a spring force of 157 lb. The latter peak value would remain constant indefinitely since we have not included damping in the present example. The time interval between successive positive (or negative) peaks is exactly equal to the natural period of the system. After the load becomes constant at 0.14 sec, the spring force varies in sinusoidal fashion, the positive and negative peaks being equidistant above and below the value of the external load.

The student is urged to study the computation shown in Table 1.1 carefully and to make similar computations on his own. It would be advisable to plot the displacement as the computations proceed. This exercise has two advantages: first, errors in arithmetic are quickly discovered; and second, it helps to develop an understanding of dynamic

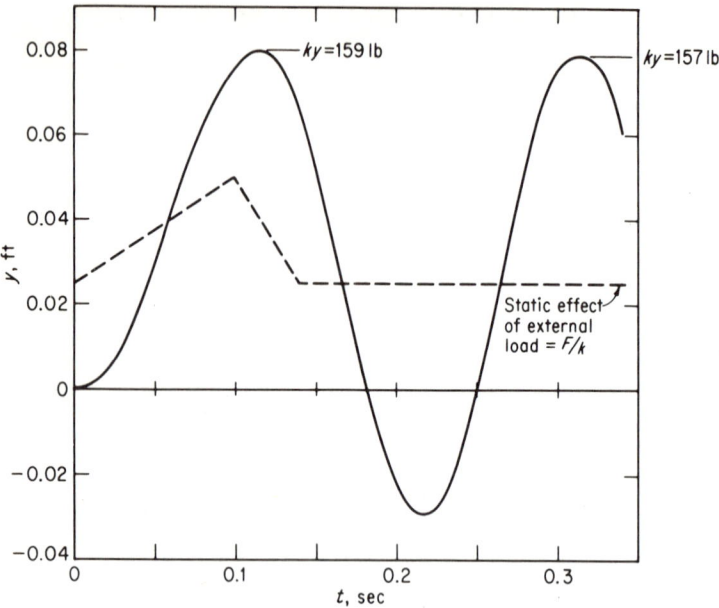

FIGURE 1.5 Example. Response of system shown in Fig. 1.4.

behavior. Although somewhat tedious, the effort is well worthwhile, since this intuitive understanding will be invaluable in later chapters.

Lest it be forgotten that the system analyzed in the previous paragraphs could represent an actual structure, consider the frame shown in Fig. 1.6a. This is a steel rigid frame to which is applied a dynamic force at the upper level. It is desired to determine the dynamic deflection of the top of the frame in the horizontal direction. Two assumptions will be made: first, the weights of the columns and walls are negligible; and second, the girder is sufficiently rigid to prevent significant rotation at the tops of the columns. These assumptions are not necessary, but will serve to simplify the problem and, in fact, are essentially correct for many actual frames of this type.

The parameters of the idealized system shown in Fig. 1.6b may be easily computed as follows:

$$W = 1000 \times 30 = 30{,}000 \text{ lb}$$
$$k = \frac{12E(2I)}{h^3} = \frac{12 \times 30 \times 10^6 \times 112.8}{(20)^3 \times 144} \quad (1.10)$$
$$= 35{,}200 \text{ lb/ft}$$

The spring constant k is simply equal to the inverse of the deflection at the top of the frame due to a unit horizontal load, and the equation

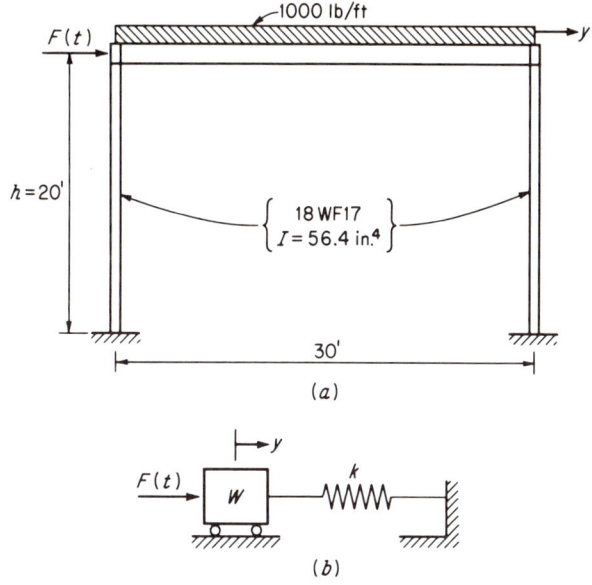

FIGURE 1.6 Rigid frame represented by one-degree system.

employed above may be easily verified by simple elastic analysis. Having these parameters, the analyst would then proceed to determine the dynamic displacements for a given load function, using the same technique as in the previous example. These displacements would be equal to the actual horizontal deflections at the top of the frame. The spring force computed for the ideal system would at all times be equal to the total shear in the two columns, and the maximum column bending moment would be given by $6EIy/h^2$. Thus the dynamic stresses at any time could be easily determined.

1.3 Two-degree Elastic Systems

In this section the analysis of two-degree systems will be discussed, applying the same basic concepts as those used in the previous section for one-degree systems. A two-degree system is defined as one in which two separate types of motion are possible, or in other words, the configuration of the system at any time is completely specified by exactly two parameters, or displacements. Such a system is shown in Fig. 1.7a, where the parameters defining the motion are the displacements of the two masses, namely, y_1 and y_2. The constants of the system are the two masses M_1 and M_2 and the two spring constants k_1 and k_2. In addition, the external forces $F_1(t)$ and $F_2(t)$, which vary with time, must

12   Introduction to Structural Dynamics

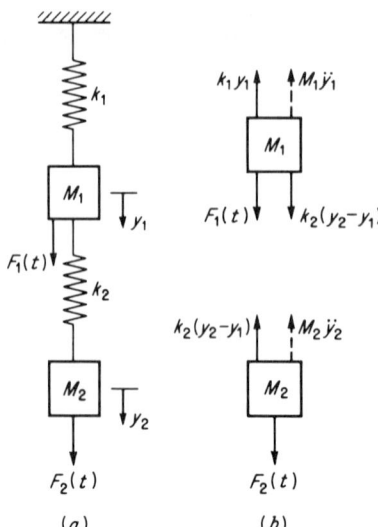

FIGURE 1.7 Two-degree system—dynamic equilibrium.

also be known. It is our purpose in this section to determine the variation of the two displacements with time for a given set of external forces having a prescribed time variation.

a. Formulation of the Problem

The equations of motion may be written by applying the concept of dynamic equilibrium as discussed in the previous section. For this purpose the two masses are isolated, as indicated in Fig. 1.7b. To each mass one must apply all the forces acting. These include (1) the internal spring forces, (2) the externally applied forces, and (3) the inertia forces. Note that the signs must always be consistent. In this example the displacements y_1 and y_2 have been taken as positive when downward. Thus the directions of the spring forces shown on the sketch are consistent with the expressions written for these forces and the positive directions of the displacements. The inertia forces are, of course, always taken as positive in a direction opposite to positive displacement.

Simply by writing the equation of equilibrium for each of the two masses shown in Fig. 1.7b we arrive at the following two differential equations of motion:

$$M_1\ddot{y}_1 + k_1 y_1 - k_2(y_2 - y_1) - F_1(t) = 0$$
$$M_2\ddot{y}_2 + k_2(y_2 - y_1) - F_2(t) = 0 \quad (1.11)$$

These two equations may be solved simultaneously to determine the time variation of the two displacements.

The numerical integration of the two differential equations of motion is carried out in exactly the same way as for a one-degree system, except that in this case two integrations, one for each equation, must be carried out simultaneously. This presents no particular difficulty since the extrapolation process yields both displacements at a particular time on the basis of the displacements at preceding time intervals. Thus the two accelerations can be determined by the application of the two equations (1.11) independently. This procedure is illustrated by the example which follows.

b. Example

As a numerical example, values are assigned to the constants of the lumped-mass two-degree system shown in Fig. 1.7. It is desired to determine the variation of the two displacements, starting at rest at $t = 0$ and continuing throughout whatever interval of time may be of interest. The following values are assumed:

$M_1 = 2$ lb-sec^2/ft $k_1 = 4000$ lb/ft $F_1(t) = 0$
$M_2 = 1$ lb-sec^2/ft $k_2 = 2000$ lb/ft $F_2(t) = 200$ lb $t > 0$

Note that zero force is applied to the mass M_1 and that the force applied to the mass M_2 is suddenly applied at $t = 0$ and remains constant indefinitely thereafter.

If these values are inserted into Eqs. (1.11), expressions for the two accelerations may be written as follows:

$$\ddot{y}_1 = 1000(y_2 - y_1) - 2000y_1 \\ \ddot{y}_2 = 200 - 2000(y_2 - y_1)$$ (1.12)

Only these two equations and the recurrence formula, Eq. (1.5), are required for the analysis. The recurrence formula applies to each of the two displacements independently.

As explained previously, a proper time interval for the numerical integration depends upon the natural period of the system. In this example there are two natural periods, since we are dealing with a two-degree system. For these particular parameters, the two natural periods are 0.20 and 0.10 sec. The former is the natural period of the first, or fundamental, mode of vibration, and the latter is the natural period of the second mode. The determination of natural periods for two-degree systems is discussed in Sec. 3.2, and will not be considered here. In order for the numerical analysis to yield accurate results, the time interval should not be larger than one-tenth of the smaller natural period. Thus, in this example, a time interval Δt of 0.01 sec will be used.

The computations for this analysis are arranged in tabular form for convenience in Table 1.2. At each time station the two displacements have been determined by the preceding computations. The procedure

Table 1.2 Numerical Integration; Undamped Elastic Two-degree System (Fig. 1.7)

t sec	$2000y_1$ lb	$1000(y_2 - y_1)$ lb	$\ddot{y}_1$ Eq. (1.12) ft/sec²	$\ddot{y}_1 (\Delta t)^2$ ft	y_1 Eq. (1.5) ft	$2000(y_2 - y_1)$ lb	$\ddot{y}_2$ Eq. (1.12) ft/sec²	$\ddot{y}_2 (\Delta t)^2$ ft	y_2 Eq. (1.5) ft	$y_2 - y_1$ ft
0	0	0	0	0	0	0	200	0.0200	0	0
0.01	0	10	10	0.0010	0.0002*	20	180	0.0180	0.0100†	0.0098
0.02	3	37	34	0.0034	0.0014	73	127	0.0127	0.0380	0.0366
0.03	12	73	61	0.0061	0.0060	145	55	0.0055	0.0787	0.0727
0.04	33	108	75	0.0075	0.0167	216	−16	−0.0016	0.1249	0.1082
0.05	70	135	65	0.0065	0.0349	269	−69	−0.0069	0.1695	0.1346
0.06	119	148	29	0.0029	0.0596	295	−95	−0.0095	0.2072	0.1476
0.07	174	148	−26	−0.0026	0.0872	296	−96	−0.0096	0.2354	0.1482
0.08	224	142	−82	−0.0082	0.1122	284	−84	−0.0084	0.2540	0.1418
0.09	258	135	−123	−0.0123	0.1290	270	−70	−0.0070	0.2642	0.1352
0.10	267	134	−133	−0.0133	0.1335	268	−68	−0.0068	0.2674	0.1339
0.11	249	139	−110	−0.0110	0.1247	278	−78	−0.0078	0.2638	0.1391
0.12	210	147	−63	−0.0063	0.1049	295	−95	−0.0095	0.2524	0.1475
0.13	158	153	−5	−0.0005	0.0788	305	−105	−0.0105	0.2315	0.1527
0.14	104	148	44	0.0044	0.0522	296	−96	−0.0096	0.2001	0.1479
0.15	60	129	69	0.0069	0.0300	258	−58	−0.0058	0.1591	0.1291
0.16	29	98	69	0.0069	0.0147	195	5	0.0005	0.1123	0.0976
0.17	13	60	47	0.0047	0.0063	119	81	0.0081	0.0660	0.0597
0.18	5	25	20	0.0020	0.0026	50	150	0.0150	0.0278	0.0252
0.19	2	4	2	0.0002	0.0009	7	193	0.0193	0.0046	0.0037
0.20	−1	1	2	0.0002	−0.0006	3	197	0.0197	0.0007	0.0013
0.21	...	...	...	...	−0.0019	...	...	...	0.0165	0.0184

* $y_1 = \frac{1}{6}(\ddot{y}_1$ at $t = 0.01)(\Delta t)^2$.
† $y_2' = \frac{1}{2}(\ddot{y}_2$ at $t = 0)(\Delta t)^2$.

is therefore as follows: (1) the two accelerations are computed by Eqs. (1.12); and (2) the displacements at the next time station are computed using the recurrence formula. For further illustration the detailed computations leading to the displacements at $t = 0.05$ are given below:

At $t = 0.03$,
$$y_1 = 0.0060, \quad y_2 = 0.0787$$
At $t = 0.04$,
$$y_1 = 0.0167, \quad y_2 = 0.1249$$
previously determined

At $t = 0.04$,
$$\ddot{y}_1 = 1000(0.1249 - 0.0167) - 2000(0.0167) = 75$$
$$\ddot{y}_2 = 200 - 2000(0.1249 - 0.0167) = -16$$

At $t = 0.05$,
$$y_1 = 2 \times 0.0167 - 0.0060 + 75(0.01)^2 = 0.0349$$
$$y_2 = 2 \times 0.1249 - 0.0787 - 16(0.01)^2 = 0.1695$$

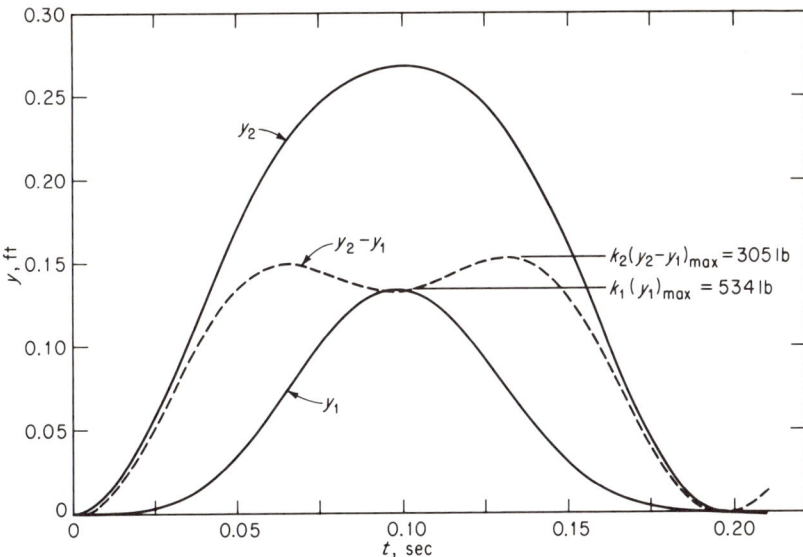

FIGURE 1.8 Example. Response of two-degree system.

All values are, of course, in units of feet and seconds. These computations are repeated for each time interval in succession.

At the first time station, y_2 is determined by the special formula (1.7), since the load is applied suddenly at $t = 0$. For y_1 at the first time station, Eq. (1.6), with $y^{(0)} = 0$, is used, since the acceleration of mass 1 begins at zero.

The result of this analysis is plotted in Fig. 1.8, where the displacements are displayed as a function of time. Inspection of such data assists one to become aware of the general characteristics of dynamic response. For example, both displacements y_1 and y_2 vary in what appears to be almost sinusoidal fashion between a minimum of zero and some maximum value. Furthermore, the period of this sinusoidal variation is approximately 0.2 sec, or in other words, the same as the first natural period of the system. Thus one concludes that the response in this particular example is primarily in the first mode and that the effect of the second mode is of less significance. On the other hand, the difference between the two displacements, which is also plotted in Fig. 1.8, distinctly shows the effect of the second mode. The reason for this behavior will become more clear when modal analysis of multidegree systems is discussed in Chap. 3.

It is also of interest to consider the maximum spring forces which result from the application of this suddenly applied load. For example,

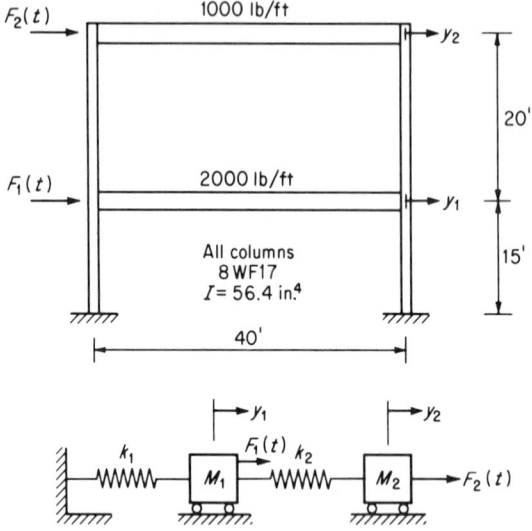

FIGURE 1.9 Two-story rigid frame.

the maximum force in spring 1 is 534 lb (Fig. 1.8), which is 2.67 times the value of the applied external load. The maximum force in spring 2 is 305 lb, which is equal to 1.53 times the maximum value of the applied load. The multipliers given are also the ratios of the maximum dynamic spring force to the force which would have occurred had the external load been applied statically. Thus one reaches the conclusion, which is generally true, that a suddenly applied force causes considerably greater stresses in a structure than would the same force applied statically.

Before leaving this subject the reader is reminded that the idealized system shown in Fig. 1.7 is a representation of an actual structure. For example, such a structure could be the two-story rigid frame subjected to horizontal dynamic forces as shown in Fig. 1.9. If it is assumed for simplicity that the girders are infinitely rigid and that the weight of the columns is negligible, the constants of the idealized spring-mass system may be easily determined as follows:

$$W_1 = 2000 \times 40 = 80{,}000 \text{ lb}$$
$$W_2 = 1000 \times 40 = 40{,}000 \text{ lb}$$
$$k_1 = \frac{12E(2I)}{h^3} = \frac{12(30 \times 10^6) \times 112.8}{(15)^3 \times 144} = 83{,}500 \text{ lb/ft}$$
$$k_2 = \frac{12(30 \times 10^6) \times 112.8}{(20)^3 \times 144} = 35{,}200 \text{ lb/ft}$$

The two lumped weights (or masses) are simply equal to the total weights at the two floor levels. The spring constant in one of the stories is merely the sum of the two column shears developed by a unit relative displacement in that story. The formula given above may be easily derived by simple elastic analysis. Having the values of these system parameters and given the two load-time functions, we could determine the horizontal displacements of the two floors by exactly the same procedure as was used for the lumped-mass system of the example. The spring-mass system shown in Fig. 1.9 is, of course, exactly equivalent to that in Fig. 1.7.

1.4 One-degree Elastic System with Damping

This section contains a very brief discussion of the effect of damping in structural dynamic systems and methods for including this effect in numerical analysis. The discussion is extended in later chapters.

a. Damping Characteristics

All structural dynamic systems contain damping to some degree, but as will be shown below, the effect may not be significant if the load duration is short and only the maximum dynamic response is of interest. On the other hand, if a continuing state of vibration is being investigated, damping may be of primary importance. In fact, if enough damping is present, vibration may be completely eliminated.

Damping in structures may be of several different forms. It is in part due to the internal molecular friction of the material. It is also due to the loss of energy associated with the slippage of structural connections either between members or between the structure and the supports. In some cases it may be due to the resistance to motion provided by air or other fluids surrounding the structure. In any case, the effect is one of forces opposing the motion, and hence the amplitude of the response is decreased.

It is generally believed that, for purposes of analysis, structural damping may be assumed to be of the viscous type; i.e., the damping force is opposite but proportional to the velocity. Although other forms of damping are usually present, this assumption provides reasonable results. Accordingly, the damping force applied to a lumped mass may be expressed by the following:

$$\text{Damping force} = -c\dot{y}$$

where c is a numerical constant, and $\dot{y}$ is the velocity of that mass. The negative sign indicates that the force is always opposed to the direction of the velocity. The magnitude of the coefficient c is extremely difficult to determine, and for this reason it is convenient to introduce the concept of *critical damping*. This is the amount of damping that

18   Introduction to Structural Dynamics

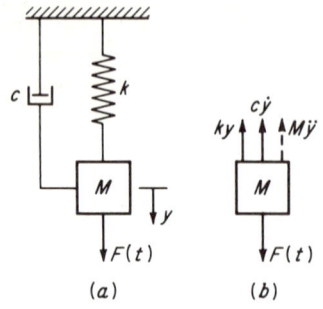

FIGURE 1.10 Damped one-degree system—dynamic equilibrium.

would completely eliminate vibration and, for a one-degree system, is given by the following:

$$c_{cr} = 2\sqrt{kM} \tag{1.13}$$

where k and M are the stiffness and mass of the system. The concept of critical damping is useful, since it is often easier to specify the amount of damping as a certain percentage of critical than it is to arrive at the numerical value for the coefficient c. The derivation of Eq. (1.13) and further discussion of this subject are contained in Sec. 2.4.

b. Formulation of the Problem

Consider the simple one-degree system shown in Fig. 1.10a and the state of dynamic equilibrium shown in Fig. 1.10b. This system differs from those in the previous examples only in the inclusion of a damping force. In Fig. 1.10b the damping force is shown upward, or opposite to the direction of positive velocity. Using the concept of dynamic equilibrium, we write the equation of motion as follows:

$$ky + M\ddot{y} + c\dot{y} - F(t) = 0 \tag{1.14}$$

Thus the equation of motion contains one additional term, and involves velocity, besides acceleration and displacement.

c. Example

To illustrate the effect of damping the analysis of Sec. 1.2c will be repeated, but in this instance 10 percent of critical damping will be assumed. The parameters of the system and the load-time function are given in Fig. 1.4. The damping coefficient is equal to one-tenth of the critical coefficient as given by Eq. (1.13):

$$c = 0.1 \times 2\sqrt{kM} = 0.2\sqrt{2000 \times 2} = 12.7 \frac{\text{lb}}{\text{ft/sec}}$$

Note that damping coefficients must have units of force divided by velocity. By substituting the proper numerical values into Eq. (1.14) the following expression for acceleration is obtained:

$$\ddot{y} = \tfrac{1}{2}F(t) - 1000y - 6.35\dot{y} \tag{1.15}$$

Although numerical integration, as in previous sections, will be used, it must be modified, since it is now necessary to determine the velocity at each time station. The velocity at time station s will be approximately expressed by the following:

$$\dot{y}^{(s)} = \frac{y^{(s)} - y^{(s-1)}}{\Delta t} + \ddot{y}^{(s)} \frac{\Delta t}{2} \tag{1.16}$$

where the first term is the average velocity in the preceding time interval, and the second term is an estimate of the amount by which that average must be increased to give the velocity at the next time station. This procedure is consistent with the concept of acceleration pulses on which Eq. (1.5) is based and in addition approximates $\dot{y}^{(s)}$ as the mean of the average velocities in the adjacent time intervals [Eq. (1.4)].

By substituting Eq. (1.16) into Eq. (1.14) and rearranging, the following is obtained:

$$\ddot{y}^{(s)} = \frac{F^{(s)} - ky^{(s)} - c(y^{(s)} - y^{(s-1)})/\Delta t}{M + c\Delta t/2} \tag{1.14a}$$

If we substitute the numerical values of the problem at hand and use a time interval of 0.02 sec, Eq. (1.15) becomes

$$\ddot{y}^{(s)} = \frac{F^{(s)}}{2.127} - 940 y^{(s)} - 5.96 \left(\frac{y^{(s)} - y^{(s-1)}}{0.02} \right) \tag{1.15a}$$

The computations for this analysis, which are based on Eqs. (1.5) and (1.15a), are shown in Table 1.3. This table differs from previous computation tables only in the additional columns, which permit the calculation of the velocity at each time station. At $t = 0$, y is obtained by Eq. (1.15) rather than (1.15a), since the velocity is known to be zero and the latter equation is not applicable. As in previous examples, the displacement at the end of the first interval is computed by Eq. (1.7) rather than (1.5).

The result of these computations is plotted in Fig. 1.11. For comparison, the solution obtained in Sec. 1.2c, which differed only in that damping was not included, is also plotted. It should be noted that the effect of damping on the first peak of response is not great. However, this effect becomes considerable at the second positive peak, and will become greater as time increases. Since the damping in most actual structural systems does not exceed 10 percent of critical, it may be generally concluded that damping is not of great importance with respect to the

20   *Introduction to Structural Dynamics*

Table 1.3   Numerical Integration; Damped Elastic One-degree System (Fig. 1.10)

t sec	$\dfrac{y^{(s)} - y^{(s-1)}}{\Delta t}$ fps	$5.96\left(\dfrac{y^{(s)} - y^{(s-1)}}{\Delta t}\right)$ ft/sec^2	$\dfrac{F^{(s)}}{2.127}$ ft/sec^2	$940 y^{(s)}$ ft/sec^2	$\ddot{y}^{(s)}$ Eq. (1.15a) ft/sec^2	$\ddot{y}^{(s)} (\Delta t)^2$ ft	$y^{(s)}$ Eq. (1.5) ft
0	0	0	25.0*	0	25.0	0.0100	0
0.02	0.25	1.5	28.2	4.7	22.0	0.0088	0.0050†
0.04	0.69	4.1	32.9	17.7	11.1	0.0044	0.0188
0.06	0.91	5.4	37.6	34.8	−2.6	−0.0010	0.0370
0.08	0.86	5.1	42.3	50.9	−13.7	−0.0055	0.0542
0.10	0.58	3.5	47.0	61.9	−18.4	−0.0074	0.0659
0.12	0.21	1.3	35.3	66.0	−32.0	−0.0128	0.0702
0.14	−0.42	−2.5	23.5	58.0	−32.0	−0.0128	0.0617
0.16	−1.06	−6.3	23.5	38.0	−8.2	−0.0033	0.0404
0.18	−1.23	−7.3	23.5	14.8	16.0	0.0064	0.0158
0.20	−0.91	−5.4	23.5	−2.2	31.1	0.0124	−0.0024
0.22	−0.29	−1.7	23.5	−7.7	32.9	0.0132	−0.0082
0.24	0.37	2.2	23.5	−0.8	22.1	0.0088	−0.0008
0.26	0.81	4.8	23.5	14.5	4.2	0.0017	0.0154
0.28	0.89	5.3	23.5	31.3	−13.1	−0.0052	0.0333
0.30	0.63	3.7	23.5	43.2	−23.4	−0.0094	0.0460
0.32	0.16	1.0	23.5	46.3	−23.8	−0.0095	0.0493
0.34	......	.....	.....	....	......	....	0.0431

* $F^{(s)}/2$ since $\dot{y}^{(s)} = 0$.
† $y(t = 0.02) = \frac{1}{2}\ddot{y}(t = 0)(\Delta t)^2$.

maximum stress in the structure, which usually occurs with the first peak of response. It should be noted, however, that for certain situations this statement may not be true. Examples of such exceptions are multidegree systems in which higher modes are important and one-degree systems with irregular load-time functions.

1.5   One-degree Elasto-plastic Systems

Up to this point only linear elastic systems have been considered; i.e., the resistance function has been a straight line with slope k and without upper limit. In many practical cases this function is nonlinear (the slope is not constant) and/or inelastic (when the spring is unloaded the resistance does not return to zero by the same path). These conditions can easily be handled by numerical analysis provided only that the resistance is a unique function of displacement. Attention here will be focused on that type of inelastic behavior which is normally assumed in structural design.

Considered below is the dynamic response of a structure which extends through the elastic and into the plastic range. Because most structures

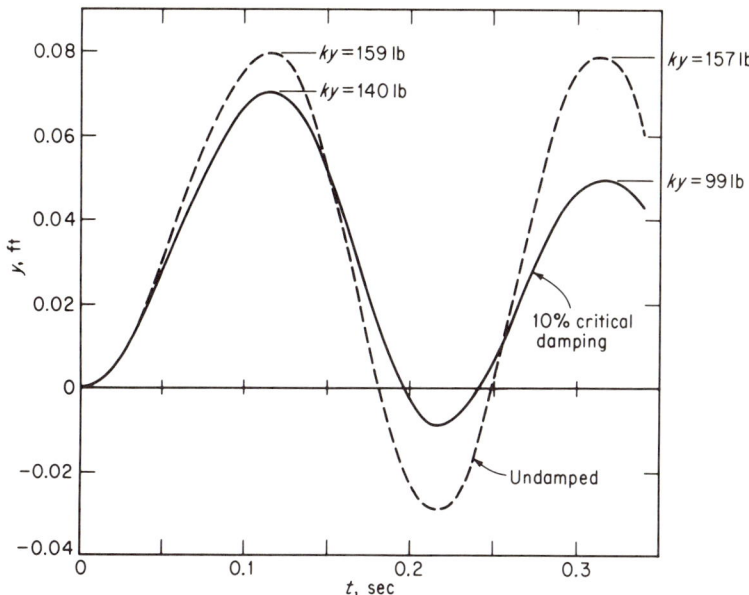

FIGURE 1.11 Example. Response of damped and undamped systems.

have considerable ductility, this type of behavior is entirely feasible. Although plastic behavior is not generally permissible under continuous operating conditions, it is quite appropriate for design when the structure is subjected to a severe dynamic loading only once or at most a few times during its life. Among other examples which might be cited, plastic behavior is normally anticipated in the design of blast-resistant structures and at least implied in the design of structures for earthquake. This concept is of considerable importance in structural design for dynamic loads because a much greater portion of the energy-absorbing capacity of the structure is utilized thereby. The economy of the design can be considerably increased by taking this fact into account. In some respects plastic behavior is similar to damping since it disrupts the harmonic motions which are characteristic of elastic vibrations.

a. Resistance Function

Consider the one-degree system shown in Fig. 1.12a, the spring of which is assumed to have the resistance function shown in Fig. 1.12b. The latter is usually called a bilinear resistance function since all the lines making up that function lie in one of two directions. As the displacement increases from zero, the resistance increases linearly with a slope of k, the spring constant. The linearity continues until the elastic-

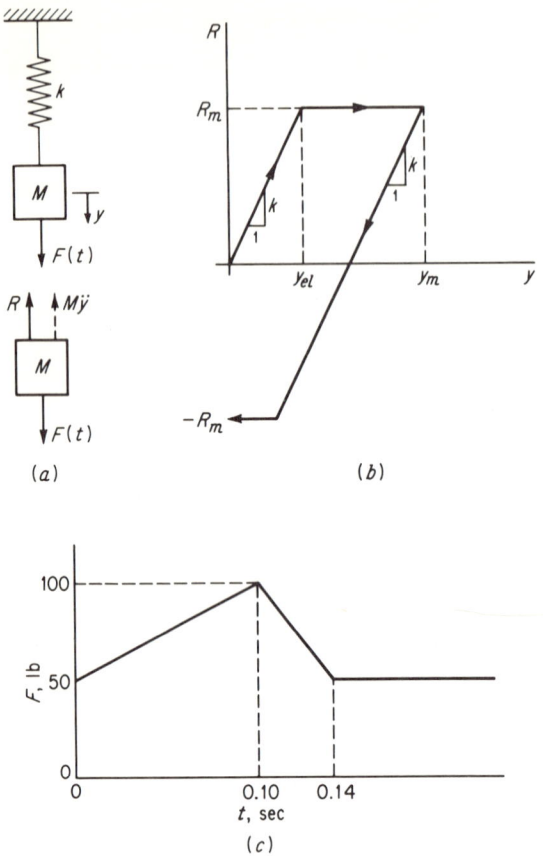

FIGURE 1.12 (a) Elasto-plastic system; (b) resistance function; (c) load-time function.

limit displacement y_{el} is reached, at which point the maximum spring force R_m has been attained. As the displacement increases further, the resistance is assumed to remain constant at R_m. The latter value will be maintained until the ductility limit of the structure is reached. However, if the displacement reaches a maximum before that limit and then decreases, the structure is said to "rebound." During rebound the resistance is assumed to decrease along a line parallel to the initial elastic slope. This decrease will continue with decreasing displacement until a spring force equal to $-R_m$ is attained.

The resistance function described above is an idealization of that of an actual structure. A real structure will have a curved transition in the region of y_{el} rather than the sharp break shown in Fig. 1.12b. This curvature occurs even when only one plastic hinge is necessary to develop the full plastic strength of the structure. If the structure is such that more than one hinge is required, the transition range will be even wider. Nevertheless, the resistance function shown in Fig. 1.12b is an adequate representation for the majority of structures. Strain hardening such as exhibited by steel may be ignored since this occurs only at very large structural deformation, which is usually beyond the range of interest. These points are further discussed in Sec. 5.2.

b. Formulation of the Problem

By isolating the mass of the one-degree system as shown in Fig. 1.12a and applying the concept of dynamic equilibrium, the equation of motion may be written as in previous examples. For this inelastic case, it is convenient to represent the spring force, or resistance, by the more general notation R since the expression for resistance changes in the various ranges of the resistance function. The equations of motion are as follows:

$$
\begin{aligned}
&(a) \quad M\ddot{y} + R - F(t) = 0 \\
&(b) \quad M\ddot{y} + ky - F(t) = 0 \qquad 0 < y < y_{el} \\
&(c) \quad M\ddot{y} + R_m - F(t) = 0 \qquad y_{el} < y < y_m \\
&(d) \quad M\ddot{y} + R_m - k(y_m - y) - F(t) = 0 \qquad (y_m - 2y_{el}) < y < y_m
\end{aligned}
\qquad (1.17)
$$

where Eq. (a) is the general equation of motion, (b) applies in the original elastic range, (c) applies in the plastic region, and (d) applies during the elastic behavior after y_m has been attained. Additional equations could be written, but this is unnecessary since in most situations rebound does not extend into the negative plastic range. The reader can easily verify these equations since (b), (c), and (d) follow directly from (a) and the included expressions for R become obvious upon inspection of the resistance function in Fig. 1.12b.

c. Example

For purposes of illustration, the following numerical values are given to the parameters of the one-degree system shown in Fig. 1.12:

$$
\begin{aligned}
M &= 2 \text{ lb-sec}^2/\text{ft} \\
k &= 2000 \text{ lb/ft} \\
R_m &= 110 \text{ lb}
\end{aligned}
$$

and therefore, $y_{el} = R_m/k = 0.055$ ft

It may be noted that the elastic limit y_{el} can be computed directly from the maximum resistance and the spring constant. The load-time function is shown in Fig. 1.12c. It is our purpose once again to determine the response of the system in terms of displacement versus time. The

Table 1.4 Numerical Integration; Undamped Elasto-plastic One-degree System (Fig. 1.12)

t sec	$\frac{1}{2}F(t)$ ft/sec²	$\frac{1}{2}R = 1000y,$ or 55 ft/sec²	$\ddot{y}$ Eq. (1.18) ft/sec²	$\ddot{y}\,(\Delta t)^2$ ft	t Eq. (1.5) ft
0	25	0	25.0	0.0100	0
0.02	30	5.0	25.0	0.0100	0.0050
0.04	35	20.0	15.0	0.0060	0.0200
0.06	40	41.0	−1.0	−0.0004	0.0410
0.08	45	55.0	−10.0	−0.0040	0.0616
0.10	50	55.0	−5.0	−0.0020	0.0782
0.12	37.5	55.0	−17.5	−0.0070	0.0928
0.14	25	55.0	−30.0	−0.0120	0.1004
0.16	25	50.6*	−25.6	−0.0102	0.0960
0.18	25	36.0*	−11.0	−0.0044	0.0814
0.20	25	17.0*	8.0	0.0032	0.0624
0.22	25	1.2*	23.8	0.0095	0.0466
0.24	25	−5.1*	30.1	0.0120	0.0403
0.26	25	0.6*	24.4	0.0098	0.0460
0.28	25	16.1*	8.9	0.0036	0.0615
0.30	25	35.2*	−10.2	−0.0040	0.0806
0.32	25	50.3*	−25.3	−0.0101	0.0957
0.34	.....	......	......	....	0.1007

* $\frac{1}{2}R = 55 - 1000(0.1004 - y)$.

problem stated is exactly the same as that solved in Sec. 1.2c, except that, in the latter case, no plastic limit was placed upon the spring resistance.

By inserting the numerical values of the parameters into Eq. (1.17), the following expressions are obtained for acceleration:

$$
\begin{aligned}
&(a) && \ddot{y} = \tfrac{1}{2}F(t) - \tfrac{1}{2}R && \\
&(b) && \ddot{y} = \tfrac{1}{2}F(t) - 1000y && 0 < y < 0.055 \\
&(c) && \ddot{y} = \tfrac{1}{2}F(t) - 55 && 0.055 < y < y_m \\
&(d) && \ddot{y} = \tfrac{1}{2}F(t) - 55 + 1000(y_m - y) && (y_m - 0.11) < y < y_m
\end{aligned}
\qquad (1.18)
$$

The last three equations correspond to the last three of Eqs. (1.17) and apply to the same ranges of the resistance function. Since these ranges are defined in terms of displacement, the analyst has no difficulty in selecting the correct equation for use.

The numerical analysis of the problem, which is shown in Table 1.4, is conducted in exactly the same manner as in previous examples. The only difference lies in the computation of the resistance R. It will be

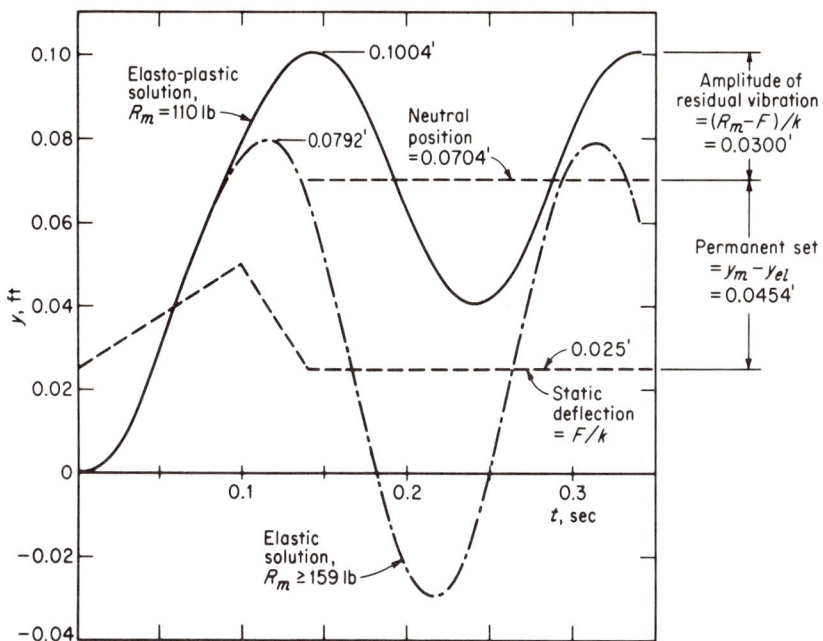

FIGURE 1.13 Example. Responses of elastic system ($R_m \geq 158$ lb) and elasto-plastic system ($R_m = 100$ lb).

noted that, in Eqs. (1.18), (b) was used up to $t = 0.06$, (c) was used from 0.08 to 0.14, and (d) was used from $t = 0.16$ to the end of the computation.

The results of the analysis are plotted in Fig. 1.13. Also shown in the same figure, for the sake of comparison, is the analysis of the completely elastic but otherwise similar system discussed in Sec. 1.2c. It will be recalled that, in the latter case, the maximum spring force developed was 159 lb compared with the 110-lb maximum resistance specified in the present example.

Several interesting comments may be made in connection with Fig. 1.13. First, the maximum displacement of the elasto-plastic system, although greater than that of the purely elastic system, is not excessively so in spite of the fact that only about two-thirds as much structural resistance is available. On the other hand, the residual vibration of the plastic structure is considerably less than that of the elastic structure. Note that, in either case, the residual vibration would continue indefinitely in the absence of damping and that the periods are the same. In the elasto-plastic system the amplitude of the residual vibration is given by $(R_m - F)/k$ since the numerator of this fraction is the amount by which the spring force must decrease in order to reach a point corresponding to static equilibrium. The neutral position for the axis of vibration may be computed as the maximum deflection minus the half-amplitude of the residual vibration.

There is, of course, a permanent set, or distortion, in the structure of this example. This may be computed as $y_m - y_{el}$ since y_{el} is the amount by which the displacement would be reduced if all load were removed. Thus, for zero stress in the spring, there would still be a permanent distortion of $y_m - y_{el}$. As indicated in Fig. 1.13, the permanent set is also equal to the distance between the neutral position and the static elastic deflection. If the structure were again loaded into the plastic range, there would of course be additional permanent distortion.

The foregoing discussion was intended to illustrate the manner in which elasto-plastic systems may be analyzed and also to point out some of the distinguishing characteristics of plastic behavior. The advantage of permitting plastic behavior should now be obvious. If the designer can accept some permanent distortion in the structure, the amount of resistance, and hence structural materials and cost, can be appreciably reduced. In the example above, a reduction in the strength of about one-third resulted in an increase in deflection of only about 25 percent.

In order to relate the above to a real structure, consider the rigid frame shown in Fig. 1.6 and discussed in Sec. 1.2c. The maximum resistance of the frame to horizontal loading is given by $4\mathfrak{M}_P/h$, where $\mathfrak{M}_P$ is the ultimate, or plastic, bending strength of the columns. For the 8WF17 column section, this value is 43.4 kip-ft ($\sigma_y = 33$ ksi), and hence

$$R_m = 8.68 \text{ kips}$$

The latter value plus the W and k as given by Eqs. (1.10) would permit an elasto-plastic analysis of the frame.

The procedure given in this section can in general be applied to multi-degree systems as illustrated in later chapters. The discussion above, together with that in Sec. 1.3, should make it apparent that no unusual difficulties are to be encountered.

1.6 Alternative Methods of Numerical Analysis

The constant-velocity procedure presented in Sec. 1.2 and used in subsequent sections is an extremely simple method which yields any desired precision provided that the time interval is taken sufficiently small. There are many other methods[1-3]* of numerical analysis, some of which are presented below. These have the advantage that greater precision

---

* Superior numbers correspond to References at the end of the text.

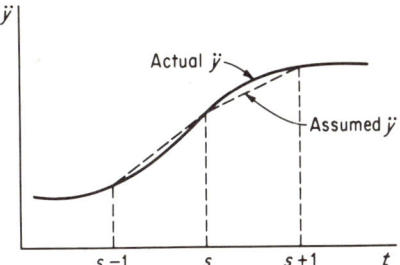

FIGURE 1.14 Numerical integration—linear-acceleration method.

is achieved for a given time interval, but the disadvantage that more complex computation is required within each interval. Conversely, for a given precision, these methods permit a larger time interval than does the constant-velocity method. The choice depends to some extent upon the computational device being used. Although the point is debatable, the author believes that more efficient solution generally results from use of a less complex recurrence formula (e.g., the constant-velocity method) together with a somewhat smaller interval.

a. Linear-acceleration Method

An obvious refinement of the constant-velocity procedure is to assume a continuous function for acceleration rather than a step function (Fig. 1.3b). Suppose, for example, that the acceleration were assumed to vary linearly between time stations as indicated in Fig. 1.14. The acceleration between stations s and $s + 1$ would then be approximated by

$$\ddot{y} = \ddot{y}^{(s)} + \frac{\ddot{y}^{(s+1)} - \ddot{y}^{(s)}}{\Delta t} (t - t^{(s)}) \tag{1.19}$$

The velocity at any time within the same interval may be obtained by

$$\dot{y} = \dot{y}^{(s)} + \int_{t^{(s)}}^{t} \ddot{y} \, dt$$

or $$\dot{y} = \dot{y}^{(s)} + \ddot{y}^{(s)}(t - t^{(s)}) + \frac{\ddot{y}^{(s+1)} - \ddot{y}^{(s)}}{2 \, \Delta t} (t - t^{(s)})^2 \tag{1.20a}$$

which, at station $s + 1$, becomes

$$\dot{y}^{(s+1)} = \dot{y}^{(s)} + \frac{\Delta t}{2} (\ddot{y}^{(s+1)} + \ddot{y}^{(s)}) \tag{1.20b}$$

The displacement at $s + 1$ is given by

$$y^{(s+1)} = y^{(s)} + \int_{t^{(s)}}^{t^{(s+1)}} \dot{y} \, dt$$

$$= y^{(s)} + \dot{y}^{(s)} \Delta t + \frac{(\Delta t)^2}{6} (2\ddot{y}^{(s)} + \ddot{y}^{(s+1)}) \tag{1.21}$$

where, from Eq. (1.20a),

$$\dot{y}^{(s)} = \dot{y}^{(s-1)} + \frac{\Delta t}{2}(\ddot{y}^{(s)} + \ddot{y}^{(s-1)}) \tag{1.22}$$

Equations (1.21) and (1.22) are the basis for the *linear-acceleration method* of numerical integration.

At a particular stage in the computations $\ddot{y}^{(s)}$ is known, and hence $\dot{y}^{(s)}$ can be computed directly. However, in order to obtain $y^{(s+1)}$, one must first obtain $\ddot{y}^{(s+1)}$, which depends upon $y^{(s+1)}$. If the analysis is for a one-degree system, the acceleration at $s+1$ is given by

$$\ddot{y}^{(s+1)} = \frac{F(t^{(s+1)})}{M} - \frac{k}{M}y^{(s+1)} = \frac{F(t^{(s+1)})}{k}\omega^2 - \omega^2 y^{(s+1)}$$

Substituting this expression into Eq. (1.21) and rearranging, we obtain

$$y^{(s+1)} = \frac{y^{(s)} + \dot{y}^{(s)}\Delta t + \dfrac{(\Delta t)^2}{3}\ddot{y}^{(s)} + \dfrac{(\Delta t)^2}{6}\dfrac{F(t^{(s+1)})}{k}\omega^2}{1 + \dfrac{(\Delta t)^2}{6}\omega^2} \tag{1.23}$$

This equation, together with Eq. (1.22), permits a direct solution in each interval; that is, $y^{(s+1)}$ may be computed directly from $y^{(s)}$ (and hence $\ddot{y}^{(s)}$) and $\dot{y}^{(s)}$.

If the system has more than one degree of freedom, the acceleration at one point depends not only upon the displacement at that point, but also upon the other displacements of the system. Therefore the equation comparable with Eq. (1.23) is much more complicated. In such cases it is usually advisable to adopt an iterative procedure in which values of $\ddot{y}^{(s+1)}$ are assumed and used in Eq. (1.21) to obtain $y^{(s+1)}$. New values of $\ddot{y}^{(s+1)}$ are then computed by the equations of motion, and the process repeated to convergence. This additional complication in the analysis of multidegree systems is not encountered in the constant-velocity procedure.

An example of the linear-acceleration method is given in Table 1.5, where the response of a one-degree system subjected to a suddenly applied constant force is computed. The time interval used is approximately one-tenth the natural period, which, although smaller than that required for sufficient precision by this method, is selected to afford a comparison with the constant-velocity procedure. For the given parameters of the problem, Eqs. (1.22) and (1.23) reduce to the simple forms shown in Table 1.5. Note that this procedure is self-starting; i.e., the recurrence formula may be used without modification in the first step.

Shown in Table 1.5 are comparative results provided by an exact solution and also by the constant-velocity method using the same time

Table 1.5 Numerical Integration by Linear-acceleration Method; Comparison with Constant-velocity Method and Exact Solution

$T = 0.1047$ sec; $\omega = 60$ rad/sec
Suddenly applied constant force F_1; $F_1/k = 1$
$\ddot{y} + \omega^2 y = F_1/M = F_1\omega^2/k$; $\ddot{y} = 3600(1 - y)$
Eq. (1.22): $\dot{y}^{(s)} = \dot{y}^{(s-1)} + 0.005(\ddot{y}^{(s)} + \ddot{y}^{(s-1)})$

Eq. (1.23): $y^{(s+1)} = \dfrac{y^{(s)} + 0.01\dot{y}^{(s)} + (\ddot{y}^{(s)}/30{,}000) + 0.06}{1.06}$

t sec	$\ddot{y}^{(s)}$ ft/sec²	$\dot{y}^{(s)}$ fps	Linear-acceleration method $y^{(s)}$ ft	Exact solution $y^{(s)}$ ft	Constant-velocity method $y^{(s)}$ ft
0	3600	0	0	0	0
0.01	2990	32.95	0.1698	0.1747	0.1800
0.02	1362	54.71	0.6217	0.6376	0.6552
0.03	−728	57.88	1.2021	1.2272	1.2555
0.04	−2570	41.39	1.7138	1.7374	1.7638
0.05	−3539	10.85	1.9830	1.9901	1.9971
0.06	−3306	−23.38	1.9183	1.8968	1.8714
0.07	−1950	−49.66	1.5416	1.4903	1.4320
0.08	+68	−59.07	0.9811	0.9125	0.8371
0.09	2062	−48.42	0.4271	0.3653	0.3008
0.10	3357	−21.33	0.0675	0.0398	0.0162
0.11	.....	......	0.0246	0.0497	0.0858

interval. At first glance it may appear that both numerical methods result in considerable error. However, comparison at discrete time stations is misleading since much of the apparent error is due to a slight phase shift which is generally of little consequence. In other words, there is little error in the peak responses.

As expected, the constant-velocity method is slightly less accurate. On the other hand, the linear-acceleration method requires considerably more computation, which is significant whether the calculations are being done by hand or by electronic computer. Actually, the difference in the accuracy of the two methods is not great in this case. However, the difference may be appreciable if the load varies with time in irregular fashion. It should be emphasized that this comparison of accuracy is given only for illustration. In a particular problem the desired accuracy is independent of the method used and the choice is one of time interval. In general, an interval of $\frac{1}{6}$ the natural period is small enough for the linear-acceleration method, while the constant-velocity method requires a ratio of about $\frac{1}{10}$.

b. Newmark β Method[4]

A versatile method developed by Newmark, which can be adjusted to suit the particular problem, is embodied in the equations

$$\dot{y}^{(s+1)} = \dot{y}^{(s)} + \frac{\Delta t}{2}(\ddot{y}^{(s)} + \ddot{y}^{(s+1)}) \tag{1.24a}$$

and $\quad y^{(s+1)} = y^{(s)} + \dot{y}^{(s)}\Delta t + (\tfrac{1}{2} - \beta)\ddot{y}^{(s)}(\Delta t)^2 + \beta\ddot{y}^{(s+1)}(\Delta t)^2 \quad$ (1.24b)

Within certain limits β may be selected at will and in effect is an assumption regarding the variation of acceleration within the time interval. The value selected affects the rate of convergence within each step (if iteration is being used), the stability of the analysis, and the amount of error. The effect of the β value is of course also related to the time interval. Thorough investigations of the method have been made,[5] and in general it may be stated that best results are obtained if β is taken within the range $\tfrac{1}{6}$ to $\tfrac{1}{4}$ and Δt at about $\tfrac{1}{6}$ to $\tfrac{1}{5}$ of the shortest natural period. Inspection of Eqs. (1.24) reveals that $\beta = \tfrac{1}{6}$ corresponds exactly to the linear-acceleration method previously given.

c. Finite-difference Methods

A large variety of recurrence formulas may be derived by finite-difference techniques.[1] These are generally of the same form as those discussed previously. To illustrate, it must be recalled that the *first backward difference* is

$$\nabla(y^{(s)}) = y^{(s)} - y^{(s-1)}$$

and the *second backward difference* is

$$\nabla^2(y^{(s)}) = \nabla(y^{(s)}) - \nabla(y^{(s-1)})$$
$$= y^{(s)} - 2y^{(s-1)} + y^{(s-2)}$$

and higher orders are given by

$$\nabla^{r+1}(y^{(s)}) = \nabla^r(y^{(s)}) - \nabla^r(y^{(s-1)})$$

Since the second difference divided by the square of the time interval approximates the second derivative at the central time station, we may write

$$(\Delta t)^2 \ddot{y}^{(s-1)} = \nabla^2(y^{(s)}) = y^{(s)} - 2y^{(s-1)} + y^{(s-2)}$$

or $\qquad y^{(s)} = 2y^{(s-1)} - y^{(s-2)} + \ddot{y}^{(s-1)}(\Delta t)^2$

or $\qquad y^{(s+1)} = 2y^{(s)} - y^{(s-1)} + \ddot{y}^{(s)}(\Delta t)^2$

which is exactly the same as the recurrence formula (Eq. 1.5) for the constant-velocity method.

As examples of formulas giving greater precision, there are those embodied in the Adams method,

$$y^{(s+1)} = y^{(s)} + \Delta t(1 + \tfrac{1}{2}\nabla + \tfrac{5}{12}\nabla^2 + \tfrac{3}{8}\nabla^3 + \cdots)\dot{y}^{(s)} \quad (1.25a)$$

or the modified Adams method,

$$y^{(s+1)} = y^{(s)} + \Delta t(1 - \tfrac{1}{2}\nabla - \tfrac{1}{12}\nabla^2 - \tfrac{1}{24}\nabla^3 - \cdots)\dot{y}^{(s+1)} \quad (1.25b)$$

Equation (1.25a) is of the *open type* (since no knowledge is required at $s+1$), and Eq. (1.25b) is of the *closed type*. As many different terms as desired may be retained in these equations, but since the successive terms diminish in magnitude, only the first two or three are necessary in practice. For example, if Eq. (1.25b) is truncated after the first difference, we obtain

$$\begin{aligned} y^{(s+1)} &= y^{(s)} + \Delta t(1 - \tfrac{1}{2}\nabla)\dot{y}^{(s+1)} \\ &= y^{(s)} + \frac{\Delta t}{2}(\dot{y}^{(s+1)} + \dot{y}^{(s)}) \end{aligned} \quad (1.26)$$

We may also increase the order of the derivatives in the last equation to obtain

$$\dot{y}^{(s+1)} = \dot{y}^{(s)} + \frac{\Delta t}{2}(\ddot{y}^{(s+1)} + \ddot{y}^{(s)}) \quad (1.27)$$

Equations (1.26) and (1.27) together provide a possible method of numerical integration. Equation (1.27) is the same as the velocity equation (1.24) of the Newmark β method. Furthermore, if Eq. (1.27) is substituted into (1.26),

$$y^{(s+1)} = y^{(s)} + \dot{y}^{(s)}\Delta t + \frac{(\Delta t)^2}{4}(\ddot{y}^{(s+1)} + \ddot{y}^{(s)}) \quad (1.28)$$

which is the Newmark method with $\beta = \tfrac{1}{4}$.

Other finite-difference equations can be written which do not involve velocity. Examples are those sometimes referred to as Stormer's method:

$$y^{(s+1)} = 2y^{(s)} - y^{(s-1)} + (\Delta t)^2(1 + \tfrac{1}{12}\nabla^2 + \tfrac{1}{12}\nabla^3 + \cdots)\ddot{y}^{(s)}$$

and

$$y^{(s+1)} = 2y^{(s)} - y^{(s-1)} + (\Delta t)^2(1 - \nabla + \tfrac{1}{12}\nabla^2 - \tfrac{1}{240}\nabla^4 + \cdots)\ddot{y}^{(s+1)}$$

Note that, if only the first term in the parentheses of the first equation is retained, the result is identical with that obtained by the constant-velocity method [Eq. (1.5)].

Problems

1.1 Compute by numerical integration the response of a one-degree system to the load-time function shown in Fig. 1.15. $W = 200$ lb, and $k = 300$ lb/in. Use the

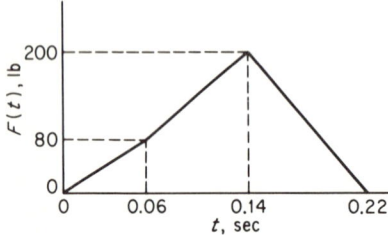

FIGURE 1.15 Problem 1.1. Load-time function.

lumped-impulse, or constant-velocity, procedure and a time interval of 0.02 sec. Plot deflection versus time up to maximum response.
Answer
$y_{max} = 0.972$ in.

1.2 *a.* Compute by numerical analysis ($\Delta t = 0.1$ sec) the horizontal deflection of the frame in Fig. 1.6 due to the load-time function shown in Fig. 1.16. Plot the result up to maximum response.

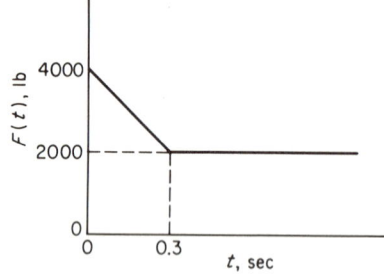

FIGURE 1.16 Problems 1.2 and 1.4. Load-time function.

Answer
$y_{max} = 0.148$ ft
b. Compute the maximum column bending stress.
Answer
$\sigma_{max} = 22.2$ ksi

1.3 A one-degree system having a natural period of 1 sec is subjected to the loading shown in Fig. 1.17. The peak force (5000 lb) if applied statically would cause a deflection of 2 in. Write the equation for $\ddot{y}$ (in terms of t and y only) to be used in a numerical analysis up to $t = 0.50$ sec.

1.4 A two-degree system (Fig. 1.7) is defined by the parameters

$$M_1 = 2.0 \text{ lb-sec}^2/\text{in.} \quad k_1 = 3000 \text{ lb/in.}$$
$$M_2 = 0.5 \text{ lb-sec}^2/\text{in.} \quad k_2 = 1500 \text{ lb/in.}$$

Determine the displacement of each mass for the load-time function shown in Fig. 1.16, with the time 0.3 sec changed to 0.05 sec, $F_1(t)$ being plus and $F_2(t)$ being minus the force values given. Use the lumped-impulse procedure and a time interval of 0.01 sec. Plot y_1 and y_2 up to $t = 0.1$ sec.

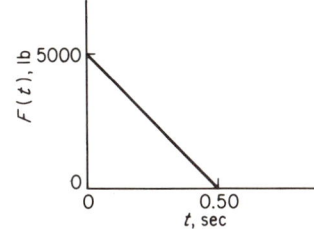

FIGURE 1.17 Problems 1.3 and 1.5. Load-time function.

1.5 Plot the horizontal displacement of both floors of the rigid frame shown in Fig. 1.9 due to $F_1(t)$ as shown in Fig. 1.17 and $F_2(t)$ with the same duration but a peak value of 1000 lb. Use a time interval of 0.08 sec.
Answer
$(y_1)_{max} = 0.060$ ft
$(y_2)_{max} = 0.113$ ft

1.6 Repeat Prob. 1.1, including 8 percent of critical damping.

1.7 Make an elasto-plastic analysis of the system and loading given in Prob. 1.1, except that in this case the maximum plastic spring resistance is 140 lb. Plot the results up to maximum displacement. Note that the two analyses are identical up to the elastic limit.
Answer
$y_{max} = 1.612$ in.

1.8 Make an elasto-plastic analysis of the frame given in Prob. 1.2, except that the magnitude of load is now twice that shown in Fig. 1.16. The ultimate bending capacity of the columns may be taken to be 43.4 ft-kips.
Answer
$y_{max} = 0.315$ ft

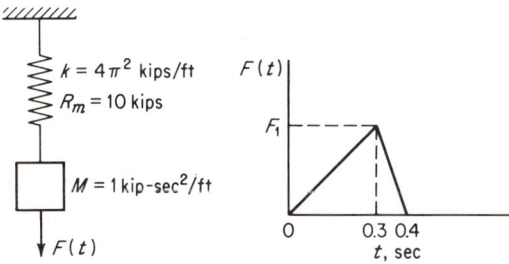

FIGURE 1.18 Problem 1.9. Dynamic system and load function.

1.9 *a.* Using a numerical analysis, compute for the one-degree system and loading shown in Fig. 1.18 the value of F_1 which would cause the spring force to reach R_m but remain elastic. Use $\Delta t = 0.05$ sec.
Answer
$F_1 = 9.1$ kips.
b. If F_1 were greater and the spring went plastic, what would be the amplitude of the residual elastic vibration?
Answer
0.254 ft

2

Rigorous Analysis of One-degree Systems

2.1 Introduction

In Chap. 1 simple dynamic systems were analyzed, using numerical methods. Rigorous, or closed, solutions will be discussed in this chapter. In other words, the solution will be obtained in the form of an equation giving the displacement as a function of time. Such solutions are obviously preferable, but, as mentioned previously, are possible only for simple systems subjected to mathematically simple time variations of load. Unfortunately, such situations do not usually occur in practice. However, a study of rigorous solutions for one-degree systems subjected to typical load-time functions is worthwhile for the following reasons: (1) it enables the student to identify certain types of response with certain types of load-time functions and to isolate the effects of the various parameters involved; and (2) many practical cases may be idealized into one of the simple forms to be discussed, thus making possible an approximate solution of the actual case. With regard to the first reason given, examples of the type of phenomena which should be understood are the effect of a suddenly applied load as compared with one which builds up to its maximum value in a finite time, the effect of a decaying load as compared with one which remains constant with time, the importance of the duration of the applied load relative to the natural period of the system, and the general effect of damping. An understanding of these effects enables the analyst to idealize an actual system as mentioned in

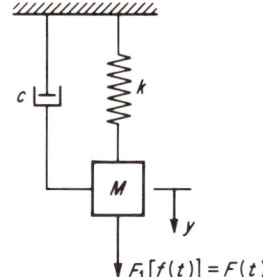

FIGURE 2.1 One-degree damped system.

reason 2 given above. Since such approximations are frequently made, it is convenient to have nondimensional charts giving the response of one-degree systems to certain standardized loads. Such charts are discussed and presented in this chapter.

All attention in this chapter will be focused on the simple system shown in Fig. 2.1, for which the equation of motion as it has been derived in Chap. 1 is

$$M\ddot{y} + ky + c\dot{y} = F_1[f(t)] \tag{2.1}$$

where F_1 is a constant-force value which may be arbitrarily chosen, and $f(t)$ is a nondimensional time function. The right side of Eq. (2.1) is thus the load-time variation. The purpose of this chapter is to develop closed solutions of the differential equation for various load-time functions. The treatment is restricted to linear elastic systems, except in the case of Secs. 2.7 and 2.8, where bilinear elasto-plastic systems are considered.

2.2 Undamped Systems
a. Free Vibration

Consider, first, the elementary case where F equals zero and there is no damping; that is, $c = 0$. Motion will occur only if the system is given an initial disturbance, which may take the form of an initial displacement y_o (imagine that the mass is displaced and then released at $t = 0$) or an initial velocity (which might be produced by an impulse or impact) or a combination of the two. The resulting motion, unaffected by any external force, is called *free vibration*. The equation of motion for this situation is simply

$$\ddot{y} + \frac{k}{M}y = 0 \tag{2.2}$$

and the solution * of this equation is

$$y = C_1 \sin \sqrt{\frac{k}{M}} t + C_2 \cos \sqrt{\frac{k}{M}} t$$

or, letting $\sqrt{k/M} = \omega$,

$$y = C_1 \sin \omega t + C_2 \cos \omega t \tag{2.3}$$

The constants C_1 and C_2 depend upon the initial conditions of the problem, and may be evaluated by substitution into the solution, with time taken as zero. For example, if the initial velocity is $\dot{y}_o$ and the initial displacement is y_o, the constants are determined as follows. Equation (2.3) at $t = 0$ may be written as

$$y_o = C_1 \sin \omega(0) + C_2 \cos \omega(0)$$
Therefore $\quad C_2 = y_o$

Differentiating Eq. (2.3) and substituting $t = 0$, we obtain

$$\dot{y}_o = C_1 \omega \cos \omega(0) - C_2 \omega \sin \omega(0)$$
Therefore $\quad C_1 = \dfrac{\dot{y}_o}{\omega}$

Substituting these expressions for the constants into Eq. (2.3), we obtain the solution for zero external load as

$$y = \frac{\dot{y}_o}{\omega} \sin \omega t + y_o \cos \omega t \tag{2.4}$$

The displacement given by Eq. (2.4) is plotted in Fig. 2.2 for the cases of initial velocity and initial displacement taken separately.

b. Natural Period and Frequency

The free vibration discussed above is said to be harmonic; that is, y varies sinusoidally with t. The motion is completely repetitive if there is no damping in the system. The harmonic motion is defined by an amplitude (y_o or $\dot{y}_o/\omega$ in Fig. 2.2) and a natural period, which is the time required for the motion to go through one complete cycle. The initial conditions affect only the amplitude of the vibration.

The parameter ω (Sec. 2.2a) is called the natural circular frequency.

$$Natural\ circular\ frequency = \omega = \sqrt{\frac{k}{M}} \quad \text{rad/sec} \tag{2.5}$$

---

* The solution of differential equations will not be discussed herein since it is assumed that the student is familiar with such procedures. The solutions given may be easily verified by substitution into the original differential equation.

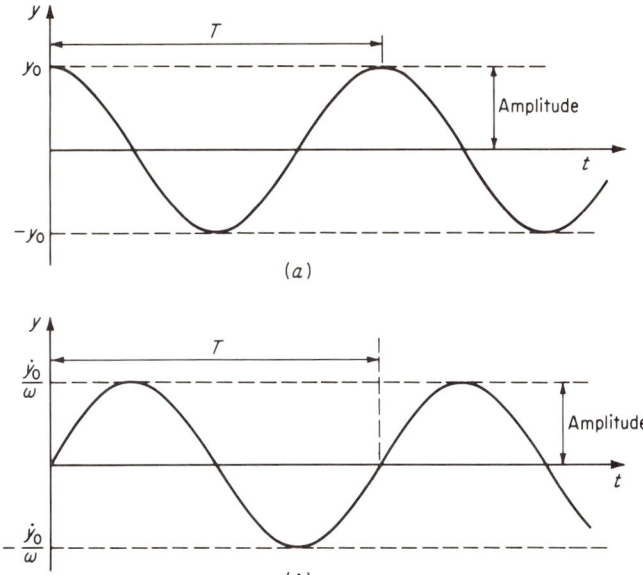

FIGURE 2.2 Free vibration of one-degree undamped system. (a) Initial displacement; (b) initial velocity.

Since one complete cycle occurs for each angular increment $\omega t = 2\pi$, the natural period of the system is given by

$$\text{Natural period} = T = \frac{2\pi}{\omega} = 2\pi \sqrt{\frac{M}{k}} \quad \text{sec} \quad (2.6)$$

Note that the natural period and frequency are characteristics of the system and depend only upon the mass and the spring constant. The natural frequency (not circular frequency) is defined as the inverse of the natural period, or the number of cycles per unit of time.

$$\text{Natural frequency} = f = \frac{1}{T} = \frac{1}{2\pi} \sqrt{\frac{k}{M}} \quad \text{cps} \quad (2.7)$$

c. Forced Vibration

Consider now a case in which the motion is the result of an applied force $F(t)$. It will be assumed that the system begins at rest; i.e., both the velocity and displacement are zero at $t = 0$. Obviously, this is not a necessary condition, and solutions could be obtained for the combination of the two effects.

To begin with a simple case, assume that $F(t)$ has a constant magnitude F_1 which is suddenly applied and remains constant indefinitely. For

38 Introduction to Structural Dynamics

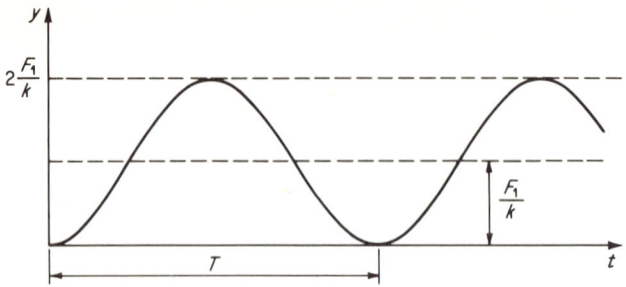

FIGURE 2.3 Response of undamped one-degree system to suddenly applied constant force.

this situation Eq. (2.1) becomes

$$\ddot{y} + \frac{k}{M} y = \frac{F_1}{M} \qquad (2.8)$$

The solution of this equation is

$$y = C_1 \sin \omega t + C_2 \cos \omega t + \frac{F_1}{k} \qquad (2.9)$$

where once again the constants C_1 and C_2 are determined by the initial conditions. Substituting into Eq. (2.9), $y = 0$ and $t = 0$, one obtains

$$0 = C_1 \sin (0) + C_2 \cos (0) + \frac{F_1}{k}$$

Therefore $\qquad C_2 = -\dfrac{F_1}{k}$

In order to obtain C_1, Eq. (2.9) is differentiated, and into the resulting equation is substituted $\dot{y} = 0$ and $t = 0$ as follows:

$$\dot{y} = C_1 \omega \cos (0) - C_2 \omega \sin (0) = 0$$

Therefore $\qquad C_1 = 0$

If these values of C_1 and C_2 are substituted into Eq. (2.9), the final solution is obtained:

$$y = \frac{F_1}{k} (1 - \cos \omega t) \qquad (2.10)$$

This solution for a suddenly applied constant load is plotted in Fig. 2.3.

It will be observed that the solution just obtained is very similar to the previous solution for free vibration (Fig. 2.2). The only difference is that the axis of the vibration has been shifted by an amount equal

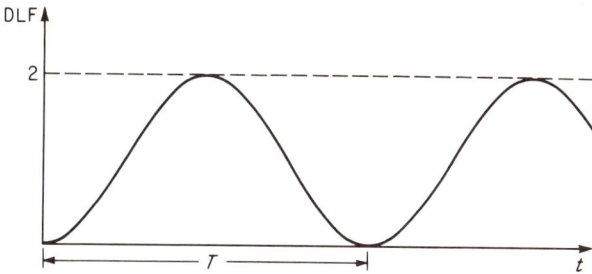

FIGURE 2.4 Dynamic load factor (DLF) for an undamped one-degree system subjected to a suddenly applied constant force.

to F_1/k. It will also be noticed that the maximum displacement $2F_1/k$ is exactly twice the displacement which would occur if the load F_1 were applied statically. Thus we reach an elementary but very important conclusion: If a constant force is suddenly applied to a linear elastic system, the resulting displacement is exactly twice that for the same force applied statically. The same observation is true regarding the dynamic force in the spring, which is proportional to the displacement. Furthermore, since the spring-mass system represents an actual structure, the same statement may be made regarding both dynamic deflections and stresses in that structure.

d. Dynamic Load Factor

It is now convenient to introduce the concept of the dynamic load factor (DLF). This factor is defined as the ratio of the dynamic deflection at any time to the deflection which would have resulted from the static application of the load F_1, which is used in specifying the load-time variation. Since deflections, spring forces, and stresses in the structure are all proportional, the dynamic load factor may be applied to any of these in order to obtain the ratio of dynamic to static effects.

In the preceding example, which involved a suddenly applied constant load, the static deflection is F_1/k. Thus the dynamic load factor is given by

$$\text{DLF} = \frac{y}{y_{st}} = \frac{y}{F_1/k} = \frac{ky}{F_1} \qquad (2.11)$$

Substituting Eq. (2.10) for y,

$$\text{DLF} = 1 - \cos \omega t \qquad (2.12)$$

Thus the dynamic load factor for this case is as shown in Fig. 2.4. It is apparent that the dynamic load factor is nondimensional and inde-

40   Introduction to Structural Dynamics

pendent of the magnitude of load. It is because of this fact that its use is convenient.

In many structural problems only the maximum value of the DLF is of interest. In the case just considered, this maximum is 2, which immediately indicates that all maximum displacements, forces, and stresses due to the dynamic load are twice the values that would be obtained from a static analysis for the load F_1.

In cases where the applied load is not constant, F_1, upon which the DLF is based, is some arbitrarily selected value of the load. This load value is usually, but not necessarily, taken as the maximum which occurs at any time during the period of interest.

2.3 Various Forcing Functions (Undamped Systems)

a. Generalized Linear-systems Theory

Before discussing responses for various load-time functions, it is convenient to obtain a general solution applicable to any such function. First, however, let us recall the concept of *impulse*, which is defined as the area under the load-time curve.

Suppose that a system at rest is subjected to a constant force F with a duration t_d. The mass of the system, having an initial acceleration $\ddot{y} = F/M$, will begin to move. If t_d is a very short time relative to the natural period, little spring resistance will be developed during the time t_d. If such resistance is negligible compared with F, the acceleration can be considered constant and the net effect will be a velocity imparted to the mass. The value of this velocity at time t_d will be

$$\dot{y} = \ddot{y} t_d = \frac{F}{M} t_d = \frac{i}{M} \qquad (2.13)$$

where i is the applied impulse equal to the area under the load-time curve. If the assumption stated above and implied by Eq. (2.13) is valid, i is said to be a *pure impulse*. To give a quantitative feeling for this concept, it may be said that the error in Eq. (2.13) is negligible if t_d is smaller than about one-tenth of the natural period. Obviously, in such cases, the actual shape of the load-time function during the time t_d is of no importance.

Turning now to a general load function such as shown in Fig. 2.5, consider the area in the element of time $d\tau$ to be a pure impulse. This causes an increment of velocity at τ equal to $F_1 f(\tau) \, d\tau / M$, which may be considered as an initial velocity imparted to a system at rest. The displacement at a later time due to this single element of impulse is given by Eq. (2.4) if $\dot{y}_o$ is the initial velocity just defined and if y_o is taken as zero (since there is no initial displacement corresponding to the effect

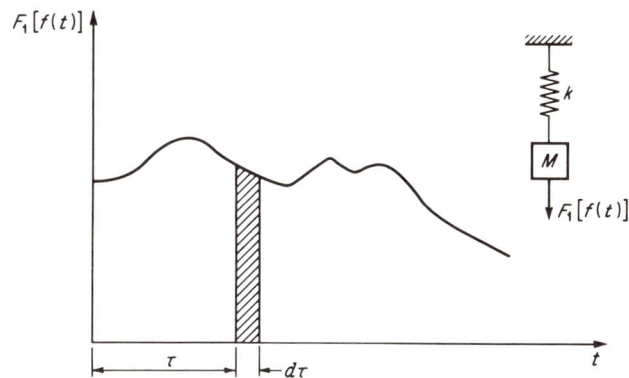

FIGURE 2.5 Linear-system theory—impulse element.

of this impulse). Thus we obtain

$$\frac{F_1 f(\tau) \, d\tau}{M\omega} \sin \omega(t - \tau)$$

which is the displacement at time t due to the load applied during $d\tau$. Since the system is linear, superposition may be employed and the total displacement at t is the sum of the effects of all elements of impulse between zero and t. Thus

$$y = \int_0^t \frac{F_1 f(\tau)}{M\omega} \sin \omega(t - \tau) \, d\tau \qquad (2.14a)$$

Since the static deflection (due to F_1) may be represented by

$$y_{st} = \frac{F_1}{k} = \frac{F_1}{\omega^2 M}$$

Eq. (2.14a) may also be written as

$$y = y_{st}\omega \int_0^t f(\tau) \sin \omega(t - \tau) \, d\tau \qquad (2.14b)$$

To make the equation even more general, the effects of initial displacement and velocity may be included by superimposing Eqs. (2.4) and (2.14b):

$$y = y_o \cos \omega t + \frac{\dot{y}_o}{\omega} \sin \omega t + y_{st}\omega \int_0^t f(\tau) \sin \omega(t - \tau) \, d\tau \qquad (2.15)$$

where y_o and $\dot{y}_o$ are the displacement and velocity (if any) at $t = 0$.

Equation (2.15) is a perfectly general expression for the response of an undamped, linearly elastic one-degree system subjected to any load function and/or initial conditions. A closed solution is of course possible only if the integral can be evaluated. Applications of Eq. (2.15) are illustrated below.

b. Rectangular-pulse Load

Consider first the case of a suddenly applied constant load with a limited duration t_d as shown in Fig. 2.6a. The system starts at rest, and there is no damping. Up to time t_d Eq. (2.10) applies, and at that time we have

$$y_{t_d} = \frac{F_1}{k}(1 - \cos \omega t_d)$$

$$\dot{y}_{t_d} = \frac{F_1}{k}\omega \sin \omega t_d$$

For the response after t_d we may apply Eq. (2.15), taking as the initial conditions the velocity and displacement at t_d. Replacing t by $t - t_d$ and y_o and $\dot{y}_o$ by y_{t_d} and $\dot{y}_{t_d}$ and noting that $f(\tau) = 0$, we obtain

$$\begin{aligned} y &= \frac{F_1}{k}(1 - \cos \omega t_d)\cos \omega(t - t_d) + \frac{F_1}{k}\sin \omega t_d \sin \omega(t - t_d) \\ &= \frac{F_1}{k}[\cos \omega(t - t_d) - \cos \omega t] \end{aligned} \quad (2.16a)$$

Since F_1/k is the static deflection and the dynamic load factor is given by y/y_{st}, we may write

$$\begin{aligned} \text{DLF} &= 1 - \cos \omega t = 1 - \cos 2\pi \frac{t}{T} & t \leqslant t_d \\ \text{DLF} &= \cos \omega(t - t_d) - \cos \omega t & t \geqslant t_d \\ &= \cos 2\pi \left(\frac{t}{T} - \frac{t_d}{T}\right) - \cos 2\pi \frac{t}{T} \end{aligned} \quad (2.16b)$$

It is often convenient to nondimensionalize the time parameter as indicated in Eqs. (2.16b), where T is the natural period. This also serves to emphasize the fact that the ratio of duration to natural period, rather than the actual value of either quantity, is the important parameter.

Two typical responses are plotted in Fig. 2.6a, and it is easy to visualize the response for an intermediate value of t_d/T. The maximum dynamic load factor obtained by maximizing Eqs. (2.16b) is plotted in Fig. 2.7a. Obviously, as the duration approaches zero, the maximum deflection, or stress, also diminishes to zero. A less trivial observation is that, if $(t_d/T) > 0.5$, the maximum response of the system is the same as if the load duration had been infinite.

Charts such as Fig. 2.7 (and also Figs. 2.8 and 2.9) are extremely useful for design purposes, as illustrated in Chap. 5. For a given load function one need know only the natural period in order to read from the chart the maximum DLF and hence the ratio of maximum dynamic to static stress. Also given in Fig. 2.7b (and Figs. 2.8b and 2.9b) is the time at which the maximum stress, or deflection, occurs. This time is often a matter of considerable importance. In the derivation of these charts no damping has been included because it would have no significant effect. The maximum dynamic load factor usually corresponds to the first peak of response, and the amount of damping normally encountered in structures is not sufficient to decrease appreciably this value.

c. Triangular Load Pulses

Consider next a system initially at rest and subjected to a force F which has an initial, suddenly applied value of F_1 and decreases linearly to zero at time t_d (Fig. 2.6b). The response may be computed by Eq. (2.15) in two stages. For the first stage,

$\tau \leqslant t_d$:

$$y_o = 0 \qquad \dot{y}_o = 0 \qquad f(\tau) = 1 - \frac{\tau}{t_d}$$

Substituting these values in Eq. (2.15) and integrating,

$$y = \frac{F_1}{k}(1 - \cos \omega t) + \frac{F_1}{kt_d}\left(\frac{\sin \omega t}{\omega} - t\right) \qquad (2.17a)$$

or
$$\mathrm{DLF} = 1 - \cos \omega t + \frac{\sin \omega t}{\omega t_d} - \frac{t}{t_d} \qquad (2.17b)$$

which defines the response before t_d. For the second stage, from Eq. (2.17a),

$\tau \geqslant t_d$:
$$y_o = \frac{F_1}{k}\left(\frac{\sin \omega t_d}{\omega t_d} - \cos \omega t_d\right)$$

$$\dot{y}_o = \frac{F_1}{k}\left(\omega \sin \omega t_d + \frac{\cos \omega t_d}{t_d} - \frac{1}{t_d}\right)$$

$$f(\tau) = 0$$

Substituting these values in Eq. (2.15), replacing t by $t - t_d$, and simplifying,

$$y = \frac{F_1}{k\omega t_d}[\sin \omega t_d - \sin \omega(t - t_d)] - \frac{F_1}{k}\cos \omega t \qquad (2.18a)$$

or
$$\mathrm{DLF} = \frac{1}{\omega t_d}[\sin \omega t_d - \sin \omega(t - t_d)] - \cos \omega t \qquad (2.18b)$$

which gives the response after t_d.

44 *Introduction to Structural Dynamics*

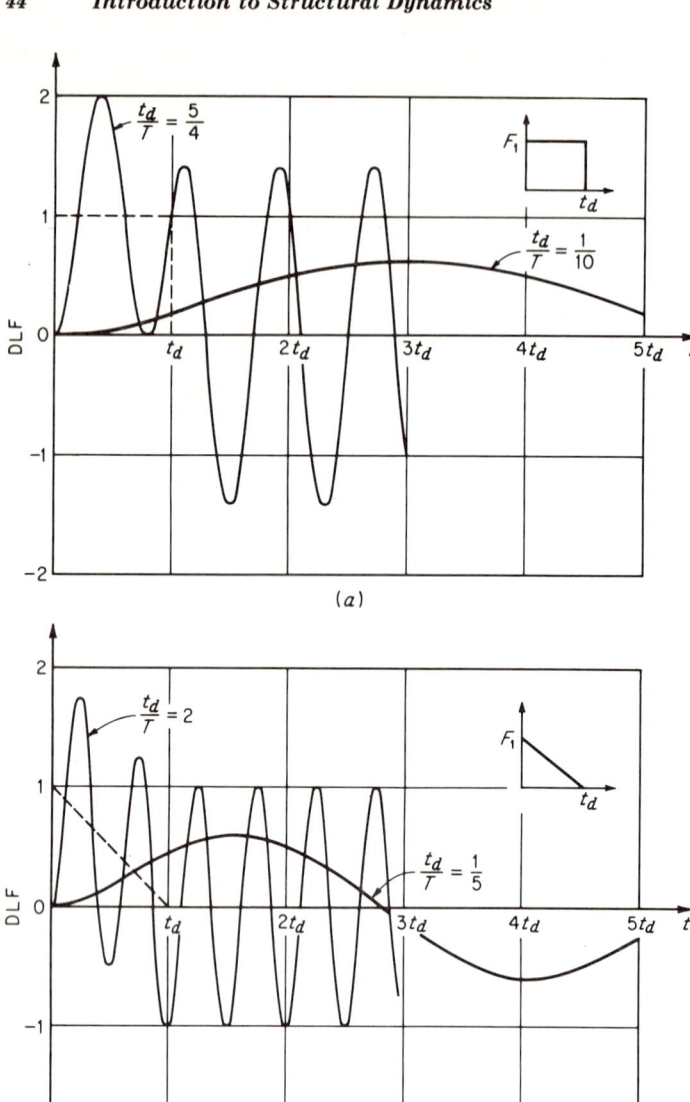

FIGURE 2.6 Typical responses of one-degree elastic systems. (a) Rectangular pulse; (b) suddenly applied triangular pulse; (c) symmetrical triangular pulse; (d) constant force with finite rise time.

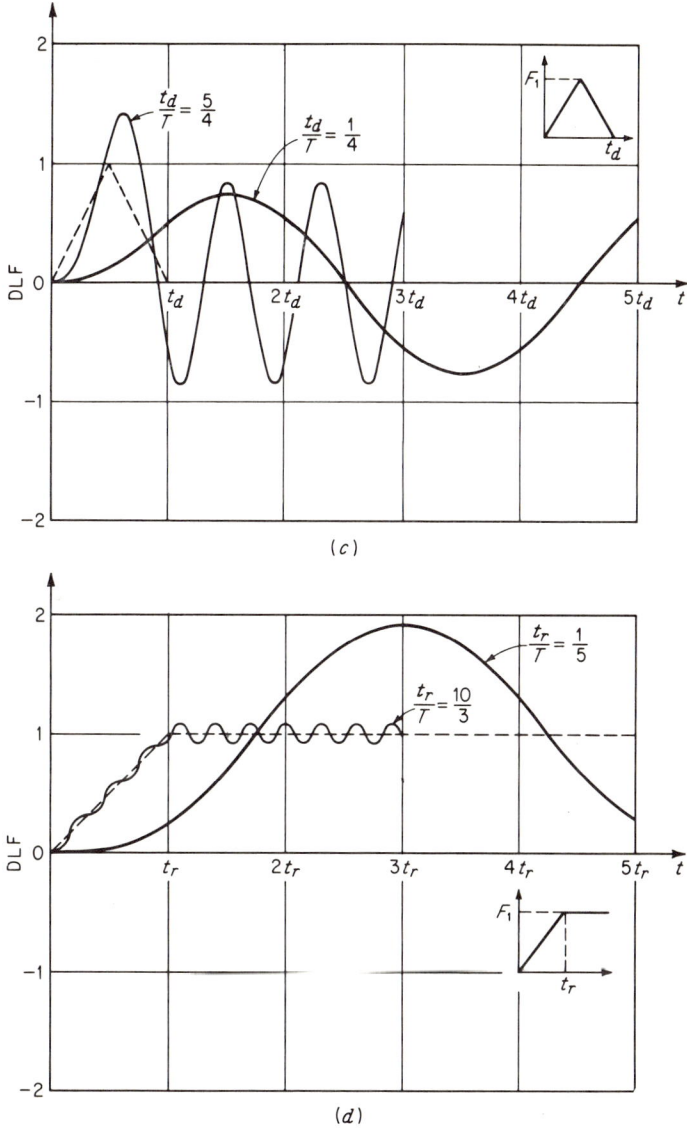

FIGURE 2.6 (*Continued*)

Typical responses for this type of forcing function are shown in Fig. 2.6b. Maximum dynamic load factors and the time of that maximum response are given in Fig. 2.7. As would be expected, $(\text{DLF})_{\max} \to 2$ as t_d/T becomes large, or in other words, the effect of the decay in force is negligible in the time required for the response to reach the first peak.

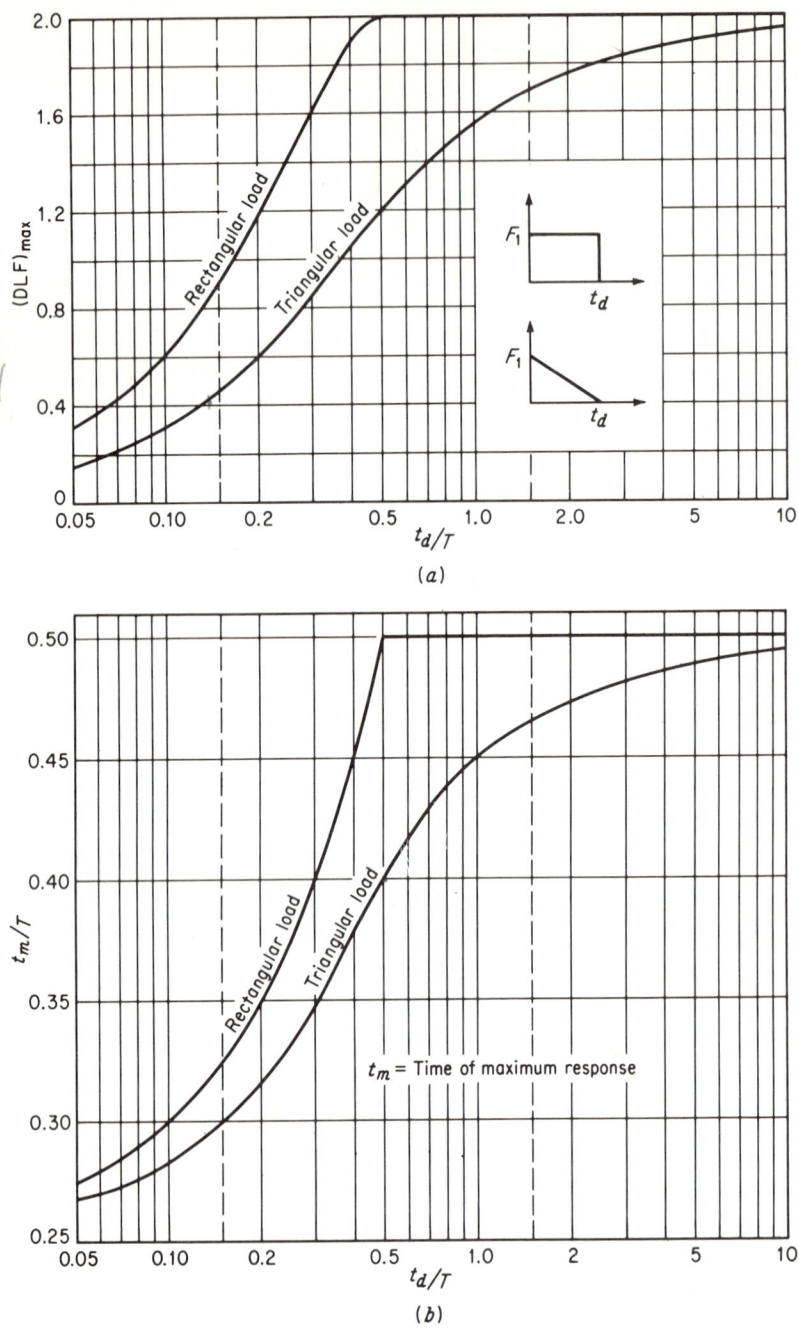

FIGURE 2.7 Maximum response of one-degree elastic systems (undamped) subjected to rectangular and triangular load pulses having zero rise time. (*U.S. Army Corps of Engineers.*[10])

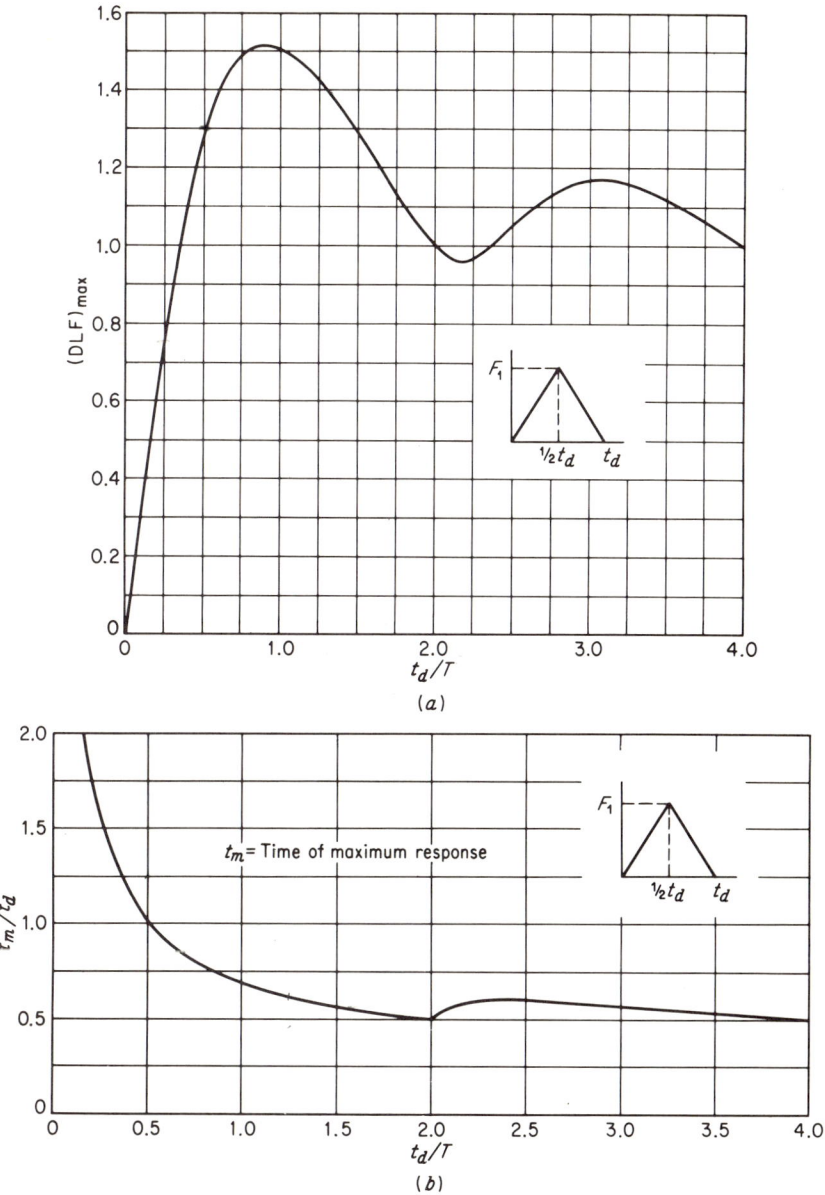

FIGURE 2.8 Maximum response of one-degree elastic systems (undamped) subjected to equilateral triangular load pulse.

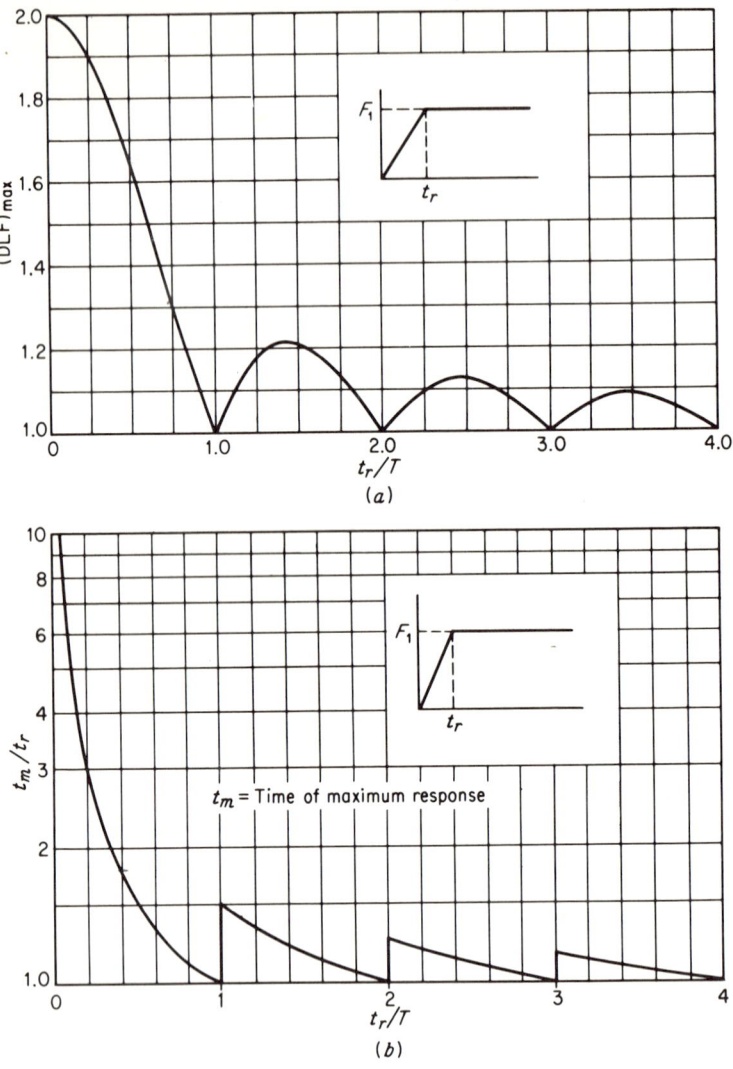

FIGURE 2.9 Maximum response of one-degree elastic systems (undamped) subjected to constant force with finite rise time. (*U.S. Army Corps of Engineers.*[10])

Consider now a symmetrical triangular pulse which starts at zero and reaches a maximum at one-half the total duration (Fig. 2.6c). In this case, Eq. (2.15) must be applied in three stages, taking for the initial conditions of each stage the final velocity and displacement of the preceding stage. The time functions to be used are as follows:

$$f(\tau) = 2\frac{\tau}{t_d} \qquad 0 \leqslant \tau \leqslant \tfrac{1}{2}t_d$$

$$f(\tau) = 2\left(1 - \frac{\tau}{t_d}\right) \qquad \tfrac{1}{2}t_d \leqslant \tau \leqslant t_d$$

$$f(\tau) = 0 \qquad t_d \leqslant \tau$$

Following the same procedure as in previous cases, we obtain

$$\text{DLF} = \frac{2}{t_d}\left(t - \frac{\sin \omega t}{\omega}\right) \qquad\qquad 0 \leqslant t \leqslant \tfrac{1}{2}t_d$$

$$\text{DLF} = \frac{2}{t_d}\left\{t_d - t + \frac{1}{\omega}\left[2 \sin \omega \left(t - \frac{t_d}{2}\right) - \sin \omega t\right]\right\} \qquad \tfrac{1}{2}t_d \leqslant t \leqslant t_d$$

(2.19)

$$\text{DLF} = \frac{2}{\omega t_d}\left[2 \sin \omega \left(t - \frac{t_d}{2}\right) - \sin \omega t - \sin \omega (t - t_d)\right] \qquad t_d \leqslant t$$

Typical responses are plotted in Fig. 2.6c, and maximum response as a function of t_d/T is given by Fig. 2.8. It may be observed that, for this load-time function, the maximum dynamic effect occurs when the duration of loading is approximately equal to the natural period of the system.

d. Constant Force with Finite Rise Time

Since, in reality, a force can never be applied instantaneously, it is of interest to investigate a loading which has a finite rise time but remains constant thereafter as shown in Fig. 2.6d. In this case the load-time functions are

$$f(\tau) = \frac{\tau}{t_r} \qquad \tau \leqslant t_r$$
$$f(\tau) = 1 \qquad \tau \geqslant t_r$$

where t_r is the rise time. Proceeding as in previous cases,

$$\text{DLF} = \frac{1}{t_r}\left(t - \frac{\sin \omega t}{\omega}\right) \qquad\qquad t \leqslant t_r$$
(2.20)
$$\text{DLF} = 1 + \frac{1}{\omega t_r}[\sin \omega (t - t_r) - \sin \omega t] \qquad t \geqslant t_r$$

Typical responses are shown in Fig. 2.6d, from which it may be deduced that, if t_r is large relative to T, the response simply follows the applied load and the dynamic effect is negligible. In Fig. 2.9, (DLF)$_{\max}$ and the time of maximum response are plotted. Here the effect of rise time is apparent. If t_r is less than about one-quarter of the natural period, the effect is essentially the same as for a suddenly applied load. This observation is of significance in practical design since it indicates that smaller rise times may be ignored. A peculiarity of this type of load

50   Introduction to Structural Dynamics

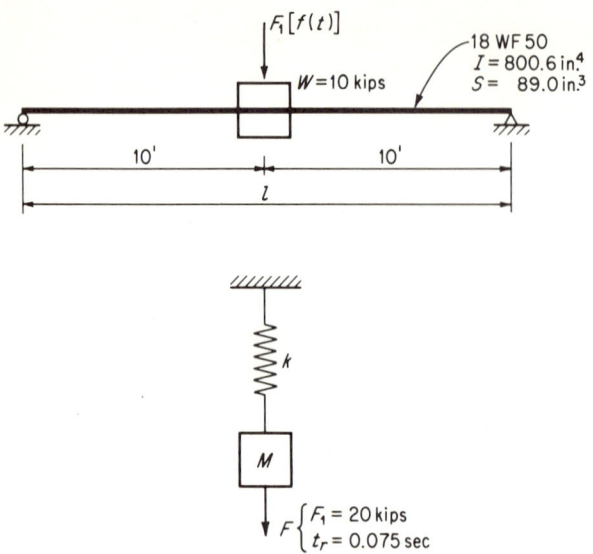

FIGURE 2.10 Example. Response of a beam having negligible mass.

pulse is the fact that, if the rise time is a whole multiple of the natural period, the response is the same as if F_1 had been applied statically.

e. Example

To illustrate the use of the charts just discussed, consider the case of a steel beam supporting a concentrated mass (Fig. 2.10) and subjected to a dynamic load of the type shown in Fig. 2.6d with a rise time t_r of 0.075 sec and a maximum value F_1 of 20 kips. If the weight of the beam is considered negligible, the system has only one degree of freedom. We wish to determine the maximum dynamic bending stress.

First the natural period of the one-degree system must be computed. The spring constant k is defined as the force at midspan necessary to cause a unit deflection at the same point. Thus

$$k = \frac{48EI}{l^3} = \frac{48 \times 30 \times 10^3 \times 800.6}{(20 \times 12)^3} = 83.4 \text{ kips/in.}$$

$$T = 2\pi \sqrt{\frac{M}{k}} = 2\pi \sqrt{\frac{10}{386 \times 83.4}} = 0.111 \text{ sec}$$

Entering Fig. 2.9 with $t_r/T = 0.075/0.111 = 0.68$, we obtain

$$(\text{DLF})_{\max} = 1.38$$
$$\frac{t_m}{t_r} = 1.23$$

The maximum dynamic stress equals the static stress due to the 20-kip force multiplied by $(DLF)_{max}$.

$$\text{max stress} = \sigma_{max} = (DLF)_{max} \times \sigma_{st}$$
$$= (DLF)_{max} \times \frac{F_1 l/4}{S} = 1.38 \times \frac{20 \times 240/4}{89.0} = 18.6 \text{ ksi}$$

The time at which this maximum stress occurs is

$$t_m = \left(\frac{t_m}{t_r}\right) t_r = 1.23 \times 0.075 = 0.092 \text{ sec}$$

2.4 Damped Systems

As mentioned previously, if one is interested in a continuing state of vibration rather than merely the first peak of response, the effect of damping must be included. For the one-degree system shown in Fig. 2.1, with viscous damping, the equation of motion as derived in Sec. 1.4 is

$$M\ddot{y} + ky + c\dot{y} = F(t) \tag{2.21}$$

Damping, indicated in Fig. 2.1 by the conventional dashpot, produces a force $c\dot{y}$ which opposes the motion and dissipates some of the energy of the system.

a. Free Vibration

We consider first the case in which there is no external force and the system is subjected to an initial disturbance. The solution of Eq. (2.21) with the right side equal to zero is

$$y = e^{-\beta t}(C_1 \sin \omega_d t + C_2 \cos \omega_d t) \tag{2.22}$$

where $\beta = c/2M$, and $\omega_d = \sqrt{\omega^2 - \beta^2}$. β is a measure of the amount of damping present, and ω_d is the natural frequency of the damped system, which is somewhat different from that of the undamped system ω. Equation (2.22) applies only when $\beta < \omega$. The solution for cases in which this condition is not met (overdamping) is of limited importance and will not be considered here.[11]

If the system is subjected to initial displacement and velocity, y_o and $\dot{y}_o$, the constants of Eq. (2.22) are determined by substituting $t = 0$ into that equation and its derivative as follows:

$$y_o = e^0[C_1 \sin(0) + C_2 \cos(0)]$$
Therefore $\qquad C_2 = y_o$

$$\dot{y} = e^{-\beta t}[-(\beta C_1 + \omega_d C_2) \sin \omega_d t + (\omega_d C_1 - \beta C_2) \cos \omega_d t]$$
$$\dot{y}_o = e^0[-(\beta C_1 + \omega_d C_2) \sin(0) + (\omega_d C_1 - \beta C_2) \cos(0)]$$

Therefore
$$C_1 = \frac{\dot{y}_o + \beta y_o}{\omega_d}$$

The total response resulting from a combination of initial displacement and velocity is therefore given by

$$y = e^{-\beta t}\left(\frac{\dot{y}_o + \beta y_o}{\omega_d}\sin \omega_d t + y_o \cos \omega_d t\right) \qquad (2.23)$$

The responses due to each of the two initial conditions taken separately are shown in Fig. 2.11a and b, where it may be observed that the exponential term in Eq. (2.23) with an appropriate multiplier forms an envelope for the decaying harmonic motion.

The condition $\beta = \omega$ creates a case of special interest. As $\beta \to \omega$, the frequency $\omega_d \to 0$, and hence $\cos \omega_d t \to 1$ and $\sin \omega_d t \to \omega_d t$. As a result, Eq. (2.23) reduces to

$$y = e^{-\omega t}[\dot{y}_o t + (1 + \omega t)y_o] \qquad (2.24)$$

From this equation it is apparent that the motion is no longer periodic, or in other words, there is no vibration in the usual sense of the word. As given by Eq. (2.24), the response to an initial displacement (zero initial velocity) is as shown in Fig. 2.11c. Rather than vibrating, the system merely creeps back to the neutral position.

The amount of damping which removes all vibration as described above ($\beta = \omega$) is known as *critical damping*. Although this case is of little importance in itself, the critical coefficient of damping is a convenient reference. For example, observations indicate that typical structures have between 5 and 10 percent of critical damping. Since, for critical damping,

$$\omega = \beta = \frac{c_{cr}}{2M}$$

the critical coefficient is given by

$$c_{cr} = 2M\omega = 2\sqrt{kM} \qquad (2.25)$$

As noted above, damping affects the natural frequency of the system. To illustrate the significance of this effect, consider a system with 10 percent of critical damping, or $\beta = 0.1\omega$. The damped frequency is then

$$\omega_d = \sqrt{\omega^2 - \beta^2} = \sqrt{0.99\omega^2} = 0.995\omega$$

which is only slightly different from the undamped natural frequency. It may be concluded, therefore, that the decrease in natural frequency due to damping may for practical purposes be ignored.

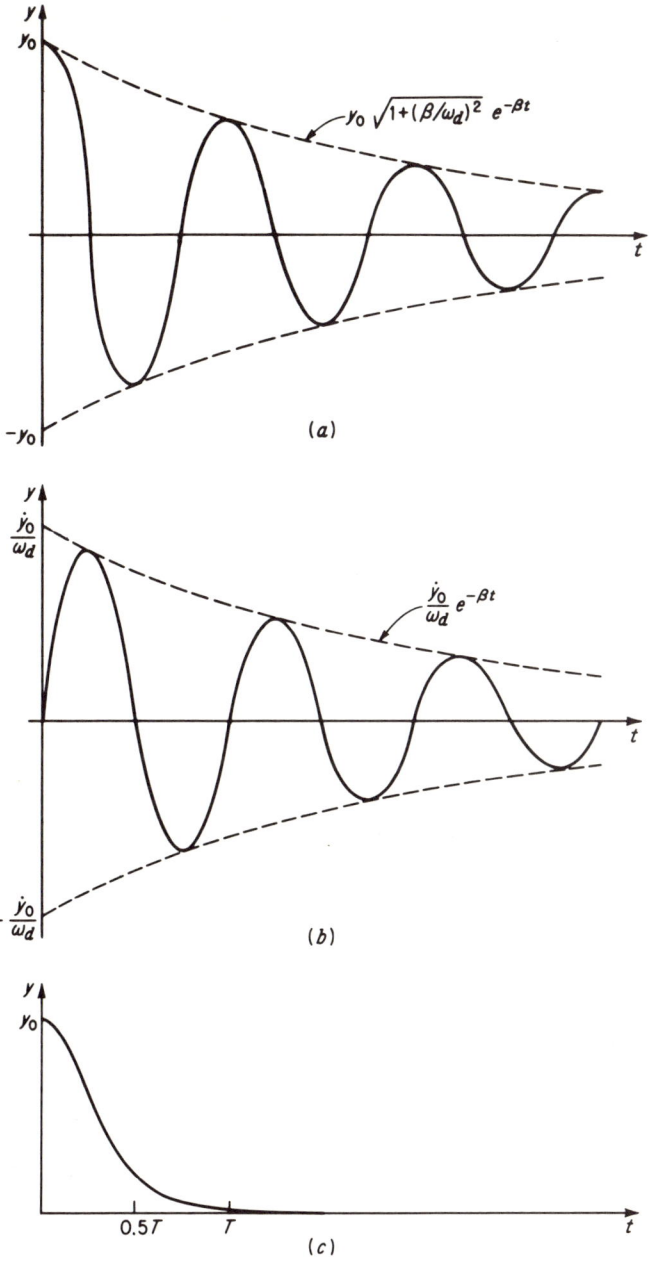

FIGURE 2.11 Free vibration with viscous damping. (a) Initial displacement; (b) initial velocity; (c) initial displacement and critical damping.

54   Introduction to Structural Dynamics

Another concept of importance is that of the *logarithmic decrement*, defined as the difference between the logarithms of two consecutive peaks in the free vibration, or identically, the logarithm of the ratio of two consecutive peaks. An expression for logarithmic decrement may be obtained by taking the logarithm of two values given by Eq. (2.23), the first for a time t and the second for a time $(t + T_d)$, where T_d is the damped natural period. Thus

$$\begin{aligned}
\text{Logarithmic decrement} &= \ln \frac{y(t=t)}{y(t=t+T_d)} \\
&= \ln \frac{e^{-\beta t}}{e^{-\beta(t+T_d)}} \\
&= \ln e^{\beta T_d} = \beta T_d \\
&= \beta \frac{2\pi}{\omega} \qquad \left(T_d \approx T = \frac{2\pi}{\omega}\right)
\end{aligned} \qquad (2.26)$$

Therefore, if a system had 10 percent of critical damping ($\beta = 0.1\omega$), the logarithmic decrement would be 0.2π, which indicates that the ratio of successive peaks would be $e^{0.2\pi}$, or 1.87. Inverting this quantity, it could be said that each and every peak would have a value 0.534 times that of the preceding peak. This is obviously a convenient way to visualize the effect of damping.

b. Forced Vibration

A generalized solution for the forced vibration of a damped system may be obtained in the same manner as was used for an undamped system in Sec. 2.3a. For the damped case, the response due to an element of impulse is given by (Fig. 2.5)

$$\frac{F_1 f(\tau) \, d\tau}{M \omega_d} e^{-\beta(t-\tau)} \sin \omega_d(t-\tau)$$

The total response obtained by summing the effects of all elements of impulse and superimposing the effects of initial conditions is

$$y = e^{-\beta t}\left(\frac{\dot{y}_o + \beta y_o}{\omega_d} \sin \omega_d t + y_o \cos \omega_d t\right)$$
$$+ y_{st} \frac{\omega^2}{\omega_d} \int_0^t f(\tau) e^{-\beta(t-\tau)} \sin \omega_d(t-\tau) \, d\tau \qquad (2.27)$$

This equation is comparable with Eq. (2.15) for the undamped system and is identical when $\beta = 0$.

Consider now a system initially at rest and subjected to a suddenly applied constant force F_1. The response may be determined by direct

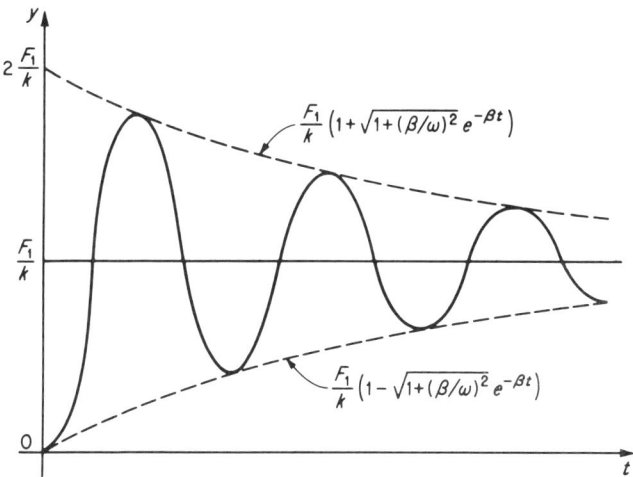

FIGURE 2.12 Response of damped one-degree system to a suddenly applied constant force.

solution of Eq. (2.21) with $F(t) = F_1$ or by Eq. (2.27) with $f(\tau) = 1$ and is given by

$$y = \frac{F_1}{k}\left[1 - e^{-\beta t}\left(\cos \omega t + \frac{\beta}{\omega}\sin \omega t\right)\right] \quad (2.28)$$

where the difference between ω and ω_d has been ignored. The response indicated by Eq. (2.28) is plotted in Fig. 2.12. It is apparent that this response is very similar to that due to an initial displacement as shown in Fig. 2.11a. If the initial displacement y_o had been equal to $-F_1/k$, the only difference would have been a shift in the neutral position by an amount equal to F_1/k.

It was stated in Sec. 2.3 that damping had little effect on the first peak of response. The validity of that statement may now be investigated by further consideration of the case just presented. With little error it may be assumed that the first peak occurs when $\omega t = \pi$, for which Eq. (2.28) gives

$$y_{\max} = \frac{F_1}{k}(1 + e^{-\pi\beta/\omega})$$

Assuming for illustration that $\beta = 0.1\omega$ (10 percent of critical damping), we obtain $y_{\max} = 1.73 F_1/k$. For an undamped system the response will be $2F_1/k$, and hence the reduction due to this amount of damping is 13.5 percent. Since the damping assumed is relatively high, this percentage effect may be considered an upper limit for most typical structures subjected to loads which are fairly rapid in application.

56 *Introduction to Structural Dynamics*

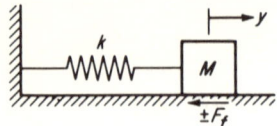

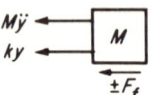

FIGURE 2.13 One-degree system with Coulomb damping.

c. Coulomb Damping

Up to this point we have concentrated our attention on systems with viscous damping, and will continue to do so hereafter, because this type is the most commonly assumed for structural analysis. However, in this particular section a different form, namely, Coulomb, or constant, damping is considered. This would apply to a system such as shown in Fig. 2.13, where the mass slides on a surface such that the resistance to motion is provided by simple friction. The magnitude of this friction force F_f is constant and depends only upon the coefficient of friction and the weight of the body. However, the direction of the force depends upon the velocity of the mass, which it always opposes. Thus, for free vibration, the equation of motion is

$$M\ddot{y} + ky \pm F_f = 0 \tag{2.29}$$

where the positive sign before F_f applies when and only when the velocity is positive. Because of the changing sign, any solution of Eq. (2.29) would apply only during a time interval in which the sign of the velocity remained unchanged.

As an illustration, consider a mass which is given an initial displacement y_o with zero initial velocity. During the first half-cycle of response, the velocity is negative and Eq. (2.29) becomes

$$M\ddot{y} + ky = +F_f$$

Thus the situation under consideration is equivalent to a system with a suddenly applied constant force and an initial displacement. The response is therefore the superposition of these two effects as given by Eqs. (2.4) and (2.10).

$$\begin{aligned} y &= y_o \cos \omega t + \frac{F_f}{k}(1 - \cos \omega t) \\ &= \left(y_o - \frac{F_f}{k}\right) \cos \omega t + \frac{F_f}{k} \end{aligned} \tag{2.30}$$

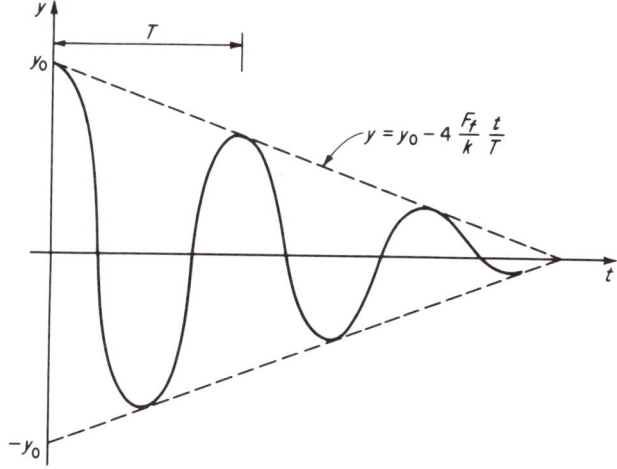

FIGURE 2.14 Free vibration with Coulomb damping.

The first negative peak given by Eq. (2.30) is

$$y\left(t = \frac{\pi}{\omega}\right) = -\left(y_o - 2\frac{F_f}{k}\right)$$

In the second half-cycle the velocity is positive and the equation of motion is

$$M\ddot{y} + ky = -F_f$$

The response indicated by this equation is the same as Eq. (2.30) if the sign of the force term is changed and if time is measured from the first negative peak ($t = \pi/\omega$), the amplitude of which is taken as the initial displacement. Thus, for the second half-cycle,

$$y = \left(-y_o + 3\frac{F_f}{k}\right)\cos\omega\left(t - \frac{\pi}{\omega}\right) - \frac{F_f}{k} \qquad (2.31)$$

The next peak (positive) occurs at $\omega t = 2\pi$, and Eq. (2.31) indicates a displacement of

$$y\left(t = \frac{2\pi}{\omega}\right) = y_o - 4\frac{F_f}{k}$$

It may now be deduced that successive positive peaks are given by $y_o - (4F_f/k)n$, where n is an integer representing the number of complete cycles, or multiples of the natural period. The complete response is shown in Fig. 2.14, where the damping envelope is formed by a pair of straight lines, and each half-cycle is a pure cosine function. The response

58 Introduction to Structural Dynamics

is completely damped out at $t = (kT/4F_f)y_o$, where T is the natural period.

2.5 Response to a Pulsating Force

In this section we shall consider the classical solution for the response of a one-degree system to a pulsating force of the form

$$F = F_1 \sin \Omega t \tag{2.32}$$

The primary reason for interest in this case is the fact that Eq. (2.32) may represent the dynamic force applied by a rotating machine to its support. The slightest imbalance of the rotating part produces this type of force. F_1 is proportional to the unbalanced weight, and Ω is the frequency, or speed, of the machine. Consideration of this problem will introduce the concept of *resonance*, traditionally a matter of interest to engineers, which occurs when the natural frequency of the supporting structure is close to the frequency of the machine.

a. Undamped System with Sinusoidal Force

For an undamped one-degree system subjected to a sinusoidal forcing function [Eq. (2.32)] of indefinite duration, the equation of motion is

$$M\ddot{y} + ky = F_1 \sin \Omega t \tag{2.33}$$

The solution is of the form

$$y = C_1 \sin \omega t + C_2 \cos \omega t + \frac{F_1}{M} \frac{\sin \Omega t}{\omega^2 - \Omega^2}$$

where ω is the natural frequency of the system. If the system starts at rest, the constants are determined by the following:

$$y_o = 0 = C_1 \sin (0) + C_2 \cos (0) + \frac{F_1}{M} \frac{\sin (0)}{\omega^2 - \Omega^2}$$

$$\dot{y}_o = 0 = C_1 \omega \cos (0) - C_2 \omega \sin (0) + \frac{F_1 \Omega}{M} \frac{\cos (0)}{\omega^2 - \Omega^2}$$

Solving for C_1 and C_2 and substituting in the general solution, we obtain the final result as

$$y = \frac{F_1}{M}\left(\frac{\sin \Omega t}{\omega^2 - \Omega^2} - \frac{\Omega}{\omega}\frac{\sin \omega t}{\omega^2 - \Omega^2}\right) \tag{2.34a}$$

or

$$\text{DLF} = \frac{1}{(1 - \Omega^2/\omega^2)}\left[\sin \Omega t - \frac{\Omega}{\omega}\sin \omega t\right] \tag{2.34b}$$

It may be observed that the response consists of two parts, the *free part*, having the natural frequency of the system, and the *forced part*, having

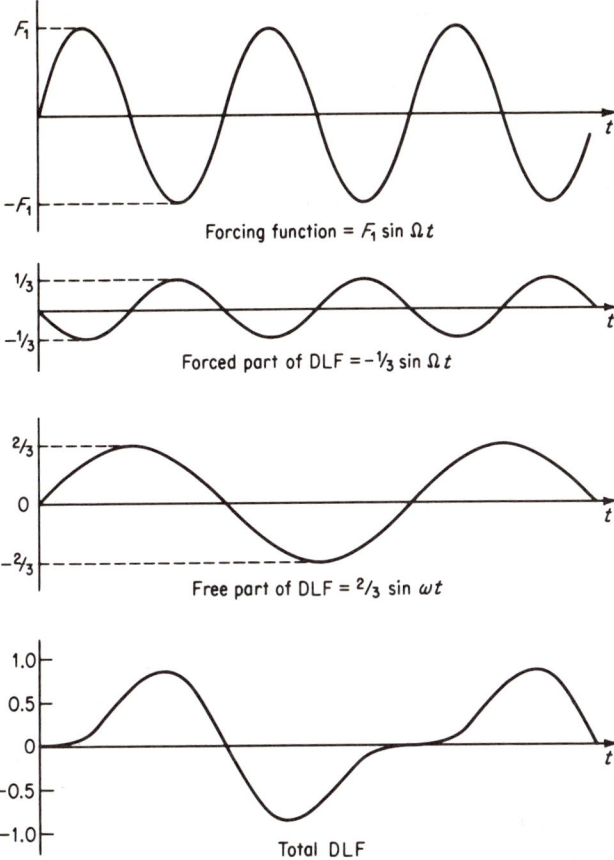

FIGURE 2.15 Response to sinusoidal force in terms of dynamic load factor. $\Omega = 2\omega$.

the frequency of the forcing function. A typical response is shown in Fig. 2.15, where the free and forced parts are separated.

The maximum DLF can be determined by differentiating and setting equal to zero Eq. (2.34b), solving for t_m, or the time of maximum response, and substituting the latter back into Eq. (2.34b). Mathematically, this procedure is rather difficult, but by plotting the free and forced sine functions separately, one can at least estimate the maximum DLF. An upper limit, which for practical purposes is sufficiently close to the actual value, may be obtained by assuming, in Eq. (2.34b), that at some time $\sin \Omega t = 1$ and $\sin \omega t = -1$. Substitution of these numerical values into Eq. (2.34b) leads directly to

$$(\text{DLF})_{\max} = \pm \frac{1}{1 - \Omega/\omega} \qquad (2.35)$$

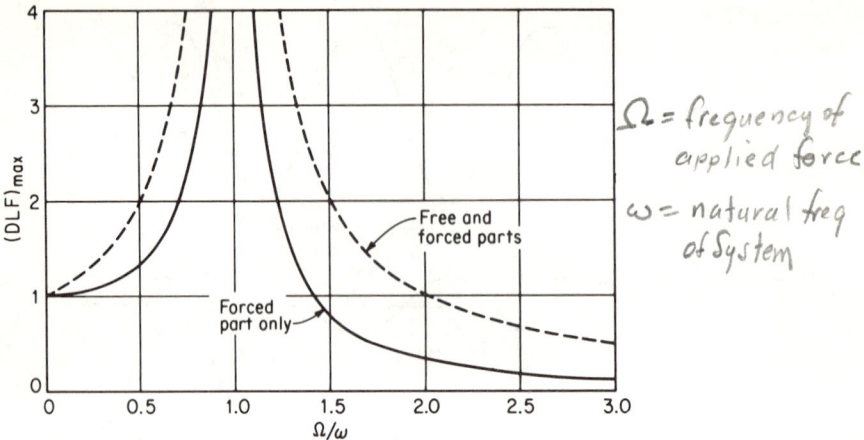

FIGURE 2.16 Maximum dynamic load factor for sinusoidal load $F_1 \sin \Omega t$, undamped systems.

This expression is plotted in Fig. 2.16 as the dashed line, and it is noted that the maximum response so determined contains both the free and forced parts.

For practical applications, Eq. (2.35) overestimates the maximum response, since even a small amount of damping quickly eliminates the free vibration. Our concern here being with a continuing state of vibration rather than with the first few cycles, it is reasonable to assume that the free vibration has been completely damped out. If the free term is removed from Eq. (2.34b), the maximum response obviously occurs when $\sin \Omega t = 1$, and is given by

$$(DLF)_{max} = \frac{1}{1 - \Omega^2/\omega^2} \qquad (2.36)$$

This solution implies that the damping is so small that the forced vibration is not affected even though the free part is completely eliminated. More detailed consideration of damping is given in Sec. 2.5b. Equation (2.36) is also plotted in Fig. 2.16. This solution is often referred to as the *steady-state* response, while that including the free part is the *transient* response.

In Fig. 2.16 it may be observed that, for either of the solutions given, $(DLF)_{max}$ approaches unity as $\Omega/\omega \to 0$ and approaches zero as $\Omega/\omega \to \infty$. Physically, this simply means that, if Ω is relatively small, the load pulsates very slowly and the mass of the systems "rides" along without vibration about the neutral position corresponding to the instantaneous load value. In other words, the effect is the same as for a static load.

Rigorous Analysis of One-degree Systems   61

On the other hand, if Ω is relatively very large, the mass cannot follow the rapid fluctuations in load and simply remains stationary. Therefore DLF = 0 at all times.

Next consider the condition of resonance which occurs when $\Omega \approx \omega$. As indicated in Fig. 2.16, this situation results in very large displacements, which theoretically become infinite if $\Omega = \omega$ [Eqs. (2.35) and (2.36)]. Actually, this is an oversimplification, as will now be demonstrated.

If in Eqs. (2.34) Ω is made equal to ω, the result is $y = 0/0$, or in other words, the response is indeterminate. The displacement can, however, be obtained by the application of L'Hospital's rule, which states that the limit of Eq. (2.34b) as $\Omega \to \omega$ is the derivative of the term in brackets with respect to Ω divided by the same derivative of the term in parentheses. Thus

$$(\text{DLF})_{\lim \Omega \to \omega} = \frac{t \cos \Omega t - (1/\omega) \sin \omega t}{-2\Omega/\omega^2}$$

With $\Omega = \omega$, this becomes

$$(\text{DLF})_{\Omega=\omega} = \tfrac{1}{2}(\sin \omega t - \omega t \cos \omega t) \tag{2.37}$$

From the last equation it is apparent that DLF does indeed become infinite at resonance, but only after an infinite time. Equations (2.35) and (2.36) are therefore valid when $\Omega = \omega$, but only after many cycles of vibration. It should be noted, however, that "many" cycles may occur in a short period of absolute time.

In practice, exact resonance does not really occur, because systems are never completely linear. As distortions become large, the characteristics of the system change because of plastic deformation and other effects. The question as to whether displacements become infinite is of course of academic interest only. The important engineering conclusion is that, at or near resonance, the deflections of the structure become very large and hence intolerable.

It is sometimes of interest to determine the amplitude of response after a limited number of cycles of pulsating load at resonant frequency. Equations (2.34) cannot be used for this purpose since it is indeterminate at this point. However, we may obtain the response by the application of Eq. (2.37). Limiting the solution to peak values, we first obtain the times of peak responses by maximizing DLF as follows:

$$\frac{d}{dt}(\text{DLF})_{\Omega=\omega} = 0 = \tfrac{1}{2}(\omega \cos \omega t + \omega^2 t \sin \omega t - \omega \cos \omega t)$$

Therefore

$$\sin \omega t = 0 \qquad \omega t = \pi, 2\pi, 3\pi, \ldots$$

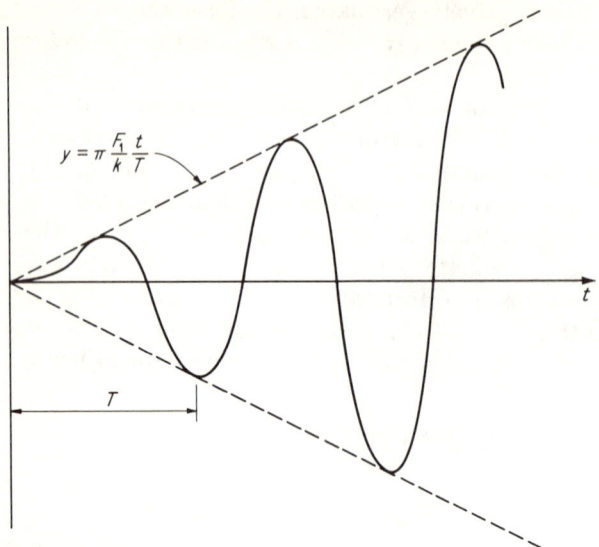

FIGURE 2.17 Initial stage of response to sinusoidal load, $F_1 \sin \Omega t$, at resonance.

Substituting these values of ωt in Eq. (2.37), we find

$$(\text{DLF})_{\max, \Omega=\omega} = \tfrac{1}{2}(\pi, -2\pi, 3\pi, - \ldots)$$
$$|(\text{DLF})_{\max, \Omega=\omega}| = \tfrac{1}{2}n\pi \qquad n = 1, 2, 3, \ldots \qquad (2.38)$$

where n is the number of half-cycles after the beginning of response. Equation (2.38) states that the maximum deflection of an undamped one-degree system subjected to sinusoidal loading is $\pi/2$ times the static deflection F_1/k after $\tfrac{1}{2}$ cycle of loading, π times y_{st} after 1 cycle, $3\pi/2$ times y_{st} after $1\tfrac{1}{2}$ cycles, etc. From the above discussion it may be deduced that the initial stages of response to resonant sinusoidal loading are as shown in Fig. 2.17.

b. Damped System with Sinusoidal Force

In this case the equation of motion is

$$M\ddot{y} + ky + c\dot{y} = F_1 \sin \Omega t \qquad (2.39)$$

for which the solution is of the form

$$y = e^{-\beta t}(C_1 \sin \omega_d t + C_2 \cos \omega_d t)$$
$$+ \frac{(F_1/k)[(1 - \Omega^2/\omega^2) \sin \Omega t - 2(\beta\Omega/\omega^2) \cos \Omega t]}{(1 - \Omega^2/\omega^2)^2 + 4(\beta\Omega/\omega^2)^2} \qquad (2.40a)$$

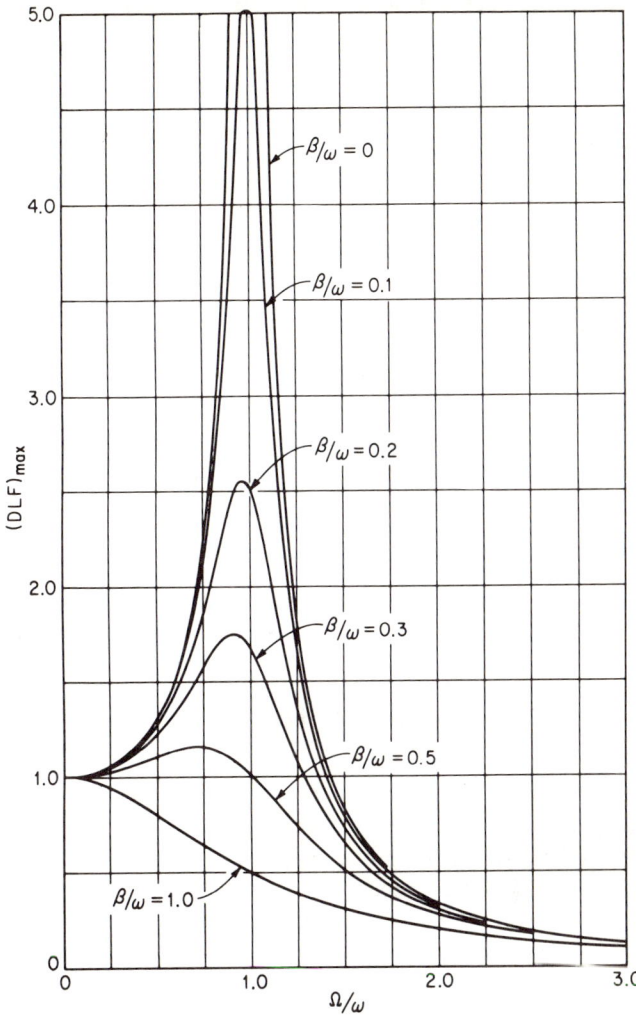

FIGURE 2.18 Maximum dynamic load factor for sinusoidal load, $F_1 \sin \Omega t$, damped systems.

where $\beta = c/2M$, and $\omega_d = \sqrt{\omega^2 - \beta^2}$, the natural frequency of the damped system.[6] As discussed previously, the contribution of the free part becomes negligible after a few cycles of response, and therefore we need consider only the steady-state, or forced, part of the response given by the second term in Eq. (2.40a). This term may be rewritten as follows:

$$y = \frac{(F_1/k)[(1 - \Omega^2/\omega^2)^2 + 4(\beta\Omega/\omega^2)^2]^{1/2} \sin(\Omega t + \theta)}{(1 - \Omega^2/\omega^2)^2 + 4(\beta\Omega/\omega^2)^2} \qquad (2.40b)$$

where θ is merely a phase angle. It is apparent that this expression is a maximum when the sine is unity, and therefore

$$(\text{DLF})_{\max} = \frac{1}{\sqrt{(1 - \Omega^2/\omega^2)^2 + 4(\beta\Omega/\omega^2)^2}} \qquad (2.41)$$

The last expression is often called the *dynamic magnification factor*, and is plotted in Fig. 2.18 for various values of β/ω, which is the ratio of actual to critical damping.

It is apparent in Fig. 2.18 that, even with small damping, theoretically infinite amplitudes do not occur at resonance. In the extreme case of critical damping ($\beta/\omega = 1$), the maximum resonant deflection is only one-half the static deflection. The curve shown for zero damping is of course the same as that shown in Fig. 2.16 for forced vibration only. As a further simplification, the following may be derived from Eq. (2.41) for the resonant condition:

$$(\text{DLF})_{\max, \Omega=\omega} = \frac{\omega}{2\beta} \qquad (2.42)$$

A development such as that leading to Eq. (2.37) could also be shown for the damped case, thus indicating that the maximums given by Eq. (2.42) are attained only after many cycles of vibration.

c. Undamped System with Step Force

It should be emphasized that a sinusoidal forcing function is not a requirement for large displacements near resonance. For example, consider the alternating step force shown in Fig. 2.19, which has the same period as the responding system. To investigate the response, we proceed as follows, recognizing that, for a suddenly applied constant force, $y = (F_1/k)(1 - \cos \omega t)$, and for an initial displacement, $y = y_o \cos \omega t$. In the first half-cycle,

$$y = \frac{F_1}{k}(1 - \cos \omega t)$$

At $\omega t = \pi$, $\qquad y_1 = 2\dfrac{F_1}{k} \qquad \dot{y}_1 = 0$

In the second half-cycle,

$$y_o = y_1 \qquad \dot{y}_o = 0$$

$$y = 2\frac{F_1}{k}\cos \omega(t - \pi) - \frac{F_1}{k}[1 - \cos \omega(t - \pi)]$$

At $\omega t = 2\pi$, $y_2 = -4\dfrac{F_1}{k}$

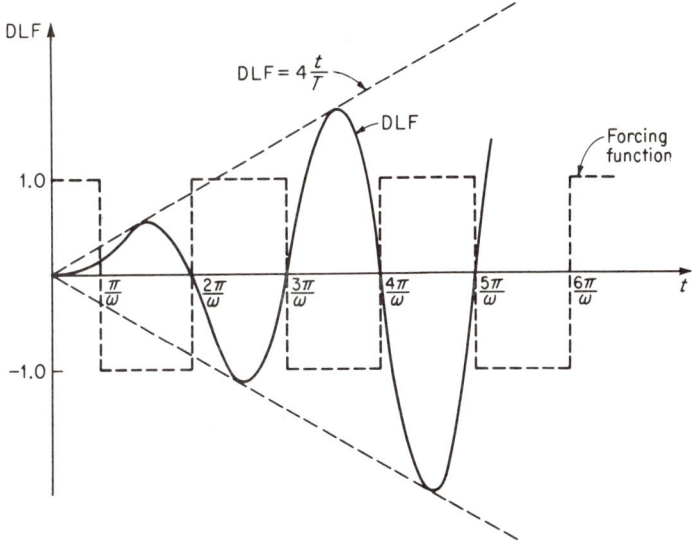

FIGURE 2.19 Initial stage of response to alternating step load at resonance.

It is therefore apparent that successive peaks (positive or negative) of the vibration have the displacements

$$|y| = 2n\frac{F_1}{k}$$

and
$$|(\text{DLF})_{\max}| = 2n \tag{2.43}$$

where n is the number of half-cycles from the starting point. The response is plotted in Fig. 2.19. Comparison of Eq. (2.43) with Eq. (2.38) indicates that, at resonance, the amplitudes resulting from an alternating step force increase even more rapidly than those due to a sinusoidal force.

2.6 Support Motions

An important class of problems is the determination of response due to movement of the support of the system rather than the application of external force. Perhaps the foremost example is the analysis and design of structures for earthquake effects. It is shown below that, with only slight modification, the preceding solutions for applied forces can also be used for the case at hand.

a. Undamped Systems

Suppose that the system shown in Fig. 2.20 is subjected to a support motion y_s defined by $y_s = y_{so}f(t)$, where y_{so} is some arbitrary magnitude

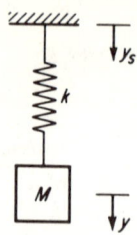

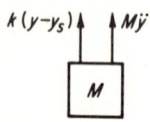

FIGURE 2.20 One-degree system with support motion.

of support displacement and $f(t)$ is the time function describing the variation of y_s with time. The equation of motion is

$$M\ddot{y} + k(y - y_s) = 0$$

or
$$M\ddot{y} + ky = ky_s = ky_{so}f(t) \tag{2.44}$$

Comparison of Eq. (2.44) with those previously written for external force functions $[F_1 f(t)]$ reveals that they are identical, except that F_1 has been replaced by ky_{so}. Thus previous solutions are valid if this simple substitution is made. For example, the general solution given by Eq. (2.14b) becomes

$$y = y_{so}\omega \int_0^t f(\tau) \sin \omega(t - \tau)\, d\tau \tag{2.45}$$

To illustrate the above, we consider a system the support of which is displaced suddenly by an amount y_{so}, and then remains fixed in that position. This imposed condition corresponds to a suddenly applied constant force, and the response may be obtained by substituting ky_{so} for F_1 in Eq. (2.10). This produces

$$y = y_{so}(1 - \cos \omega t) \tag{2.46}$$

for the absolute displacement and

$$u = y - y_s = -y_{so} \cos \omega t \tag{2.47}$$

where u is the relative displacement of the mass with respect to the support. The force in the spring is of course ku, and the negative sign in Eq. (2.47) indicates that the spring is initially in compression if y_{so} is positive.

As a second example, let the support motion be $y_s = y_{so} \sin \Omega t$. Using the previously given solution for a sinusoidal force, ky_{so} is substituted for F_1 in Eq. (2.34a) to obtain

$$y = y_{so}\omega^2 \left(\frac{\sin \Omega t}{\omega^2 - \Omega^2} - \frac{\Omega}{\omega} \frac{\sin \omega t}{\omega^2 - \Omega^2} \right)$$
$$= y_{so}(\text{DLF}) \qquad (2.48)$$

Since DLF is the same as that expressed by Eq. (2.34b), the maximum DLF for sinusoidal support motion is also given by Fig. 2.16 without modification and $y_{\max} = y_{so}(\text{DLF})_{\max}$. The relative motion, which is the same as the spring distortion, may be obtained from Eq. (2.48):

$$u = y - y_s = y_{so}\Omega^2 \left(\frac{\sin \Omega t}{\omega^2 - \Omega^2} - \frac{\omega}{\Omega} \frac{\sin \omega t}{\omega^2 - \Omega^2} \right)$$
$$= y_{so}(\text{DLF})_r \qquad (2.49)$$

where $(\text{DLF})_r$ is the dynamic load factor for relative displacement. Equation (2.49), together with a consideration of the discussion of maximum DLF in Sec. 2.5a, leads to the conclusion that Fig. 2.16 can also be used to obtain maximum relative displacement if Ω and ω are simply interchanged. Thus, at the extreme limits, we conclude:

$$\text{As } \frac{\Omega}{\omega} \to 0, \ y = y_s \text{ and } u = 0$$

$$\text{As } \frac{\Omega}{\omega} \to \infty, \ y = 0 \text{ and } u = -y_s$$

This applies throughout the response, as well as to the maximum values.

From the above discussion it should be apparent that the charts of $(\text{DLF})_{\max}$ for various forcing functions, previously given (Figs. 2.7 to 2.9), apply equally well to the cases of support motion having the same time functions; that is, $y_{\max} = y_{so}(\text{DLF})_{\max}$. However, this fact is of limited usefulness since the maximum spring force which is proportional to relative displacement cannot be obtained directly in this manner.

An alternative approach which is often useful involves representation of the input in terms of support acceleration rather than displacement. For this purpose it is convenient to change the variable to the relative displacement of the mass with respect to the support, which is identified by $u = y - y_s$. Since $\ddot{u} = \ddot{y} - \ddot{y}_s$, Eq. (2.44) becomes

$$M(\ddot{u} + \ddot{y}_s) + ku = 0$$
or
$$M\ddot{u} + ku = -M\ddot{y}_s = -M\ddot{y}_{so}f_a(t) \qquad (2.50)$$

where $f_a(t)$ is the time function for support acceleration, and $\ddot{y}_{so}$ is some arbitrary (usually maximum) value of support acceleration. This equation is identical with those for forcing functions if F_1 is replaced by $-M\ddot{y}_{so}$.

Therefore the general solution for the relative motion is

$$u = -\frac{\ddot{y}_{so}}{\omega} \int_0^t f_a(\tau) \sin \omega(t - \tau) \, d\tau \qquad (2.51)$$

if it is assumed that the initial support velocity is zero. Thus results given elsewhere for external forcing functions may be used to determine directly relative distortions due to support motions provided that the latter are given in terms of acceleration. Examples of this procedure are to be found in Chap. 6.

b. Damped Systems

When damping is involved, it is generally more convenient to employ the approach represented by Eq. (2.50), in which the support motion is specified in terms of acceleration rather than displacement. This is true because the damping force is usually proportional to relative rather than absolute velocity. If damping is included, Eq. (2.50) becomes

$$M\ddot{u} + ku + c\dot{u} = -M\ddot{y}_s = -M\ddot{y}_{so}f_a(t) \qquad (2.52)$$

The general solution obtained by replacing y_{st} by $-M\ddot{y}_{so}/k$ in Eq. (2.27) is therefore

$$u = -\frac{\ddot{y}_{so}}{\omega_d} \int_0^t f_a(\tau) e^{-\beta(t-\tau)} \sin \omega_d(t - \tau) \, d\tau \qquad (2.53)$$

if the system starts at rest.

The relative response to sinusoidal support motion ($y_s = \ddot{y}_{so} \sin \Omega t$) may be obtained from Eq. (2.40b) if F_1 is replaced by $-M\ddot{y}_{so}$ since $f(t) = f_a(t)$. Thus

$$u = \frac{(\ddot{y}_{so}/\omega^2)[(1 - \Omega^2/\omega^2)^2 + 4(\beta\Omega/\omega^2)^2]^{1/2} \sin (\Omega t + \theta)}{(1 - \Omega^2/\omega^2)^2 + 4(\beta\Omega/\omega^2)^2} \qquad (2.54)$$

and the maximum relative response is given by

$$u_{\max} = \frac{\ddot{y}_{so}}{\omega^2} \left\{ \frac{1}{[(1 - \Omega^2/\omega^2)^2 + 4(\beta\Omega/\omega^2)^2]^{1/2}} \right\} \qquad (2.55)$$

Because of the similarity between the last equation and Eq. (2.41), it is apparent that Fig. 2.18 is a plot of the bracketed term in Eq. (2.55). Thus the maximum relative displacement, and hence the maximum spring force $ku_{\max}$, can be obtained by Fig. 2.18 and Eq. (2.55); that is,

$$(u)_{\max} = (\ddot{y}_{so}/\omega^2)(\text{DLF})_{\max}$$

where $(\text{DLF})_{\max}$ is given by Fig. 2.18.

If the support motion cannot be expressed in terms of acceleration and if damping is to be considered, it is necessary to include as input both displacement and velocity. For a damped system, Eq. (2.44) becomes

$$M\ddot{y} + k(y - y_s) + c(\dot{y} - \dot{y}_s) = 0$$

or
$$M\ddot{y} + ky + c\dot{y} = ky_s + c\dot{y}_s \qquad (2.56)$$

Additional discussion of response to support motion is given in Chap. 6 in connection with analysis for earthquake.

2.7 Elasto-plastic Systems

Consider the single-degree undamped system in Fig. 2.21a, which is assumed to have the bilinear resistance function shown in Fig. 2.21b. A rigorous solution for the response due to a suddenly applied, constant load is given below. Since there are two discontinuities in the response, this involves three separate stages: (1) the elastic response up to the elastic limit y_{el}, (2) the plastic response between the elastic limit and the maximum displacement, and (3) rebound, or the elastic response which occurs after the maximum has been attained and the displacement begins to decrease. In the determination of displacements for stages 2 and 3, the initial conditions are the final displacement and velocity of the pre-

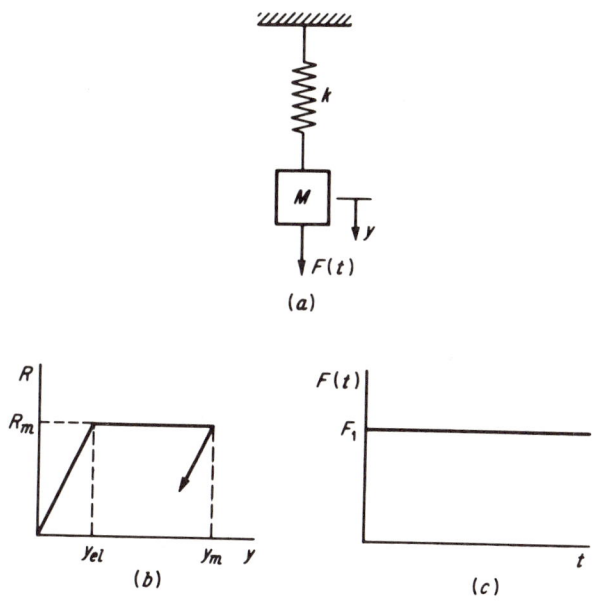

FIGURE 2.21 Elasto-plastic system.

70   Introduction to Structural Dynamics

ceding stage. If the applied force were not constant, additional stages, one for each force discontinuity, would have to be included. Obviously, this is a very laborious process, and except in the most simple cases, one would be better advised to use numerical analysis as illustrated in Sec. 1.5.

For the example indicated by Fig. 2.21, with zero initial displacement and velocity, the response in the first stage is given by (Sec. 2.2c)

$y \leq y_{el}$:
$$y = y_{st}(1 - \cos \omega t) \quad (2.57a)$$
$$\dot{y} = y_{st}\omega \sin \omega t \quad (2.57b)$$

where $y_{st} = F_1/k$, and $\omega = \sqrt{k/M}$. The time at which y_{el} is reached, t_{el}, may be obtained from Eq. (2.57a):

$$\cos \omega t_{el} = 1 - \frac{y_{el}}{y_{st}}$$

and
$$\sin \omega t_{el} = \sqrt{1 - \cos^2 \omega t_{el}} \quad (2.58)$$

Proceeding to the second stage and letting $t_1 = t - t_{el}$, we have the initial conditions for this stage:

$$y_o = y_{el}$$
$$\dot{y}_o = y_{st}\omega \sin \omega t_{el} \quad (2.59)$$

and the equation of motion

$$M\ddot{y} + R_m = F_1$$

The solution by direct integration is

$$y = \frac{1}{2M}(F_1 - R_m)t_1^2 + C_1 t_1 + C_2 \quad (2.60)$$

Making use of the initial conditions [Eqs. (2.59)] at $t_1 = 0$ to solve for C_1 and C_2 and substituting back into Eq. (2.60), we obtain the final solution for this stage:

$$y_{el} \leq y \leq y_m: \quad y = \frac{1}{2M}(F_1 - R_m)t_1^2 + y_{st}\omega t_1 \sin \omega t_{el} + y_{el} \quad (2.61)$$

By differentiating Eq. (2.61) and setting the result equal to zero, the time of maximum response is obtained as

$$t_{1m} = \frac{M\omega y_{st}}{R_m - F_1} \sin \omega t_{el} \quad (2.62)$$

The maximum displacement y_m is obtained by substituting t_{1m} into Eq. (2.61).

For the third and final stage, one could proceed as above, using a suitable equation of motion with initial conditions from the second stage. However, an easier procedure is to make use of some obvious facts regard-

Rigorous Analysis of One-degree Systems 71

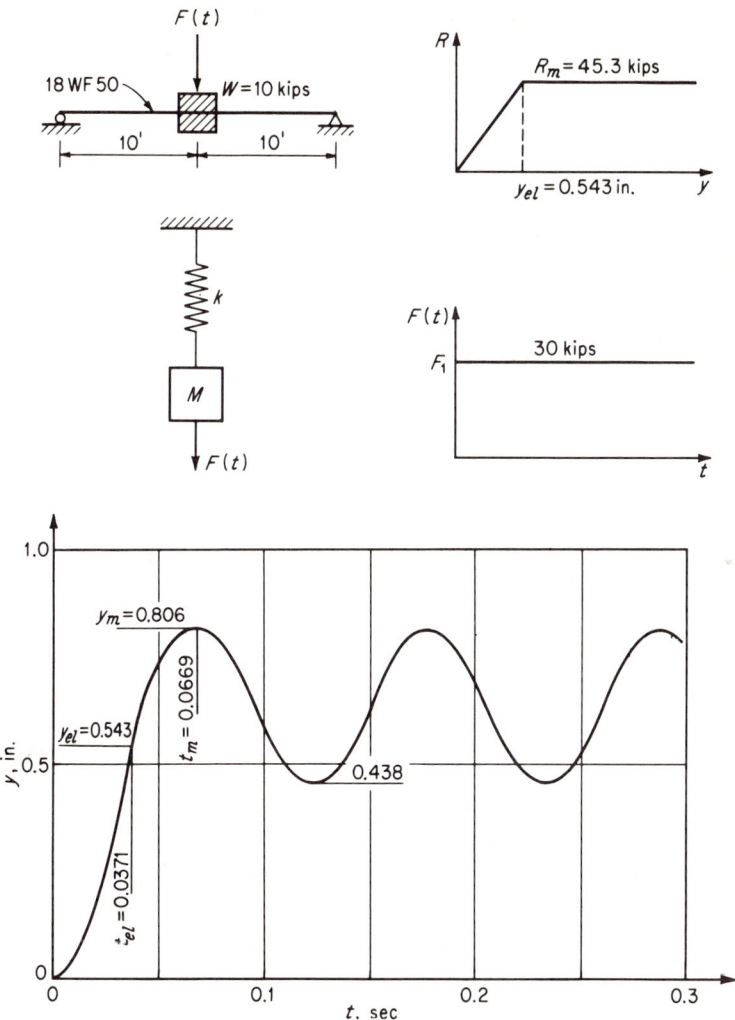

FIGURE 2.22 Example. Response of elasto-plastic one-degree structure to suddenly applied constant force.

ing the response, which were discussed in Sec. 1.5. This stage consists of a residual vibration, which is of course elastic, or harmonic. When the mass is in its neutral position, the spring force is equal to the applied load F_1. Therefore the amplitude of vibration, or the amount by which the deflection must decrease below y_m to reach the neutral position, is $(R_m - F_1)/k$. The situation is equivalent to an initial displacement of this amount on a system whose neutral position is $y_m - (R_m - F_1)/k$.

72 *Introduction to Structural Dynamics*

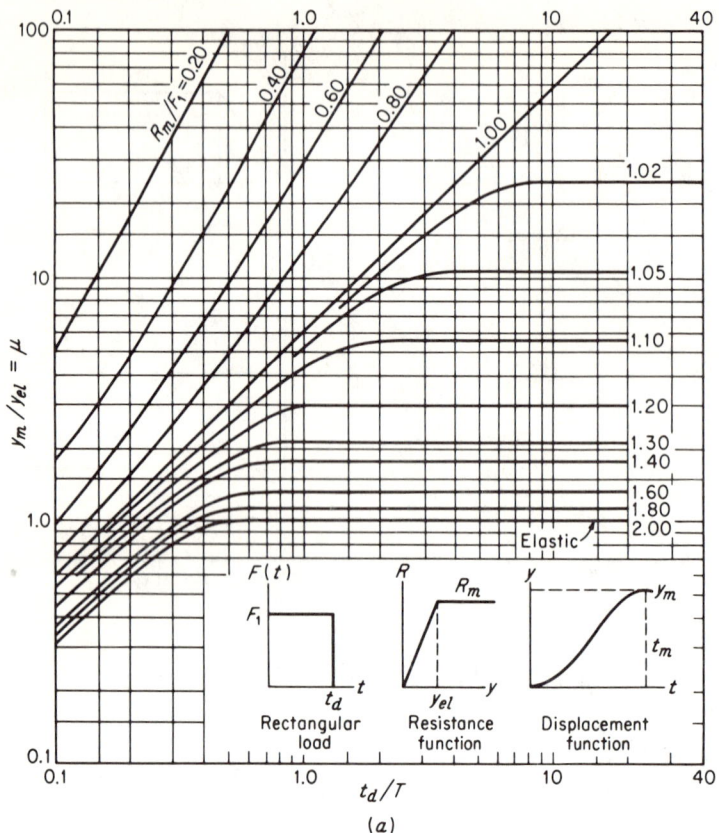

FIGURE 2.23 Maximum response of elasto-plastic one-degree systems (undamped) due to rectangular load pulses. (*U.S. Army Corps of Engineers.*[10])

Therefore the response is given by

$$y = \left(y_m - \frac{R_m - F_1}{k}\right) + \frac{R_m - F_1}{k} \cos \omega t_2 \qquad (2.63)$$

where $t_2 = t - t_{1m} - t_{el}$.

The complete response is therefore given by Eqs. (2.57a), (2.61), and (2.63), and the times at the interior boundaries of the three stages by Eqs. (2.58) and (2.62).

To illustrate rigorous elasto-plastic analysis, the response of the simple steel beam shown in Fig. 2.22 will be investigated. The elastic properties of the system were determined in Sec. 2.3e, but in addition to these, the maximum, or plastic, resistance is now needed. The total

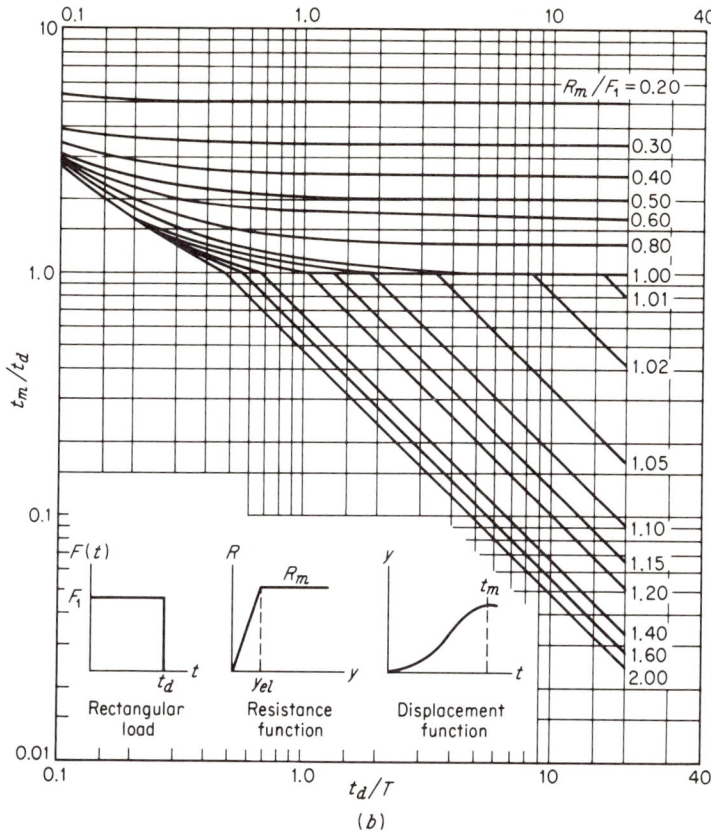

FIGURE 2.23 (*Continued*)

maximum resistance is given by

$$R_{mT} = \frac{4\mathfrak{M}_P}{l} = \frac{4 \times 3320}{20 \times 12} = 55.3 \text{ kips}$$

where $\mathfrak{M}_P$ is the ultimate bending moment (rectangular stress block) based on a yield-point stress of 33 ksi. Since the beam supports a dead weight of 10 kips, the maximum force available to resist the dynamic load is

$$R_m = 55.3 - 10 = 45.3 \text{ kips}$$

The resistance function may be assumed to be bilinear, as indicated in Fig. 2.22. The above determination of resistance of course implies that lateral buckling of the beam is prevented by some means.

74   **Introduction to Structural Dynamics**

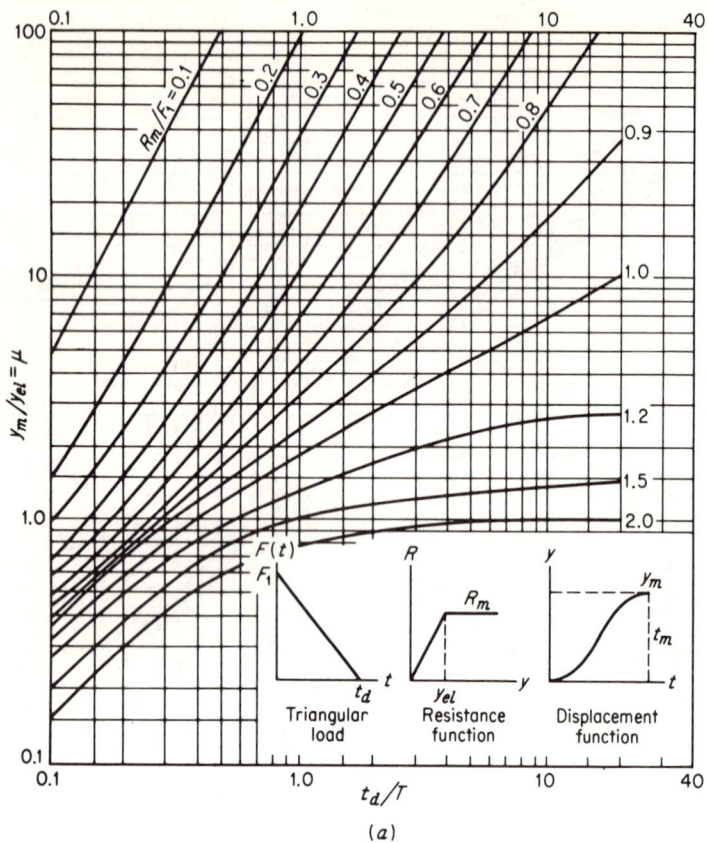

FIGURE 2.24 Maximum response of elasto-plastic one-degree systems (undamped) due to triangular load pulses with zero rise time. (*U.S. Army Corps of Engineers.*[10])

We wish to determine the complete response due to a suddenly applied, constant force of 30 kips. The parameters required for analysis of the system are as follows:

$$k = 83.4 \text{ kips/in.}$$
$$M = 0.0259 \text{ kip-sec}^2/\text{in.}$$
$$T = 0.111 \text{ sec}; \quad \omega = 56.8 \text{ rad/sec}$$
$$R_m = 45.3 \text{ kips}$$
$$y_{el} = R_m/k = 0.543 \text{ in.}$$
$$y_{st} = F_1/k = 0.360 \text{ in.}$$

It is apparent that the response will reach the plastic range since

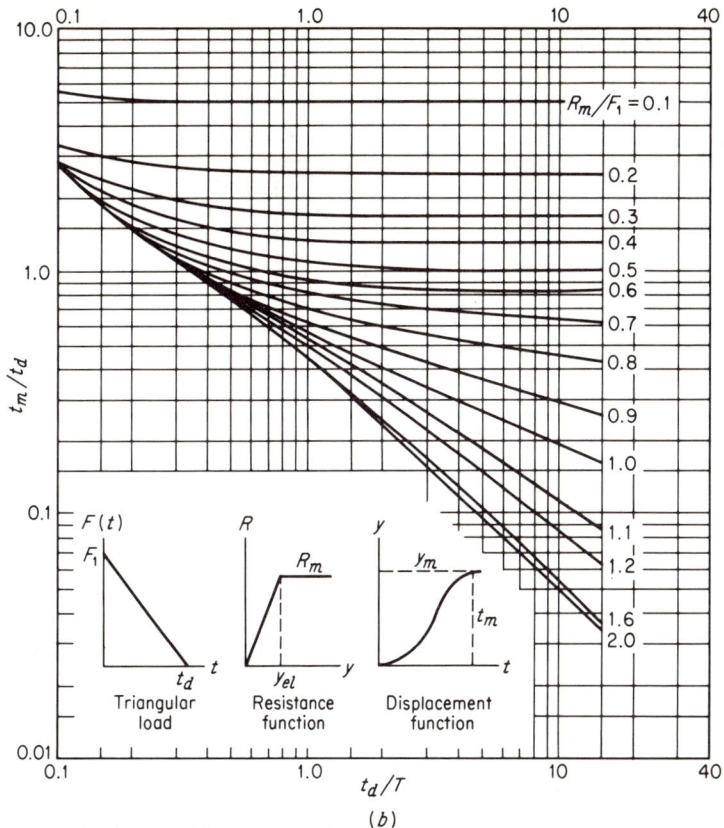

FIGURE 2.24 (*Continued*)

$R_m < 2F_1$. The total solution based on the equations derived above is as follows:

$t \leqslant t_{el}$:

$$y = 0.360(1 - \cos 56.8t) \quad \text{Eq. (2.57a)}$$
$$\cos 56.8 t_{el} = 1 - \frac{0.543}{0.360} = -0.508 \quad \text{Eq. (2.58)}$$
$$56.8 t_{el} = 2.10 \text{ rad} \quad t_{el} = 0.0371 \text{ sec}$$
$$\sin \omega t_{el} = 0.861$$

$t_{el} \leqslant t \leqslant t_m$:

$$y = -295 t_1^2 + 17.6 t_1 + 0.543 \quad \text{Eq. (2.61)}$$
$$t_{1m} = 0.0298 \text{ sec} \quad \text{Eq. (2.62)}$$
$$t_m = t_{el} + t_{1m} = 0.0669 \text{ sec}$$

76 *Introduction to Structural Dynamics*

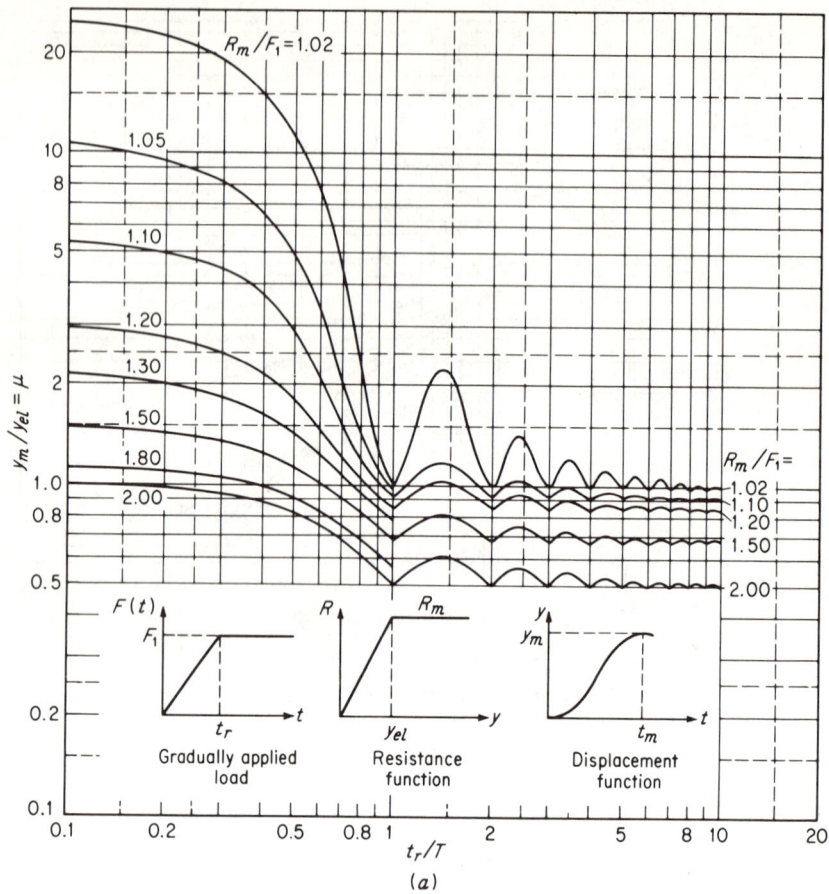

FIGURE 2.25 Maximum response of elasto-plastic one-degree systems (undamped) due to constant force with finite rise time.

Therefore
$$y_m = -295(0.0298)^2 + 17.6(0.0298) + 0.543$$
$$= 0.806 \text{ in.}$$

$t_m \leqslant t$:
$$y = 0.622 + 0.184 \cos 56.8 t_2 \qquad \text{Eq. (2.63)}$$
$$t_2 = t - 0.0669$$

The complete solution given by the foregoing is shown in Fig. 2.22.

2.8 Charted Solutions for Maximum Response of One-degree Undamped Elasto-plastic Systems

Because the analysis of elasto-plastic systems is cumbersome, it is convenient to make use of charts giving the maximum response. Usually,

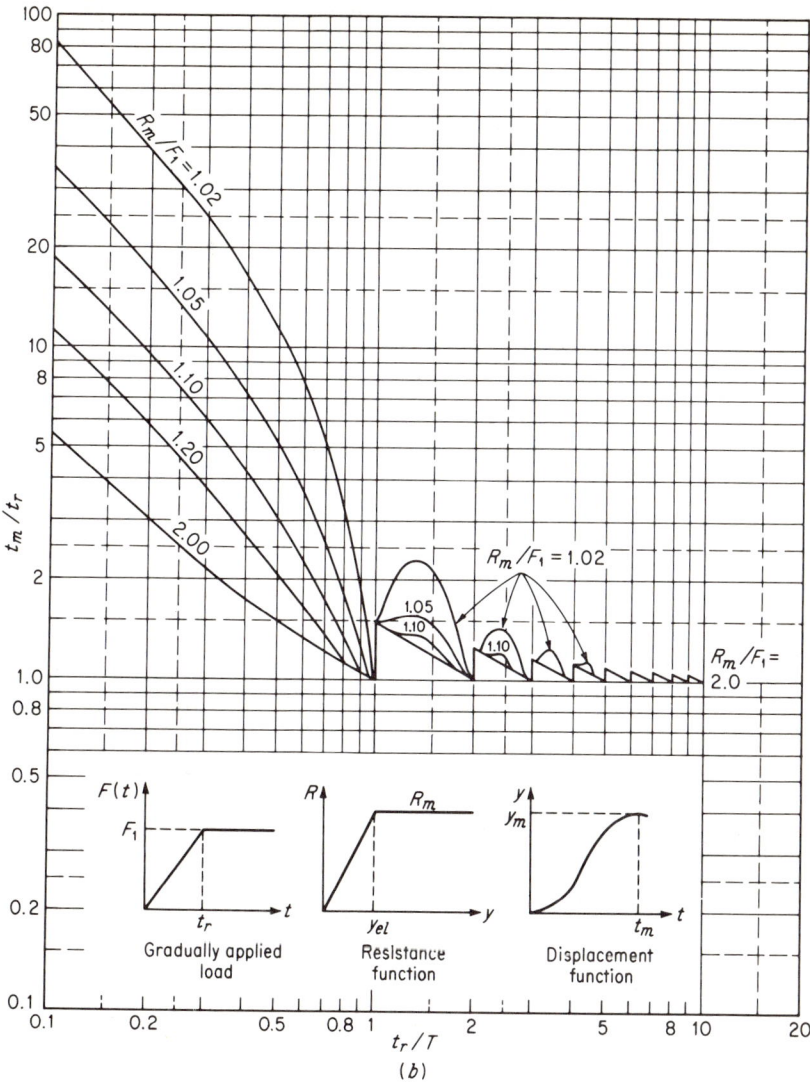

FIGURE 2.25 (Continued)

the structural designer is interested only in the maximum displacement, and therefore such charts need give only that quantity rather than the complete response as a function of time. Presented in this section are response charts for four load-time functions. It will be found in practice that many actual loading conditions can be approximated by one of these simple functions.

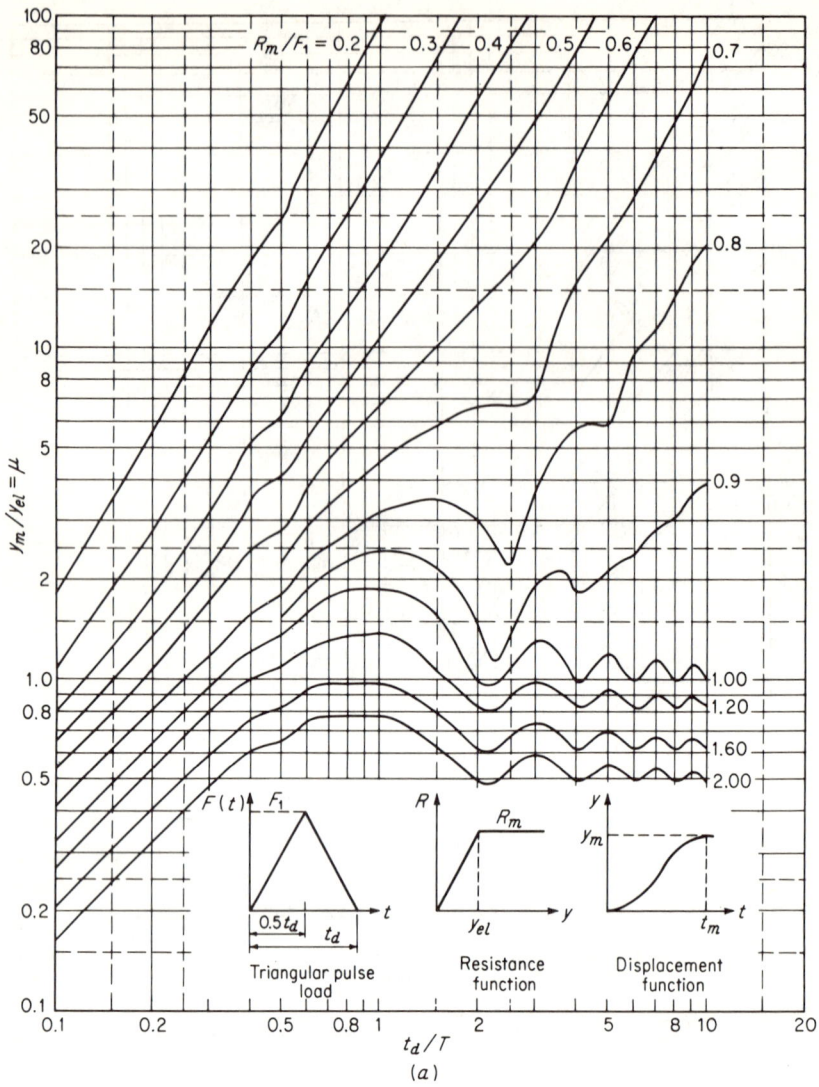

FIGURE 2.26 Maximum response of elasto-plastic one-degree systems (undamped) due to equilateral triangular load pulses.

a. Nondimensional Equations of Motion

The procedure given in Sec. 2.7 may be used to derive the desired charts, but for plotting purposes it is convenient to nondimensionalize the parameters. In the basic equation of elastic motion,

$$M\ddot{y} + ky = F_1[f(t)]$$

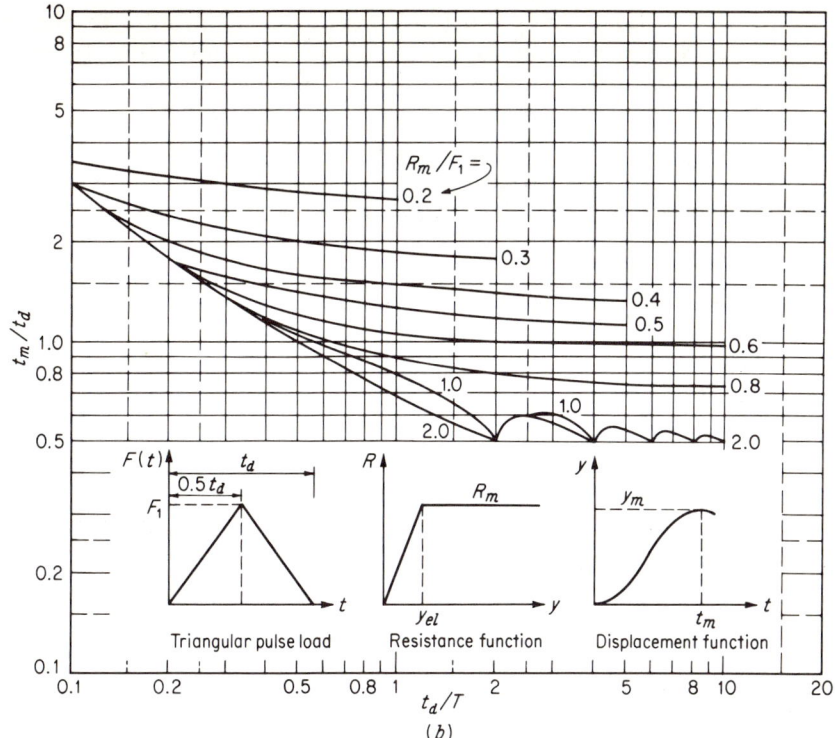

FIGURE 2.26 (*Continued*)

we transform the variables by letting $\xi = t/T$ and $\eta = y/y_{el}$.

Then
$$\ddot{y} = \frac{d^2 y}{dt^2} = \frac{y_{el}}{T^2}\frac{d^2\eta}{d\xi^2} = \frac{y_{el}}{T^2}\ddot{\eta}$$

and the equation of motion becomes

$$\frac{My_{el}}{T^2}\ddot{\eta} + ky_{el}\eta = F_1[f(\xi)]$$

or since $ky_{el} = R_m$ and $T^2 = 4\pi^2 M/k$,

$$\frac{1}{4\pi^2}\ddot{\eta} + \eta = \frac{F_1}{R_m}f(\xi) \tag{2.64}$$

In the plastic range the term ky becomes constant with a value of R_m. If the same substitutions as those leading to Eq. (2.64) are made, the equation of motion for this range becomes

$$\frac{1}{4\pi^2}\ddot{\eta} + 1 = \frac{F_1}{R_m}f(\xi) \tag{2.65}$$

Inspection of Eqs. (2.64) and (2.65) reveals that, in order to obtain the response in terms of the parameter η, one need only know the ratio F_1/R_m and the load-time function in terms of the parameter ξ. In addition to the variable itself, the latter involves only the ratio t_d/T, where t_d is the duration or some other time value characterizing the loading (see below). Thus the two parameters F_1/R_m and t_d/T are sufficient for a complete solution (if the system starts at rest and there is no damping).

b. Maximum-response Charts

The charts shown in Figs. 2.23 to 2.26 inclusive are based upon numerical solutions of Eqs. (2.64) and (2.65). The results thus provided are for undamped one-degree systems with bilinear resistance functions and without initial motion. In the case of the rectangular pulse, $f(\xi)$ is unity up to $\xi = t_d/T$ and zero thereafter. For the triangular pulse with zero rise time, $f(\xi) = 1 - \xi(T/t_d)$ up to $\xi = t_d/T$ and then zero. When the force is constant but with finite rise time, $f(\xi) = \xi(T/t_r)$ up to $\xi = t_r/T$ and unity at later times. For the triangular function of Fig. 2.26, $f(\xi) = \xi(2T/t_r)$ up to $\xi = t_r/2T$, followed by $f(\xi) = 2 - \xi(2T/t_d)$ up to $\xi = t_d/T$ and zero thereafter. The charts were constructed by inserting these expressions for $f(\xi)$ into Eqs. (2.64) and (2.65) and obtaining the maximum displacement by numerical integration for discrete values of the parameters F_1/R_m and t_d/T or t_r/T.

The values provided by the charts are the maximum nondimensional displacement, $\eta_{max} = y_m/y_{el} = \mu$, and the nondimensional time of maximum displacement, $\xi_{max} = t_m/t_d$ or t_m/t_r. It should be noted that the bottom curve in each case $(R_m/F_1 = 2)$ represents completely elastic response. If $(R_m/F_1) > 2$, the elastic-response charts (Figs. 2.7 to 2.9) should be used.

c. Example

To illustrate use of the charts presented above, let the beam cited in Sec. 2.7 (Fig. 2.22) be subjected to a suddenly applied triangular-pulse loading defined by $F_1 = 40$ kips and $t_d = 0.2$ sec. The maximum midspan deflection of the beam and the time of that deflection are obtained from Fig. 2.24 as follows:

$$\frac{R_m}{F_1} = \frac{45.3}{40} = 1.13 \qquad \frac{t_d}{T} = \frac{0.2}{0.111} = 1.80$$

From Fig. 2.24a, $\qquad \mu = 2.0$

From Fig. 2.24b, $\qquad \dfrac{t_m}{t_d} = 0.36$

Therefore

$$y_m = \mu y_{el} = 2.0 \times 0.543 = 1.09 \text{ in.}$$

$$t_m = \left(\frac{t_m}{t_d}\right) t_d = 0.36 \times 0.2 = 0.077 \text{ sec}$$

As will be illustrated in Chaps. 5 and 7, charts such as Figs. 2.23 to 2.26 are extremely useful for design purposes, provided that the load-time variation can be approximated by one of these simple functions. Similar charts can be developed for other time functions if the shapes are completely defined by two parameters, that is, F_1 and t_d.

Problems

2.1 Write the expressions for the natural frequency and natural period of the system shown in Fig. 2.27. Both the beam and the spring may be assumed massless, and rotational inertia may be neglected.

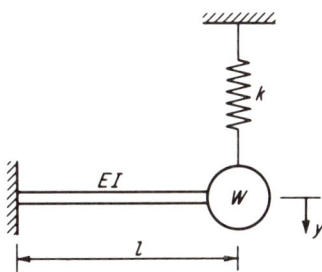

FIGURE 2.27 Problems 2.1, 2.2, 2.5, 2.6, 2.8, and 2.9.

2.2 For the system in Prob. 2.1, $l = 60$ in., $EI = 10^8$ lb-in.^{2}, $W = 2000$ lb, and $k = 2$ kips/in. If the weight has a displacement of 0.5 in. and a velocity of 10 in./sec at $t = 0$, what is the displacement and velocity at $t = 1$ sec? Assume no damping.
Answer
$y = +0.62$ in.
$\dot{y} = +3.2$ in./sec

2.3 Compute the natural frequency in the horizontal mode of the steel rigid frame shown in Fig. 2.28. The horizontal girder may be assumed infinitely rigid, and the mass of the columns may be neglected.
Answer
$\omega = 9.5$ rad/sec

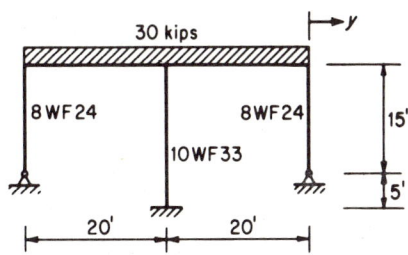

FIGURE 2.28 Problems 2.3, 2.4, and 2.7

2.4 Repeat Prob. 2.3, taking into account the girder flexibility. The horizontal member is a steel 18WF50 member.
Answer
$\omega = 9.2$ rad/sec

82   Introduction to Structural Dynamics

2.5 The weight of the system given in Prob. 2.2 is subjected to a vertical force. Compute the displacement at $t = 0.3$ sec for the following load functions, assuming no damping: (a) a force of 1000 lb applied suddenly at $t = 0$ and removed suddenly at $t = 0.2$ sec; (b) a force of 1000 lb applied suddenly at $t = 0$ and decreasing linearly to zero at $t = 0.5$ sec.

2.6 Using the appropriate charts, determine the maximum displacement and time of maximum displacement for both load cases of Prob. 2.5.

2.7 The rigid frame of Prob. 2.3 is subjected to a horizontal force applied at the girder level. The force increases linearly from zero at $t = 0$ to 4 kips at $t = 0.5$ sec and then remains constant. Neglect damping.
  a. Compute the horizontal deflection at $t = 0.7$ sec.
  b. Using the appropriate chart, determine the maximum deflection and the time of maximum deflection.

Answer
  a. $y = 0.71$ in.
  b. $y_m = 0.74$ in.
     $t_m = 0.55$ sec

2.8 Repeat Prob. 2.2, assuming that the system has 15 percent of critical damping.

2.9 Repeat Prob. 2.5a, assuming that there is 10 percent of critical damping.

2.10 For the load-time function in Fig. 2.29, derive the expression for DLF as a function of t, ω, and t_d which applies when $t > t_d$.

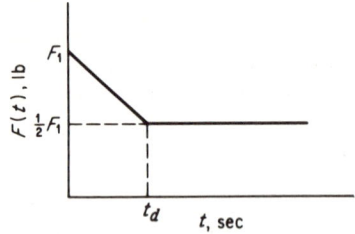

FIGURE 2.29 Problem 2.10. Load-time function.

2.11 For the dynamic system and load function shown in Fig. 2.30, compute by rigorous methods and plot the displacement versus time up to $t = 0.6$ sec.

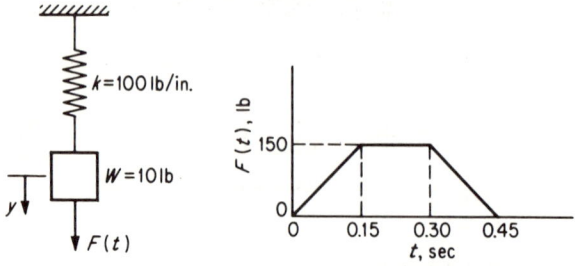

FIGURE 2.30 Problem 2.11. Dynamic system and load-time function.

2.12 It is observed experimentally that the amplitude of free vibration in the fundamental mode of a certain structure decreases from 1 to 0.6 in. in 10 cycles. What is the percentage of critical damping?
Answer
  0.815 percent

2.13 The sliding block shown in Fig. 2.13 has a natural period of 0.5 sec, and the coefficient of friction between the block and surface is 0.05.
  a. If the block is given an initial displacement of 1 in., what is the displacement after 1 cycle of vibration?
  b. If the block is given an initial velocity of 10 in./sec, what is the velocity after 1 cycle?
Answer
  a. 0.51 in.
  b. 3.85 in./sec

2.14 A simple undamped spring-mass system has a natural frequency of 10 rad/sec and is subjected to a force $F_1 \sin \Omega t$. Compute the DLF at $t = 0.4\pi$ sec if (a) $\Omega = 2\omega$, (b) $\Omega = \omega$, and (c) $\Omega = \tfrac{1}{2}\omega$. The system starts at rest.

2.15 Repeat Prob. 2.14 for the case of 5 percent of critical damping.

2.16 What would be the steady-state maximum DLF for the damped system of Prob. 2.15 after many cycles of loading?

2.17 The steel rigid frame shown in Fig. 2.31 supports a rotating machine which exerts a horizontal force at the girder level of $(1000 \sin 11t)$ lb. Assuming 4 percent of critical damping, what is the steady-state amplitude of vibration?
Answer
  0.80 in.

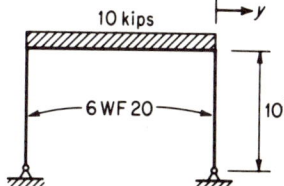

FIGURE 2.31 Problems 2.17 to 2.20.

2.18 The frame of Fig. 2.31 is subjected to horizontal support motion. Using the appropriate chart, determine the maximum absolute deflection at the top of the frame due to the support motion shown in Fig. 2.32. Assume no damping.
Answer
  $y_m = 1.51$ in.

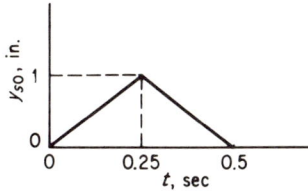

FIGURE 2.32 Problem 2.18. Support motion.

2.19 In Prob. 2.18, compute the bending stress in the columns at $t = 0.30$ sec.
Answer
  $\sigma = 11{,}800$ psi

2.20 In Prob. 2.17, compute the maximum column bending stress due to a continuous support acceleration given by $(50 \sin 11t)$ in./sec^2.

2.21 For the dynamic system and load function shown in Fig. 2.33, compute the deflection at $t = 0.15$ sec.

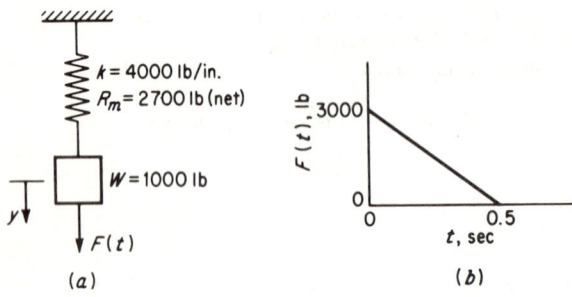

FIGURE 2.33 Problems 2.21 and 2.22. Dynamic system and load-time function.

2.22 Using the appropriate chart, determine the maximum deflection and time of maximum deflection for the system and loading of Fig. 2.33. R_m is the available resistance in excess of the dead-weight spring force.

2.23 Using the appropriate charts, determine the maximum deflection and the time of that deflection for the system of Fig. 2.33a and the load functions of Fig. 2.34a and b.

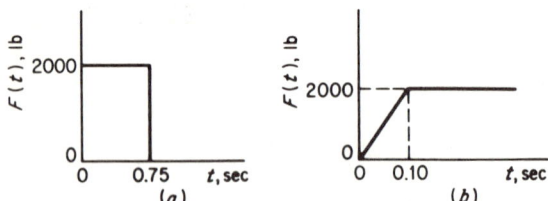

FIGURE 2.34 Problem 2.23. Load-time functions.

3

Lumped-mass Multidegree Systems

3.1 Introduction

The subject of this chapter is the analysis of discrete-parameter systems, i.e., systems consisting of a finite number of lumped masses connected by springs. The number of degrees of freedom is equal to the number of independent types of motion possible in the system. Stated differently, the number of degrees equals the number of independent coordinates necessary and sufficient to define completely the configuration of the system. To illustrate, the position of the pendulum shown in Fig. 3.1a could be defined either by y or by θ, but not by both, since the two coordinates are not independent. Therefore it is a one- rather than a two degree system. The double pendulum shown in Fig. 3.1b is, on the other hand, a two-degree system, since two coordinates (for example, y_1 and y_2 or θ_1 and θ_2) are required.

It may be stated that, for each degree of freedom, there is an independent differential equation of motion. For example, the equations for the two-degree system shown in Fig. 3.2, obtained by considering the dynamic equilibrium of the two masses, are

$$M_1\ddot{y}_1 + k_1 y_1 - k_2(y_2 - y_1) = F_1(t)$$
$$M_2\ddot{y}_2 + k_2(y_2 - y_1) = F_2(t) \tag{3.1}$$

As a somewhat different example, consider the rigid mass supported by two springs as shown in Fig. 3.3. Assuming no horizontal motion,

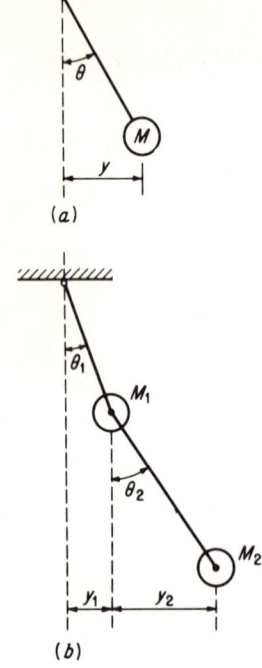

(a)

(b)

FIGURE 3.1 One- and two-degree systems.

there are two independent coordinates, y and θ, and hence it is a two-degree system. The spring forces are given by $k(y \pm \theta d)$ for small rotations, and the two equations of motion based on vertical and rotational dynamic equilibrium are

$$M\ddot{y} + 2ky = F(t)$$
$$I\ddot{\theta} + (2kd^2)\theta = \mathfrak{M}_t(t) \tag{3.2}$$

where I is the mass moment of inertia.

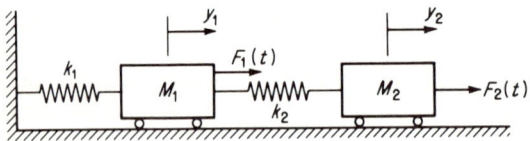

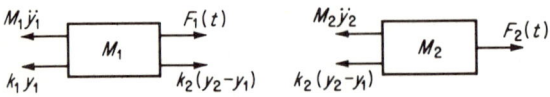

FIGURE 3.2 Two-degree system—dynamic equilibrium.

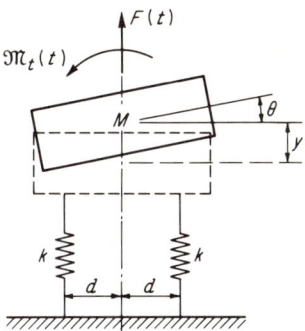

FIGURE 3.3 Uncoupled two-degree system.

An important distinction may be made between Eqs. (3.1) and (3.2). In the former, y_1 and y_2 appear in both equations, and the pair are said to be *coupled*. Determination of the response of the system therefore involves the *simultaneous* solution of two equations. Equations (3.2) are, on the other hand, *uncoupled*, and each of the two equations may be solved separately. The analysis of the system represented by the latter equations may therefore be treated as that for two independent one-degree systems. It should also be apparent from the above that the number of degrees of freedom is not necessarily equal to the number of lumped masses.

Our consideration herein will be restricted to planar systems for which there can be no more than three degrees of freedom per mass. In the most general case of three-dimensional motion, six coordinates are required to define the position of each mass.

The springs in a lumped-mass system may be arranged in different ways, depending upon the characteristics of the structure. For example, if in the three-story building frame of Fig. 3.4a, the girder rigidity approaches infinity, the system (considering only horizontal motion) may be represented as shown in Fig. 3.4b. On the other hand, if the girders are flexible, a proper representation is as shown in Fig. 3.4c. The reason for this difference may be understood if one imagines that the third floor is displaced horizontally while the second floor is held fixed. With rigid girders, no force is transmitted to the first floor, and therefore no spring is needed between the third and first. With flexible girders, however, the joints at the second level would rotate, the columns below would be distorted, and forces would therefore be applied to the first floor, causing it to displace. This interaction is represented in Fig. 3.4c by the spring k_4, and the same reasoning accounts for springs k_5 and k_6. The system in Fig. 3.4b is said to be *close-coupled*, and that in Fig. 3.4c, *far-coupled*. Although the number of springs affects the equations of motion, both

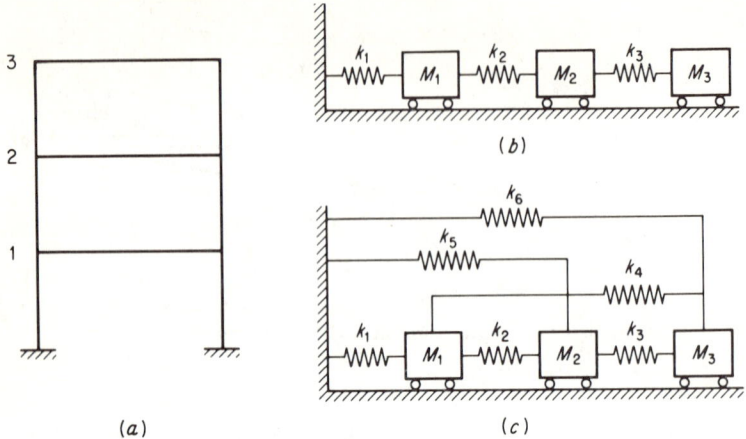

FIGURE 3.4 Close- and far-coupled systems.

systems have three degrees of freedom and the methods of analysis are identical.

Beams or other elements with significant mass at more than one point are always far-coupled systems. For example, the beam with two mass concentrations as shown in Fig. 3.5a could be represented as indicated in Fig. 3.5b. This is proper since a deflection at mass 1 (but not at mass 2) would cause a reaction at support 3. This is accounted for by spring k_4.

It should be emphasized that lumped-mass systems are not idealistic and analyses of such systems are not intended to be mere academic exercises. Many structures such as the frame in Fig. 3.4a have essentially lumped masses since the weight of the columns and walls is often negligible compared with that of the floors. Hence an analysis based on the systems shown in Fig. 3.4a or b is essentially exact. Truly distributed mass systems, e.g., a beam with uniformly distributed weight, have an infinite number of degrees of freedom. However, as will be seen in Chap. 4, any practical analysis deals with a limited number of degrees which can be represented by a lumped-mass-spring system. Thus the methods of analysis given in this chapter have a wide range of application.

It is appropriate at this point to introduce tentatively the concept of *normal modes* (or natural modes) of vibration. A system has exactly the same number of normal modes as degrees of freedom. Associated with each mode is a natural frequency and a *characteristic shape*. The distinguishing feature of a normal mode is that the system could, under certain circumstances, vibrate freely in that mode alone, and during such vibration the ratio of the displacements of any two masses is constant with time. These ratios define the characteristic shape of the mode. An extremely important fact, which is the basis of multidegree analysis, is that the complete motion of the system may be obtained by superimposing

Lumped-mass Multidegree Systems

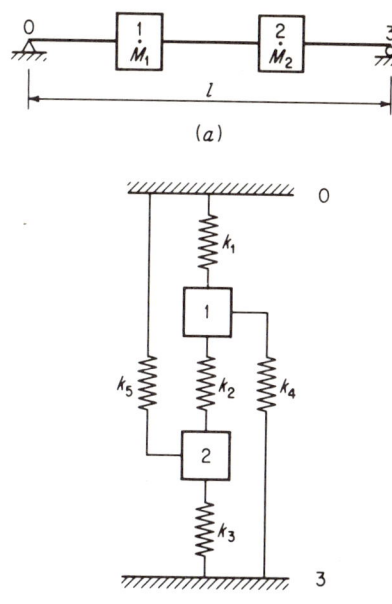

FIGURE 3.5 Beam with lumped masses —equivalent mass-spring system.

the independent motions of the individual modes. More complete definition and the great significance of normal modes will unfold in later sections of this chapter.

The next four sections deal with the determination of natural frequencies and characteristic shapes of normal modes. This emphasis is deliberate for two reasons: (1) such determination must be the first step in the dynamic analysis of the system; and (2) an experienced analyst can deduce a great deal concerning the behavior of a structure from knowledge of its normal modes.

3.2 Direct Determination of Natural Frequencies

The equations of motion for a system having N masses and N degrees of freedom but no external forces have the following form:

$$\begin{aligned} M_1\ddot{y}_1 + k_{11}y_1 + k_{12}y_2 + \cdots + k_{1N}y_N &= 0 \\ M_2\ddot{y}_2 + k_{21}y_1 + k_{22}y_2 + \cdots + k_{2N}y_N &= 0 \\ &\cdots \\ M_N\ddot{y}_N + k_{N1}y_1 + k_{N2}y_2 + \cdots + k_{NN}y_N &= 0 \end{aligned} \quad (3.3)$$

where the k's are stiffness coefficients, which are spring constants or combinations thereof, and the y's are the displacements of the lumped masses.*

---

* Matrix notation is obviously convenient when dealing with simultaneous equations of motion. However, it is not used in this text because it is believed to obscure the physical meaning of equations, and hence is not advisable for introductory material. A matrix formulation of the modal method of analysis is given in the Appendix.

It will first be shown that vibration in a normal mode must always be harmonic. As stated in the previous section, during vibration in a single mode, the displacements of the several masses are always in the same proportion; i.e., all possible positions are geometrically similar. This may be indicated by

$$y_1 = a_1 f(t), \; y_2 = a_2 f(t), \; \ldots, \; y_N = a_N f(t) \qquad (3.4)$$

where $f(t)$ is the same time function in each case, and the a's are the amplitudes of the individual motions. Substituting into Eqs. (3.3),

$$M_1 a_1 \ddot{f}(t) + k_{11} a_1 f(t) + k_{12} a_2 f(t) + \cdots + k_{1N} a_N f(t) = 0$$
$$M_2 a_2 \ddot{f}(t) + k_{21} a_1 f(t) + k_{22} a_2 f(t) + \cdots + k_{2N} a_N f(t) = 0$$
$$\cdots \cdots \cdots \cdots \cdots \cdots \cdots \cdots \cdots \cdots \cdots \cdots \cdots$$
$$M_N a_N \ddot{f}(t) + k_{N1} a_1 f(t) + k_{N2} a_2 f(t) + \cdots + k_{NN} a_N f(t) = 0$$

where $\ddot{f}(t)$ is the second derivative of $f(t)$ with respect to time. Rearranging these last equations,

$$\begin{aligned} \frac{\ddot{f}(t)}{f(t)} &= \frac{-k_{11} a_1 - k_{12} a_2 - \cdots - k_{1N} a_N}{M_1 a_1} \\ \frac{\ddot{f}(t)}{f(t)} &= \frac{-k_{21} a_1 - k_{22} a_2 - \cdots - k_{2N} a_N}{M_2 a_2} \\ &\cdots \cdots \cdots \cdots \cdots \cdots \cdots \cdots \\ \frac{\ddot{f}(t)}{f(t)} &= \frac{-k_{N1} a_1 - k_{N2} a_2 - \cdots - k_{NN} a_N}{M_N a_N} \end{aligned} \qquad (3.5)$$

Since the left sides of Eqs. (3.5) are all identical, the right sides must be equal to the same constant, which will be identified by $-\omega^2$. Thus all equations may be written as

$$\frac{\ddot{f}(t)}{f(t)} = -\omega^2$$

or
$$\ddot{f}(t) + \omega^2 f(t) = 0 \qquad (3.6)$$

Equation (3.6), when solved, yields

$$f(t) = C_1 \sin \omega t + C_2 \cos \omega t = C_3 \sin \omega(t + \alpha) \qquad (3.7)$$

Thus it has been shown that motion as defined by Eq. (3.4) is possible and, furthermore, that such motion is harmonic with a natural frequency of ω. This conclusion applies to any one of the N normal modes of the system. Note that all masses vibrate in phase with the same natural frequency.

To determine the natural frequency of the several modes, Eqs. (3.5) may be used. Substituting the constant $-\omega^2$ for the left-hand sides and rearranging,

$$\begin{aligned}
(k_{11} - M_1\omega^2)a_1 + k_{12}a_2 + \cdots + k_{1N}a_N &= 0 \\
k_{21}a_1 + (k_{22} - M_2\omega^2)a_2 + \cdots + k_{2N}a_N &= 0 \\
&\kern-4em\cdots\cdots\cdots\cdots\cdots\cdots\cdots\cdots \\
k_{N1}a_1 + k_{N2}a_2 + \cdots + (k_{NN} - M_N\omega^2)a_N &= 0
\end{aligned} \qquad (3.8)$$

These equations can be used to solve for relative values of the amplitudes $a_1 \ldots a_N$. Recalling Cramer's rule for solving such equations, we may state that nontrivial values of the amplitudes exist only if the determinant of the coefficients of a is equal to zero, because the equations are homogeneous; i.e., the right sides are zero. Since free vibration must be possible in a normal mode, we write

$$\begin{vmatrix} (k_{11} - M_1\omega^2) & k_{12} & \cdots & k_{1N} \\ k_{21} & (k_{22} - M_2\omega^2) & \cdots & k_{2N} \\ \cdots & \cdots & \cdots & \cdots \\ k_{N1} & k_{N2} & \cdots & (k_{NN} - M_N\omega^2) \end{vmatrix} = 0 \qquad (3.9)$$

All k's are presumably known, and expansion of this determinant leads to a *frequency equation* which can be solved for ω. There is one real root for each normal mode, and hence N natural frequencies are obtained.

This procedure for the determination of natural frequencies is illustrated by an example involving a two-degree system, in the following paragraph. It will become apparent that solution of the frequency equation becomes extremely cumbersome as the number of modes increases. For this reason other procedures have been devised. An iterative method is given in Sec. 3.4, and an approximate method in Sec. 3.5.

In mathematical terms the problem discussed above is known as a *characteristic-value problem* and the quantities ω^2 as *characteristic values*, or *eigenvalues*. The solution of this problem is of importance in many engineering fields, and several methods for the determination and manipulation of eigenvalues are to be found in the literature.[3,12] Attention herein will be focused on those methods which seem most useful for the type of problems considered.

a. Two-degree Systems

The equations of free motion for the far-coupled undamped two-degree system shown in Fig. 3.6 are

$$\begin{aligned} M_1\ddot{y}_1 + k_1 y_1 - k_2(y_2 - y_1) &= 0 \\ M_2\ddot{y}_2 + k_2(y_2 - y_1) + k_3 y_2 &= 0 \end{aligned} \qquad (3.10)$$

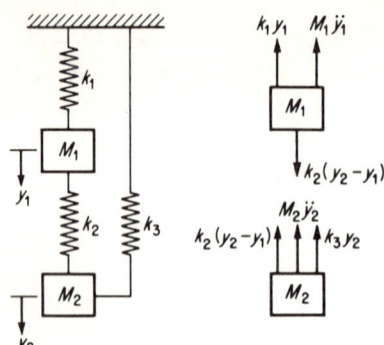

FIGURE 3.6 Two-degree system—dynamic equilibrium.

As indicated by Eq. (3.7), if the system is vibrating in a normal mode, the two displacements are harmonic and in phase, and may be expressed by

$$y_1 = a_1 \sin \omega(t + \alpha) \qquad \ddot{y}_1 = -a_1\omega^2 \sin \omega(t + \alpha)$$
$$y_2 = a_2 \sin \omega(t + \alpha) \qquad \ddot{y}_2 = -a_2\omega^2 \sin \omega(t + \alpha)$$

As an alternative to the direct use of Eqs. (3.9) for the determination of natural frequencies, we may substitute these expressions into Eqs. (3.10) to obtain

$$[-M_1\omega^2 a_1 + k_1 a_1 - k_2(a_2 - a_1)] \sin \omega(t + \alpha) = 0$$
$$[-M_2\omega^2 a_2 + k_2(a_2 - a_1) + k_3 a_2] \sin \omega(t + \alpha) = 0$$

or
$$(-M_1\omega^2 + k_1 + k_2)a_1 + (-k_2)a_2 = 0$$
$$(-k_2)a_1 + (-M_2\omega^2 + k_2 + k_3)a_2 = 0 \qquad (3.11)$$

In order for the amplitudes to have any values other than zero (a necessary condition for a normal mode), the determinant of the coefficients must be equal to zero.

$$\begin{vmatrix} (-M_1\omega^2 + k_1 + k_2) & (-k_2) \\ (-k_2) & (-M_2\omega^2 + k_2 + k_3) \end{vmatrix} = 0$$

Expanding this determinant gives the equation

$$(-M_1\omega^2 + k_1 + k_2)(-M_2\omega^2 + k_2 + k_3) - (k_2)^2 = 0$$

or $\quad (\omega^2)^2 - \left(\dfrac{k_1 + k_2}{M_1} + \dfrac{k_2 + k_3}{M_2}\right)\omega^2 + \dfrac{k_2(k_1 + k_3) + k_1 k_3}{M_1 M_2} = 0 \quad (3.12)$

This is the frequency equation for the two-degree system in Fig. 3.6. The two real roots are the squares of the natural frequencies of the two normal modes.

To illustrate further, suppose that all three spring constants are equal to k and both masses equal to M. Equation (3.12) then becomes

$$(\omega^2)^2 - \left(\frac{4k}{M}\right)\omega^2 + \frac{3k^2}{M^2} = 0$$

The two roots of this equation are

$$\omega_1{}^2 = \frac{k}{M} \qquad \omega_1 = \sqrt{\frac{k}{M}}$$
$$\omega_2{}^2 = 3\frac{k}{M} \qquad \omega_2 = 1.73\sqrt{\frac{k}{M}}$$

These are the natural circular frequencies of the two normal modes. The smaller frequency ω_1 corresponds to the *fundamental*, or *first*, *mode*, while ω_2 is the frequency of the *second mode*.

3.3 Characteristic Shapes

Having the natural frequencies of the multidegree system represented by Eqs. (3.3), the characteristic shapes of the modes may be obtained by the use of Eqs. (3.8). If the value of ω^2 for a particular mode is substituted into these N equations, there are then exactly N unknowns, namely, the *characteristic amplitudes* $a_1 \cdots a_N$ of that mode. Since the right sides of Eqs. (3.8) are zero, unique values of the a's are not obtained. However, it is possible to obtain the relative values of all amplitudes, or in other words, the ratio of any two. If an arbitrary value is given one amplitude, all others are then fixed in magnitude. A set of such arbitrary amplitudes defines the characteristic shape, since the latter is not dependent upon absolute values of amplitude. In mathematical terms, a set of modal amplitudes is known as a *characteristic vector*.

It is not surprising that unique values of the characteristic amplitudes are unobtainable. We are here dealing with free vibration, the cause of which has not been defined by either initial conditions or forcing function. The important point is that the amplitudes of a normal mode are always in the same proportion; i.e., the shape is maintained, regardless of the cause of the vibration.

To illustrate the above, consider again the two-degree system of Sec. 3.2a and Fig. 3.6, for which the natural frequencies were found to be $\sqrt{k/M}$ and $1.73\sqrt{k/M}$. Since the k's were taken to be equals, as were the M's, Eqs. (3.11) become

$$(-M\omega^2 + 2k)a_1 + (-k)a_2 = 0$$
$$(-k)a_1 + (-M\omega^2 + 2k)a_2 = 0$$

Substituting ω_1, the frequency of the first mode, into the first equation yields

$$(-k + 2k)a_{11} + (-k)a_{21} = 0$$

Therefore
$$a_{11} = a_{21}$$

which defines the characteristic shape of the first mode. The same result would have been obtained by substitution into the second equation. The notation adopted is that the first subscript on the a indicates the mass, or point on the structure at which the amplitude occurs, and the second subscript designates the mode. Substituting ω_2 into either equation yields

$$(-3k + 2k)a_{12} + (-k)a_{22} = 0$$

Therefore
$$a_{12} = -a_{22}$$

which defines the characteristic shape of the second mode. If it is desired to assign arbitrary values to the amplitudes, the two modal shapes could be indicated by

$$a_{11} = +1 \qquad a_{21} = +1$$
$$a_{12} = +1 \qquad a_{22} = -1$$

The two characteristic shapes, i.e., the motions associated with the normal modes, are indicated in Fig. 3.7. In the first mode the two masses move in the same direction and by the same amount. In the second mode they move by the same amount but in opposite directions. In both cases

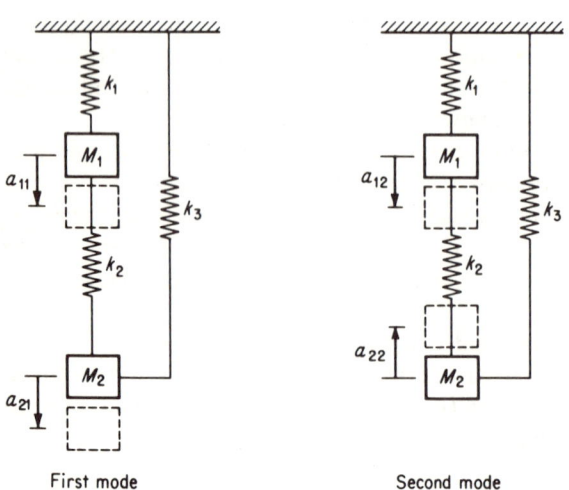

FIGURE 3.7 Characteristic shapes of normal modes.

the motions of the two masses are in phase; i.e., the maximum displacements are attained simultaneously. The neutral point of the vibration is the static dead-load position, and the a's are in reality amplitudes of the total motion. It should be intuitively obvious that the type of distortion associated with the first mode should, as we have shown, have a lower natural frequency than that associated with the second mode.

a. Orthogonality

An extremely important property of normal modes is the fact that any two modes are orthogonal. This may be expressed as follows:

$$\sum_{r=1}^{j} M_r a_{rn} a_{rm} = 0 \qquad (3.13a)$$

where n and m identify any two normal modes of the system, and the subscript r refers to the rth mass out of a total of j masses. The summation therefore indicates a series with one term for each mass of the system. The proof of Eq. (3.13a) is given below.*

When the system is vibrating in the nth mode and has attained the maximum amplitude, the masses could be placed in static equilibrium by the application of inertia forces equal to $Ma\omega_n{}^2$. This statement, which is not limited to maximum displacement but could be made for any modal position, is merely D'Alembert's principle of dynamic equilibrium. For the nth mode and the rth mass, the inertia force would be identified by $M_r a_{rn} \omega_n{}^2$. Suppose now that a virtual displacement corresponding to the mth mode is introduced. The virtual work done by the inertia forces during this process is

$$\sum_{r=1}^{j} (M_r a_{rn} \omega_n{}^2) a_{rm}$$

If the process is now reversed, i.e., virtual displacements corresponding to the nth mode are imposed on the system already in the mth mode, the virtual work is

$$\sum_{r=1}^{j} (M_r a_{rm} \omega_m{}^2) a_{rn}$$

According to the elementary principle known as Betti's law,[13] these two

---

* An equivalent statement of orthogonality is that the scalar product of two normalized characteristic vectors is zero. Various proofs may be found in textbooks on applied mathematics.[12] The one given here is purposely developed by the use of common structural concepts. An alternative proof in matrix notation is given in the Appendix.

96 *Introduction to Structural Dynamics*

virtual work quantities must be equal. Therefore

$$\omega_n^2 \sum_{r=1}^{j} M_r a_{rn} a_{rm} = \omega_m^2 \sum_{r=1}^{j} M_r a_{rm} a_{rn}$$

or

$$(\omega_n^2 - \omega_m^2) \sum_{r=1}^{j} M_r a_{rn} a_{rm} = 0$$

Since $(\omega_n^2 - \omega_m^2)$ cannot be zero if $n \neq m$,

$$\sum_{r=1}^{j} M_r a_{rn} a_{rm} = 0 \tag{3.13a}$$

which is the *orthogonality condition*. This is an extremely useful concept in the analysis of multidegree systems.

The validity of Eq. (3.13a) can be demonstrated by consideration of the two-degree example of Secs. 3.2a and 3.3, for which the characteristic shapes were found to be

$$a_{11} = +1 \quad a_{21} = +1$$
$$a_{12} = +1 \quad a_{22} = -1$$

Writing Eq. (3.13a) for a two-degree system,

$$M_1 a_{11} a_{12} + M_2 a_{21} a_{22} = 0$$

and substituting $M_1 = M_2 = M$ (as assumed in the example) and the numerical characteristic amplitudes, we obtain

$$M(+1)(+1) + M(+1)(-1) \equiv 0$$

Thus the orthogonality condition is satisfied.

Another form of orthogonality which is also useful involves spring constants rather than masses and may be developed as follows.

For a system vibrating in a normal mode, the inertia and spring forces form a set of forces in equilibrium. By the law of virtual work the net work done by such a set during a virtual distortion must be zero. For the nth mode, let the inertia force at mass r be $M_r \omega_n^2 a_{rn}$ and the force in spring g be $k \Delta_{gn}$, where Δ_{gn} is the spring distortion. Then the total work during a virtual distortion in the form of the mth mode is

$$\sum_{r=1}^{j} (M_r \omega_n^2 a_{rn}) a_{rm} + \sum_{g=1}^{s} k_g \Delta_{gn} \Delta_{gm} = 0$$

or

$$\sum_{r=1}^{j} M_r a_{rn} a_{rm} + \frac{1}{\omega_n^2} \sum_{g=1}^{s} k_g \Delta_{gn} \Delta_{gm} = 0$$

where j and s are the number of masses and springs, respectively. By Eq. (3.13a), the first series in the last equation is zero, and therefore

$$\sum_{g=1}^{s} k_g \Delta_{gn} \Delta_{gm} = 0 \tag{3.13b}$$

which is the *second orthogonality condition*.

In the two-degree example previously cited, the modal spring distortions may be determined from the modal amplitudes as follows (Fig. 3.7):

$$\Delta_{11} = a_{11} = +1 \qquad \Delta_{21} = a_{21} - a_{11} = 0$$
$$\Delta_{12} = a_{12} = +1 \qquad \Delta_{22} = a_{22} - a_{12} = -2$$
$$\Delta_{31} = a_{21} = +1 \qquad \Delta_{32} = a_{22} = -1$$

Therefore, since all spring constants are k, substitution into Eq. (3.13b) gives

$$\sum_{g=1}^{3} k_g \Delta_{gn} \Delta_{gm} = k(+1)(+1) + k(0)(-2) + k(+1)(-1) \equiv 0$$

which demonstrates the validity of the second orthogonality condition.

3.4 Stodola-Vianello Procedure for Natural Frequencies and Characteristic Shapes

Direct determination of natural frequencies and characteristic shapes as given in Sec. 3.2 is excessively tedious if there are more than two, or perhaps three, degrees of freedom. This is true because, for an N-degree system, the frequency equation [e.g., Eq. (3.12)] is of degree N and the solution is very laborious if $N > 2$ or 3. Furthermore, the expansion of the determinant [Eq. (3.9)] may be impractical in such cases. It is therefore necessary to resort to numerical, iterative (i.e., trial-and-error) procedures. The most commonly used of these is that associated with the names of Stodola and Vianello.

In the general case, this procedure involves the solution of Eqs. (3.8) by iteration, which yields both the natural frequency and the characteristic shape. The procedure is as follows: (1) assume a characteristic shape, i.e., a set of a values; (2) using one of Eqs. (3.8), solve for ω^2; (3) using the remaining $(N - 1)$ equations, obtain a new shape by solving for $(N - 1)$ a's in terms of the Nth a; and (4) using the new shape just computed as the assumed shape in the next cycle, repeat the procedure to convergence, i.e., until the computed shape is the same as the previously assumed shape. In step 3 it is usually convenient to assign a unit value to the Nth a. The rate of convergence may be increased by "overrelaxation," i.e., by using an improved estimate of the shape in step 4,

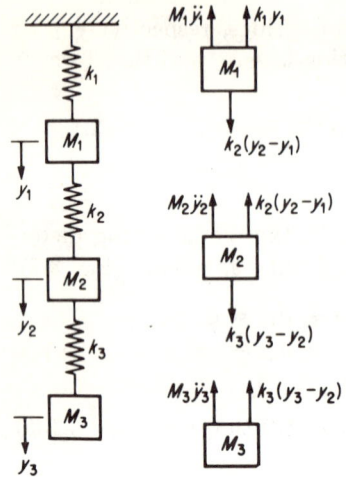

FIGURE 3.8 Three-degree system—dynamic equilibrium.

based on the trend of convergence rather than that actually computed in step 3. However, this refinement is not necessary and requires experience to be done successfully.

The procedure outlined will converge on either the highest or the lowest mode, depending upon the form of the equations of motion (Sec. 3.4a). The other modes are then obtained, using the same procedure, after having first eliminated one of the equations by use of the orthogonality condition.

The Stodola-Vianello method is best described by a numerical example. For this purpose the natural frequencies and characteristic shapes of the three-degree system shown in Fig. 3.8 will be determined. The equations of free motion are

$$M_1 \ddot{y}_1 + k_1 y_1 - k_2(y_2 - y_1) = 0$$
$$M_2 \ddot{y}_2 + k_2(y_2 - y_1) - k_3(y_3 - y_2) = 0 \quad (3.14)$$
$$M_3 \ddot{y}_3 + k_3(y_3 - y_2) = 0$$

Proceeding as before, we substitute

$$y_r = a_{rn} \sin \omega_n(t + \alpha) \quad \text{and} \quad \ddot{y}_r = -a_{rn}\omega_n^2 \sin \omega_n(t + \alpha)$$

cancel the common sine terms, and rearrange to obtain

$$(-M_1 \omega_n^2 + k_1 + k_2)a_{1n} + (-k_2)a_{2n} = 0$$
$$(-k_2)a_{1n} + (-M_2 \omega_n^2 + k_2 + k_3)a_{2n} + (-k_3)a_{3n} = 0 \quad (3.15)$$
$$(-k_3)a_{2n} + (-M_3 \omega_n^2 + k_3)a_{3n} = 0$$

where the subscript n indicates that the equations apply to any mode. For the numerical example, the following values are given:

$$M_1 = 2 \text{ lb-sec}^2/\text{in.} \quad M_2 = M_3 = 1 \text{ lb-sec}^2/\text{in.}$$
$$k_1 = 6000 \text{ lb/in.} \quad k_2 = 4000 \text{ lb/in.} \quad k_3 = 2000 \text{ lb/in.}$$

Insertion of these values in Eq. (3.15) and rearrangement leads to the following convenient form of the equations:

(a) $\quad \omega_n^2 a_{1n} = 5000 a_{1n} - 2000 a_{2n}$
(b) $\quad \omega_n^2 a_{2n} = -4000 a_{1n} + 6000 a_{2n} - 2000 a_{3n}$ $\qquad$ (3.16)
(c) $\quad \omega_n^2 a_{3n} = -2000 a_{2n} + 2000 a_{3n}$

The procedure is to substitute an assumed shape in the right sides of the equations, use one equation to compute ω_n^2, and use the two remaining equations to obtain the second trial shape. After this process has been repeated several times, the shape will converge on that of the highest, or third, mode, and therefore the analyst should begin by making an estimate of the relative amplitudes for that mode. For example, let us assume

$$a_{13} = +1 \qquad a_{23} = -1.5 \qquad a_{33} = +0.5$$

Substituting these in Eq. (3.16a),

$$\omega_3^2(+1) = 5000(+1) - 2000(-1.5) \qquad \omega_3^2 = 8000$$

Substituting this value for ω_3^2 and the a's in the right side of Eq. (3.16b),

$$8000(a_{23}) = -4000(+1) + 6000(-1.5) - 2000(+0.5)$$
Therefore $\qquad a_{23} = -1.75$

Finally, from Eq. (3.16c),

$$8000(a_{33}) = -2000(-1.5) + 2000(0.5) \qquad a_{33} = +0.5$$

Therefore the first estimate of ω_3^2 is 8000, and the next trial shape is to be taken as $a_{13} = +1$, $a_{23} = -1.75$, and $a_{33} = +0.5$. This procedure is repeated until convergence is achieved as indicated in Table 3.1. It may be seen that five cycles are required to obtain what is considered to be satisfactory agreement between the assumed and computed shapes. The last values computed are adopted as shown in the summary of Table 3.1. It is important to retain sufficient significant figures in the first mode computed since small errors here would result in large errors in the last mode computed. It should be recalled that the absolute amplitudes are indeterminate and the values used are arbitrary, since a_{13} was arbitrarily taken as unity. However, the ratios of the amplitudes are all that is required.

Proceeding to the next mode, we use the orthogonality condition to reduce the number of equations by one. If, in Eq. (3.13a), n is taken to be the third mode just computed and m the second mode now to be com-

Introduction to Structural Dynamics

Table 3.1 Stodola-Vianello Procedure (for System Shown in Fig. 3.8)

Third mode

Trial no.	Trial values			Computed values		
	a_{13}	a_{23}	a_{33}	ω_3^2 Eq. (3.16a)	a_{23} Eq. (3.16b)	a_{33} Eq. (3.16c)
1	+1	−1.50	+0.50	8000	−1.75	+0.500
2	+1	−1.75	+0.50	8500	−1.82	+0.53
3	+1	−1.82	+0.53	8640	−1.85	+0.545
4	+1	−1.85	+0.545	8700	−1.86	+0.550
5	+1	−1.86	+0.550	8720	−1.866	+0.554

Second mode

Trial no.	Trial values		Computed values		
	a_{12}	a_{22}	ω_2^2 Eq. (3.18a)	a_{22} Eq. (3.18b)	a_{32} Eq. (3.17)
1	+1	+0.5	4000	0.713	
2	+1	+0.713	3570	+0.754	
3	+1	+0.754	3440	+0.764	
4	+1	+0.764	3470	+0.765	−1.033

Summary

	ω^2	a_1	a_2	a_3	ω	f, cps
Third mode	8720	+1	−1.866	+0.554	93.4	14.86
Second mode	3470	+1	+0.765	−1.033	58.9	9.37
First mode	780	+1	+2.11	+3.50	28.0	4.45

puted, the expanded series is

$$\sum_{r=1}^{3} M_r a_{rm} a_{rn} = M_1 a_{12} a_{13} + M_2 a_{22} a_{23} + M_3 a_{32} a_{33}$$

$$= 2(a_{12})(+1) + 1(a_{22})(-1.866) + 1(a_{32})(+0.554) = 0$$

or
$$a_{32} = -3.610 a_{12} + 3.368 a_{22} \tag{3.17}$$

Substituting the last expression for a_{32} into Eqs. (3.16) results in only two independent equations, which may be written as follows:

(a) $\quad \omega_2^2 a_{12} = 5000 a_{12} - 2000 a_{22}$
(b) $\quad \omega_2^2 a_{22} = 3220 a_{12} - 736 a_{22}$ $\tag{3.18}$

We now iterate Eqs. (3.18) to obtain ω_2^2, a_{12}, and a_{22}, and then use Eq. (3.17) to obtain a_{32}. Assume $a_{12} = +1$, $a_{22} = +0.5$. By Eq. (3.18a),

$$\omega_2^2(+1) = 5000(+1) - 2000(+0.5) \qquad \omega_2^2 = 4000$$

By Eq. (3.18b),

$$4000(a_{22}) = 3220(+1) - 736(+0.5) \qquad a_{22} = +0.713$$

Subsequent cycles of iteration are given in Table 3.1, where it may be seen that four cycles are required for satisfactory results. In the last step, Eq. (3.17) is used to obtain a_{32}.

The first mode can now be computed directly from the orthogonality condition. This is always true of the last mode computed. Equation (3.13a) must be applied twice, first combining the third and first modes, and then combining the second and first modes.

$$\sum_{r=1}^{3} M_r a_{r2} a_{r1} = M_1 a_{12} a_{11} + M_2 a_{22} a_{21} + M_3 a_{32} a_{31}$$

$$= 2(+1)a_{11} + 1(+0.765)a_{21} + 1(-1.033)a_{31} = 0 \qquad (3.19)$$

$$\sum_{r=1}^{3} M_r a_{r3} a_{r1} = M_1 a_{13} a_{11} + M_2 a_{23} a_{21} + M_3 a_{33} a_{31}$$

$$= 2(+1)a_{11} + 1(-1.866)a_{21} + 1(+0.554)a_{31} = 0$$

Taking $a_{11} = +1$, these two equations are solved simultaneously to provide

$$a_{21} = +2.11 \qquad a_{31} = +3.50$$

Any one of Eqs. (3.16) may now be used to compute ω_1^2. Using Eq. (3.16a),

$$\omega_1^2(+1) = 5000(+1) - 2000(+2.11)$$
$$\omega_1^2 = 780$$

As a check on the accuracy of the solution, the frequency and shape of the last computed mode may be substituted into either Eq. (3.16b) or (3.16c). It will be found that, in most cases, slide-rule computation does not provide sufficient significant figures if there are more than three modes.

The summary in Table 3.1 gives the complete solution of the problem, i.e., the frequencies and characteristic shapes of the normal modes for the system in Fig. 3.8. The characteristic shapes are depicted graphically in Fig. 3.9.

a. Alternative Use of Flexibility and Stiffness Equations

In the preceding sections the natural frequencies and characteristic shapes were derived from stiffness equations; i.e., the equations of motion

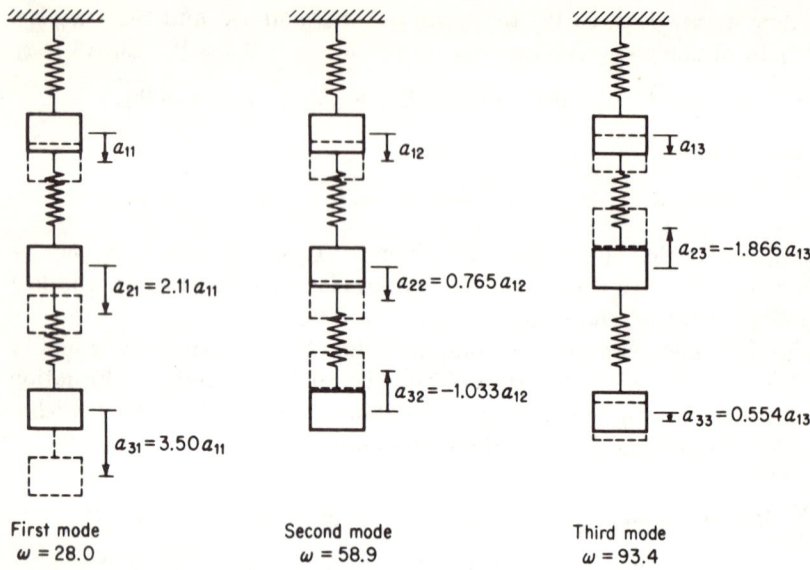

FIGURE 3.9 Characteristic shapes of normal modes.

were written in terms of stiffness coefficients k. For example, each of Eqs. (3.3) is of the form

$$M\ddot{y} + \sum ky = 0 \qquad (3.20)$$

where the k's are stiffnesses. When the Stodola-Vianello procedure is applied to such equations, the first mode obtained is the *highest mode;* e.g., in the computations based on Fig. 3.8 the third mode was first obtained. In many cases this is undesirable since the fundamental, or first, mode is usually of greatest interest. In fact, when dealing with many degrees of freedom, analysis is often based on the first few modes alone, while the higher modes are neglected.

This difficulty can be circumvented if the equations of motion are written in terms of *flexibility coefficients*. The form of the equations is then

$$y + \sum DM\ddot{y} = 0 \qquad (3.21)$$

where the D's are flexibility coefficients. If the Stodola-Vianello method is applied to equations such as (3.21), the first mode obtained, rather than the highest, is the fundamental.

To illustrate the use of flexibility coefficients, consider the beam with three concentrated masses as shown in Fig. 3.10. The coefficients are denoted by D_{ij}, which, by the usual structural convention, indicates the

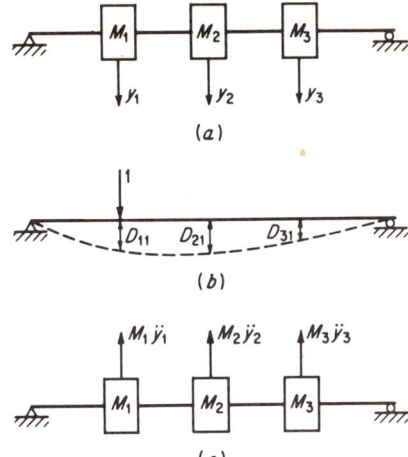

FIGURE 3.10 Beam with concentrated masses—flexibility coefficients.

deflection at point i due to a unit force applied at point j. Thus the coefficients are determined by the successive application of unit forces at the various points. For example, in Fig. 3.10b, the coefficients obtained by applying a unit force at point 1 are indicated. Since, in a normal mode, the only forces involved are inertia forces, the equations of motion are derived from the fact that, at any instant, the total deflection is the sum of the deflections due to the individual inertia forces. Considering the force system shown in Fig. 3.10c, the following may be written:

$$
\begin{aligned}
y_1 &= -M_1\ddot{y}_1 D_{11} - M_2\ddot{y}_2 D_{12} - M_3\ddot{y}_3 D_{13} \\
y_2 &= -M_1\ddot{y}_1 D_{21} - M_2\ddot{y}_2 D_{22} - M_3\ddot{y}_3 D_{23} \\
y_3 &= -M_1\ddot{y}_1 D_{31} - M_2\ddot{y}_2 D_{32} - M_3\ddot{y}_3 D_{33}
\end{aligned} \tag{3.22}
$$

As in previous discussions, we now substitute $y_r = a_{rn} \sin \omega_n(t + \alpha)$ and $\ddot{y}_r = -a_{rn}\omega_n^2 \sin \omega_n(t + \alpha)$, cancel the common sine terms, and rearrange to obtain

$$
\begin{aligned}
\left(M_1 D_{11} - \frac{1}{\omega_n^2}\right) a_{1n} + (M_2 D_{12}) a_{2n} + (M_3 D_{13}) a_{3n} &= 0 \\
(M_1 D_{21}) a_{1n} + \left(M_2 D_{22} - \frac{1}{\omega_n^2}\right) a_{2n} + (M_3 D_{23}) a_{3n} &= 0 \\
(M_1 D_{31}) a_{1n} + (M_2 D_{32}) a_{2n} + \left(M_3 D_{33} - \frac{1}{\omega_n^2}\right) a_{3n} &= 0
\end{aligned} \tag{3.23}
$$

These equations are equivalent to Eqs. (3.8) or, in the previous example given, Eqs. (3.15). The Stodola-Vianello procedure may be applied to Eqs. (3.23) in exactly the same manner as previously described for stiffness equations. The first step would be to rewrite the equations in the form

$$\frac{a_{1n}}{\omega_n^2} = M_1 D_{11} a_{1n} + M_2 D_{12} a_{2n} + M_3 D_{13} a_{3n}$$

$$\frac{a_{2n}}{\omega_n^2} = M_1 D_{21} a_{1n} + M_2 D_{22} a_{2n} + M_3 D_{23} a_{3n} \qquad (3.24)$$

$$\frac{a_{3n}}{\omega_n^2} = M_1 D_{31} a_{1n} + M_2 D_{32} a_{2n} + M_3 D_{33} a_{3n}$$

which are equivalent to Eqs. (3.16). The procedure would then be exactly the same as illustrated after Eqs. (3.16). It will be found that the fundamental mode is first obtained.

In view of the disadvantage of obtaining the highest mode first, the only reason for using stiffness equations is that, in many cases, the stiffness coefficients are more easily computed. For example, the stiffness coefficients for the rigid frame of Fig. 3.11 are obtained by introducing unit deflections at the floors. One set of coefficients would be computed by a simple moment-distribution sidesway solution as illustrated in Fig. 3.11b, where the holding forces are the desired stiffness coefficients. On the other hand, determination of a similar set of flexibility coefficients requires a more complex analysis involving the superposition of three sidesway solutions followed by deflection computations as indicated in Fig. 3.11c.

In other cases, the flexibility coefficients are more easily determined than are the stiffness coefficients. For example, for the beam of Fig. 3.10, it is easier to determine the deflections due to unit loads than to determine holding forces due to imposed unit deflections. If shearing distortions are to be included, it is impossible to obtain the stiffness coefficients directly, and one must begin with flexibilities.

It should be noted that equations of one type can, if necessary, be *inverted* to obtain the equations of the other type. For example, if Eqs.

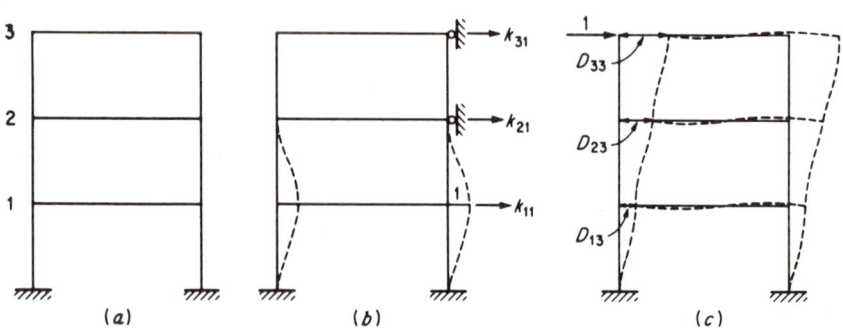

FIGURE 3.11 Rigid frame—stiffness and flexibility coefficients.

(3.22) are solved for $M_1\ddot{y}_1$, $M_2\ddot{y}_2$, and $M_3\ddot{y}_3$, one obtains the corresponding stiffness equations in the form of Eqs. (3.3). This or the reverse operation is sometimes a useful procedure.

3.5 Modified Rayleigh Method for Natural Frequencies

When the system has many degrees of freedom, the Stodola-Vianello method for determining frequencies and characteristic shapes becomes exceedingly cumbersome, if not impossible. This is true because one must deal with the complete set of equations of motion, one equation for each degree of freedom. In this section an approximate method usually attributed to Rayleigh is presented for use when one is interested in only a few of the lower modes.

It is shown in Sec. 3.5a that, by this method, the natural frequency of the fundamental mode can be obtained with considerable accuracy and yet with relative ease. The characteristic shape obtained is less accurate, but can be improved by an iterative procedure. In Sec. 3.5b, the method is extended to include higher natural modes. The method described herein is particularly useful for systems with continuous mass distribution and hence an infinite number of degrees of freedom. Applications of this sort are given in Chap. 4.

a. Fundamental Mode

The Rayleigh method is an energy procedure based on an assumed characteristic shape. Let $\phi'(x)$ be an assumed nondimensional shape, where x is a coordinate defining positions on the structure, and let the assumed displacements be $A'\phi'(x)$, where A' is an arbitrary constant. For a lumped-mass system, the assumed displacement at mass r is expressed by $A'\phi'_r$. If this were the true shape of a normal mode, the corresponding inertia force would be

$$M_r A' \phi'_r \omega^2 = \text{constant } (M_r \phi'_r)$$

Since the absolute value of the inertia force is indeterminate and of no importance, the constant may be dropped. Therefore let F_{ri} represent this force and be given by

$$F_{ri} = M_r \phi'_r$$

Suppose that all the forces F_{ri} were applied to the system and the resulting deflected shape determined. This new shape will be identified by $\phi''(x)$, the displacements by $A''\phi''(x)$, and that at mass r by $A''\phi''_r$. If these displacements represent the true modal shape, it can be stated that the kinetic energy of this system at zero displacement equals the work done by the inertia forces as the system moves from zero to maximum displacement. This statement may be further explained by the fact that the work done by the inertia forces must equal the strain energy in the system at maximum displacement and, furthermore, that this maximum strain energy must equal the maximum kinetic energy at zero displacement. Since $A''\phi_r''$ is the maximum displacement and the motion is harmonic, the maximum velocity of mass r is $A''\phi_r''\omega$. Therefore, for the complete system,

$$\mathcal{K} = \sum_{r=1}^{j} \tfrac{1}{2} M_r (A''\phi_r''\omega)^2$$

$$\mathcal{U} = \sum_{r=1}^{j} \tfrac{1}{2} F_{ri}(A''\phi_r'')$$

where $\mathcal{K}$ = total kinetic energy at zero displacement
$\mathcal{U}$ = total strain energy at maximum displacement
j = number of masses

In the expression for strain energy, the $\tfrac{1}{2}$ factor appears because the force varies linearly from zero to a maximum as the displacement increases. Equating the above energy expressions, we obtain

$$\omega^2 = \frac{\sum_{r=1}^{j} F_{ri}\phi_r''}{A'' \sum_{r=1}^{j} M_r(\phi_r'')^2} \qquad (3.25)$$

which gives the natural frequency of the fundamental mode for a lumped-mass system based on an assumed shape $\phi'(x)$. The computed shape $\phi''(x)$ is a better approximation of the characteristic shape of the mode.

To summarize, the complete procedure is as follows: (1) assume a shape $\phi'(x)$; (2) compute the corresponding inertia forces F_{ri}; (3) compute the deflections due to F_{ri}, or $A''\phi''(x)$; and (4) compute the natural frequency by Eq. (3.25). If greater accuracy is required, this procedure can be repeated using $\phi''(x)$ as the assumed shape for the next cycle.

The success of the Rayleigh method lies in the fact that accurate values of frequency are obtained even though the assumed shape is only approximately correct. The best first estimate of the fundamental-mode shape is usually that produced by the dead weight, or in other words, by the gravity forces acting on the masses of the system. This was, in fact, the basic concept of the original Rayleigh method. The complication encountered in applying the method to higher modes is that no such simple device exists for estimating the modal shapes.

To illustrate the above method, we shall determine the natural frequency and characteristic shape for the fundamental mode of the three-degree system shown in Fig. 3.12. This is the same system as that used

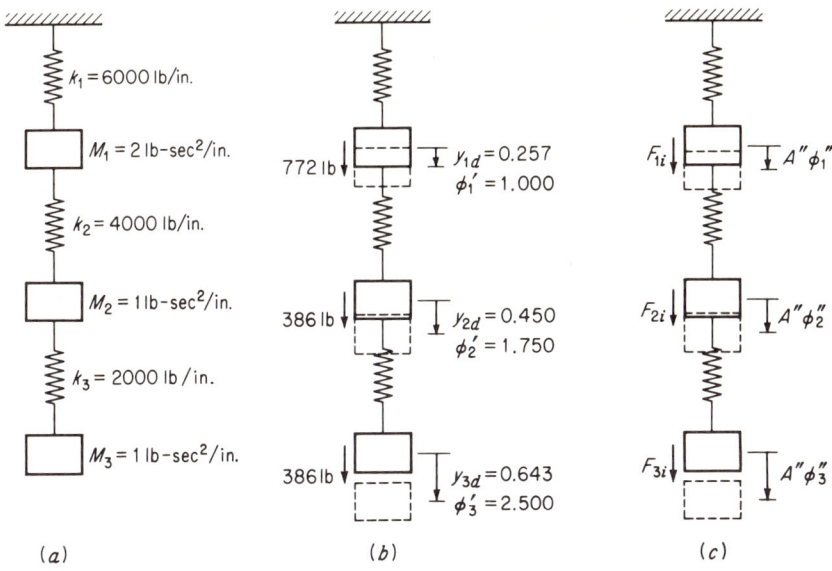

FIGURE 3.12 Determination of fundamental mode by the Rayleigh method.

for illustration of the Stodola-Vianello method in Sec. 3.4. For the Rayleigh method, the first step is to compute the dead-load deflections as indicated in Fig. 3.12b. $\phi'(x)$ is evaluated by arbitrarily letting A' equal the deflection at mass 1, and hence $\phi'_1 = 1$. The remainder of the procedure is shown in Table 3.2, where the dead-load shape is the assumed shape in the first cycle, and the computed shape of that cycle is taken to be the assumed shape of the second cycle. It may be observed that the frequency obtained in the first cycle is indeed accurate, since no later improvement is made. However, the characteristic shape is not very accurate, and three cycles are shown to indicate the rate of convergence.

Comparison with the values for the fundamental mode in Table 3.1 (the same problem done by Stodola-Vianello) reveals that the two give essentially the same result. Since the first mode was the last obtained in the Stodola-Vianello procedure, the values given in Table 3.2 are the more accurate.

It should be apparent that the Rayleigh method is extremely useful for systems having many degrees of freedom. This is particularly true when the complete set of equations of motion are not required for other purposes and when the structure is such that stiffness and flexibility coefficients are difficult to compute. By this method deflections need be computed for only one set of loads, while to write the equations needed for the Stodola-Vianello method requires a whole series of computations for stiffness or flexibility coefficients.

108   Introduction to Structural Dynamics

Table 3.2   Modified Rayleigh Method for Fundamental Mode of System in Fig. 3.12

Cycle	Mass point	Assumed shape ϕ_r'	$F_{ri} = M_r\phi_r'$	Computed deflection $A''\phi_r''$	Computed shape ϕ_r''	$F_{ri}\phi_r''$	$M_r(\phi_r'')^2$
1	1	1.00	2.00	0.001042	1.00	2.00	2.00
	2	1.75	1.75	0.002104	2.02	3.54	4.08
	3	2.50	2.50	0.003354	3.22	8.05	10.37
						13.59	16.46
2	1	1.00	2.00	0.001207	1.00	2.00	2.00
	2	2.02	2.02	0.002517	2.08	4.20	4.33
	3	3.22	3.22	0.004127	3.42	11.01	11.70
						17.21	18.03
3	1	1.00	2.00	0.001250	1.00	2.00	2.00
	2	2.08	2.08	0.002625	2.10	4.37	14.41
	3	3.42	3.42	0.004335	3.47	11.87	12.04
						18.24	18.45

Eq. (3.25):

$$\omega^2 = \frac{13.59}{0.001042 \times 16.46} = 792 \qquad \omega = 28.2 \text{ rad/sec}$$

$$\omega^2 = \frac{17.21}{0.001207 \times 18.03} = 791 \qquad \omega = 28.2 \text{ rad/sec}$$

$$\omega^2 = \frac{18.24}{0.00125 \times 18.45} = 791 \qquad \omega = 28.2 \text{ rad/sec}$$

b. Higher Modes

The Rayleigh method may be extended to obtain a higher-mode shape and frequency by the *Schmidt orthogonalization procedure*.[12] In short, this procedure is to assume a shape, "sweep out" those components of the shape associated with lower modes, and then apply the Rayleigh method based on the residual shape. The computation will always converge on the next higher mode.[3] The theoretical basis for the procedure is given below.

Any shape which might be assumed can be expressed as

$$\phi_{ra} = \sum_{m=1}^{N} \psi_m \phi_{rm} \qquad (3.26)$$

where ϕ_{ra} = assumed deflection of mass r
ϕ_{rm} = deflection coordinate for mth mode
ψ_m = *participation factor* for mth mode
N = number of modes

Lumped-mass Multidegree Systems

This equation states the known fact that any shape which might be imagined can be formed by a linear combination of modal shapes.* If both sides of Eq. (3.26) are multiplied by $M_r \phi_{rq}$, where ϕ_{rq} is the deflection coordinate for the qth mode, we obtain

$$M_r \phi_{ra} \phi_{rq} = \sum_{m=1}^{N} \psi_m M_r \phi_{rm} \phi_{rq} \qquad (3.27)$$

Summing over all masses,

$$\sum_{r=1}^{j} M_r \phi_{ra} \phi_{rq} = \sum_{r=1}^{j} \sum_{m=1}^{N} \psi_m M_r \phi_{rm} \phi_{rq} \qquad (3.28)$$

Since by the orthogonality condition

$$\sum_{r=1}^{j} M_r \phi_{rm} \phi_{rq} = 0$$

all terms on the right side of Eq. (3.28) may be eliminated, except those for $m = q$, and the equation may be written

$$\sum_{r=1}^{j} M_r \phi_{ra} \phi_{rq} = \sum_{r=1}^{j} \psi_q M_r \phi_{rq}^2$$

Therefore the participation factor for the qth mode is

$$\psi_q = \frac{\sum_{r=1}^{j} M_r \phi_{ra} \phi_{rq}}{\sum_{r=1}^{j} M_r \phi_{rq}^2} \qquad (3.29)$$

and the participation of the qth mode in the assumed deflection at mass r is $\psi_q \phi_{rq}$. If, for example, the shapes of two modes u and v had been previously determined and the assumed deflection "swept" of these modes, ϕ_{ras} would be given by

$$\phi_{ras} = \phi_{ra} - \psi_u \phi_{ru} - \psi_v \phi_{rv} \qquad (3.30)$$

This swept shape would be used in Eq. (3.25) to compute the frequency of the next higher mode after u and v. It should be recognized that the foregoing development is merely a device by which the assumed shape is adjusted so as to satisfy the orthogonality condition for all previously computed shapes.

---

* This is obviously true since any shape is defined by exactly N coordinates. Given these coordinates, Eq. (3.26) provides N simultaneous equations, which could be used to solve for the N participation factors.

Table 3.3 Modified Rayleigh Method for Second Mode of System in Fig. 3.12

ϕ_{r1} = first-mode shape, Table 3.2

$\sum_{r=1}^{3} M_r \phi_{r1}^2 = 18.45$

Cycle	Mass point	ϕ_{r1}	Assumed shape ϕ_{ra}	$M_r \phi_{ra} \phi_{r1}$	$\psi_1 \phi_{r1}$	$\phi_{ras} = \phi_{ra} - \psi_1 \phi_{r1}$	$F_{ri} = M_r \phi_{ras}$	Computed deflection $A'' \phi_r''$	Computed shape ϕ_r''	$F_{ri} \phi_r''$	$M_r (\phi_r'')^2$
1	1	1.00	1.00	2.000	−0.011	1.011	2.022	0.000280	1.000	2.022	2.000
	2	2.10	0.60	1.260	−0.023	0.623	0.623	0.000195	0.696	0.434	0.484
	3	3.47	−1.00	−3.470	−0.038	−0.962	−0.962	−0.000286	−1.021	0.982	1.042
				−0.2000						3.438	3.526

Eq. (3.29): $\psi_1 = \dfrac{-0.2000}{18.45} = -0.0108$ Eq. (3.25): $\omega^2 = \dfrac{3.438}{0.00028 \times 3.526} = 3482$

2	1	1.00	1.000	2.000	−0.004	1.004	2.008	0.000284	1.000	2.008	2.000
	2	2.10	0.696	1.462	−0.009	0.705	0.705	0.000209	0.736	0.519	0.542
	3	3.47	−1.021	−3.543	−0.015	−1.006	−1.006	−0.000294	−1.035	1.041	1.071
				−0.081						3.568	3.613

$\psi_1 = \dfrac{-0.081}{18.45} = -0.0044$ $\omega^2 = \dfrac{3.568}{0.000284 \times 3.613} = 3477$

3	1	1.00	1.000	2.000	−0.002	1.002	2.004	0.000286	1.000	2.004	2.000
	2	2.10	0.736	1.545	−0.005	0.741	0.741	0.000215	0.752	0.557	0.565
	3	3.47	−1.035	−3.591	−0.009	−1.026	−1.026	−0.000298	−1.042	1.069	1.086
				−0.046						3.630	3.651

$\psi_1 = \dfrac{-0.046}{18.45} = -0.0025$ $\omega^2 = \dfrac{3.630}{0.000286 \times 3.651} = 3476$ $\omega = 59.0$ rad/sec

To illustrate the above procedure we shall obtain the second mode of the system shown in Fig. 3.12. The computations are shown in Table 3.3, where the first mode is swept from the assumed shape in each cycle, using the characteristic shape computed in Table 3.2 for that mode. Otherwise the procedure is the same as that used for the first mode. The first-mode shape ϕ_{r1} is taken as the last computed values in Table 3.2. The assumed shape in the first cycle is merely an educated guess.

As in the case of the first mode, the natural frequency is obtained quite accurately in the first cycle of computation, but several cycles are required to obtain an accurate characteristic shape. The difference between the final shape computed (ϕ_r'' in the third cycle) with that given by the Stodola-Vianello method in Table 3.1 is due to errors in both procedures resulting from roundoff and incomplete convergence.

For a many-degree system, the above procedure could be continued for as many modes as required. The last mode, e.g., the third mode in the example, can be computed directly by the orthogonality condition. This would not occur in practice, however, since if all modes are to be obtained, the Stodola-Vianello method is probably more convenient.

3.6 Lagrange's Equation

Before proceeding to the determination of response for multidegree systems, it is convenient to develop a basic tool, namely, Lagrange's equation. This formulation, which is based upon energy concepts, is an extremely powerful device for the analysis of dynamic systems. It is useful not only for lumped-mass systems, but perhaps even more so for systems with distributed mass, and will be used extensively in Chap. 4. Several derivations of Lagrange's equation may be found in the literature.[11,12,14,15] The approach given below is based on the law of virtual work because structural engineers are familiar with this principle. More concise development is possible, but requires knowledge of certain principles of mathematics which are not presumed herein.

Consider the configuration shown in Fig. 3.13, consisting of an elastic structure (shown as a simple beam for convenience) supporting a group

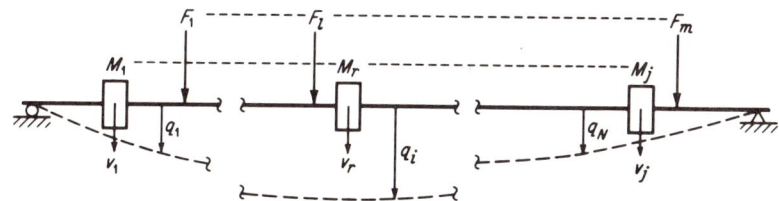

FIGURE 3.13 Lagrange's equation—notation.

of j masses, M_r, and subjected to a group of m forces, F_l. The deflected shape is defined by a set of N *generalized coordinates*, q_i. It is necessary that the shape be completely defined by N and only N coordinates. Thus the system has N degrees of freedom. The points at which these coordinates are given need not include the mass and load points, but the deflections at the latter points must be defined by the generalized coordinates.

Suppose now that a virtual distortion is introduced consisting of a small change in *one* generalized coordinate, q_i. Let this change be designated by δq_i. By the law of virtual work, the work done by the external forces during the virtual distortion must equal the change in internal strain energy. Noting that, in a dynamic system, external forces include both real loads and inertia forces, we may write the preceding statement as

$$\delta W_e + \delta W_{in} + \delta W_c = \delta \mathcal{U} \tag{3.31}$$

where δW_e = virtual work done by external loads F_l
 δW_{in} = virtual work done by inertia forces
 δW_c = virtual work done by damping forces
 $\delta \mathcal{U}$ = change in internal strain energy

Three of these terms may be expressed simply as

(a) $$\delta W_e = \frac{\partial W_e}{\partial q_i} \delta q_i$$

(b) $$\delta W_c = \frac{\partial W_c}{\partial q_i} \delta q_i \tag{3.32}$$

(c) $$\delta \mathcal{U} = \frac{\partial \mathcal{U}}{\partial q_i} \delta q_i$$

and evaluated by the partial differentiation indicated. In order for Eqs. (3.32) to produce the desired result W_e, W_c and $\mathcal{U}$ must be the work and energy due to small changes in all generalized coordinates. The partial derivatives then give the rate of change with respect to one coordinate, q_i. The fourth term, δW_{in}, requires further manipulation. This may be expressed by

$$\delta W_{in} = -\sum_{r=1}^{j} (M_r \ddot{v}_r) \frac{\partial v_r}{\partial q_i} \delta q_i$$

where v_r is the total displacement at mass r. The right side indicates the sum for all masses of the product of the inertia force and the virtual displacement at r resulting from δq_i. For reasons which will become apparent below, this is now expressed in the equivalent form

$$\delta W_{in} = -\frac{d}{dt} \sum_{r=1}^{j} M_r \dot{v}_r \frac{\partial v_r}{\partial q_i} \delta q_i + \sum_{r=1}^{j} M_r \dot{v}_r \frac{\partial \dot{v}_r}{\partial q_i} \delta q_i \tag{3.33}$$

The last is based upon the fact that

$$\frac{d}{dt}\left(\dot{v}_r \frac{\partial v_r}{\partial q_i}\right) = \ddot{v}_r \frac{\partial v_r}{\partial q_i} + \dot{v}_r \frac{\partial \dot{v}_r}{\partial q_i}$$

It is now convenient to introduce the kinetic energy $\mathcal{K}$ defined by

(a) $$\mathcal{K} = \sum_{r=1}^{j} \tfrac{1}{2} M_r \dot{v}_r^2$$

(b) $$\frac{\partial \mathcal{K}}{\partial \dot{q}_i} = \sum_{r=1}^{j} M_r \dot{v}_r \frac{\partial \dot{v}_r}{\partial \dot{q}_i} \qquad (3.34)$$

(c) $$\frac{\partial \mathcal{K}}{\partial q_i} = \sum_{r=1}^{j} M_r \dot{v}_r \frac{\partial \dot{v}_r}{\partial q_i}$$

Furthermore, since $v_r = f(q_i)$,

$$\dot{v}_r = \frac{\partial v_r}{\partial q_i} \dot{q}_i \qquad \text{and} \qquad \frac{\partial \dot{v}_r}{\partial \dot{q}_i} = \frac{\partial v_r}{\partial q_i}$$

Equation (3.34b) may therefore be written

(d) $$\frac{\partial \mathcal{K}}{\partial \dot{q}_i} = \sum_{r=1}^{j} M_r \dot{v}_r \frac{\partial v_r}{\partial q_i} \qquad (3.34)$$

Substituting Eqs. (3.34c) and (3.34d) into Eq. (3.33),

$$\delta \mathcal{W}_{in} = -\frac{d}{dt}\left(\frac{\partial K}{\partial \dot{q}_i}\right) \delta q_i + \left(\frac{\partial K}{\partial q_i}\right) \delta q_i \qquad (3.35)$$

Finally, substituting Eqs. (3.35) and (3.32) into Eq. (3.31) and canceling δq_i, we obtain

$$\frac{d}{dt}\left(\frac{\partial \mathcal{K}}{\partial \dot{q}_i}\right) - \frac{\partial \mathcal{K}}{\partial q_i} + \frac{\partial \mathcal{U}}{\partial q_i} - \frac{\partial \mathcal{W}_c}{\partial q_i} = \frac{\partial \mathcal{W}_e}{\partial q_i} \qquad (3.36)$$

which is the usual form of Lagrange's equation.

In the application of Eq. (3.36), expressions are first written for the kinetic energy $\mathcal{K}$, the strain energy $\mathcal{U}$, the work done by the damping forces $\mathcal{W}_c$, and the work done by real external forces $\mathcal{W}_e$, all in terms of the generalized coordinates $q_1 \cdots q_N$. When these expressions are differentiated as indicated and substituted into Eq. (3.36), the result is an equation of motion, there being one such equation for each coordinate q. In all cases considered herein, $\partial K/\partial q_i$ is zero, since kinetic energy is a function of velocity rather than of displacement.

114   *Introduction to Structural Dynamics*

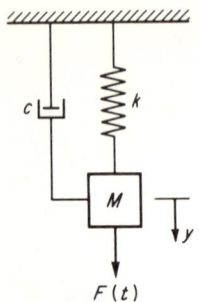

FIGURE 3.14 Example. One-degree system.

a. Examples

In order to gain familiarity with Lagrange's equation, we investigate two simple examples below. These are trivial applications of the Lagrange equation, but are presented for the purpose of emphasizing the physical meaning of the various terms and of demonstrating the validity of Eq. (3.36).

For the simple one-degree system in Fig. 3.14, there is only one coordinate; that is, $q_i = y$. The energy expressions in terms of y are as follows:

Kinetic energy $= \mathcal{K} = \frac{1}{2}M\dot{y}^2$
Strain energy $= \mathcal{U} = \frac{1}{2}ky^2$
Work by damping force $= \mathcal{W}_c = (-c\dot{y})y$
Work by external force $= \mathcal{W}_e = F(t)y$

The damping force must be taken as negative, since a positive damping force is always in a direction opposite to positive y. The necessary derivatives are

$$\frac{\partial \mathcal{K}}{\partial \dot{q}_i} = \frac{\partial \mathcal{K}}{\partial \dot{y}} = M\dot{y} \qquad \frac{d}{dt}\left(\frac{\partial \mathcal{K}}{\partial \dot{q}_i}\right) = M\ddot{y}$$

$$\frac{\partial \mathcal{K}}{\partial q_i} = 0$$

$$\frac{\partial \mathcal{U}}{\partial q_i} = ky \qquad \frac{\partial \mathcal{W}_c}{\partial \dot{q}_i} = -c\dot{y}$$

$$\frac{\partial \mathcal{W}_e}{\partial q_i} = F(t)$$

Substitution into Eq. (3.36) leads to the equation of motion

$$M\ddot{y} + ky + c\dot{y} = F(t)$$

which is obviously correct and the same as that obtained directly by consideration of dynamic equilibrium.

Second, consider the two-degree damped system in Fig. 3.15. There are now two generalized coordinates, y_1 and y_2. The energy expressions

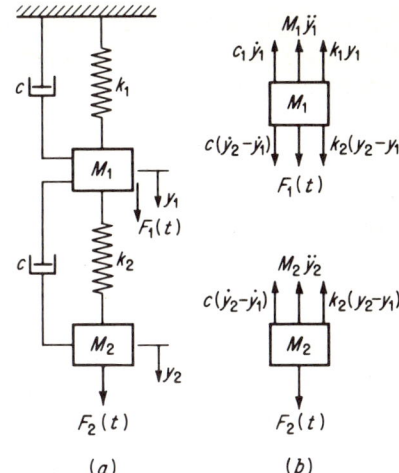

FIGURE 3.15 Example. Two-degree system.

in these terms are

$$\mathcal{K} = \tfrac{1}{2} M_1 \dot{y}_1^2 + \tfrac{1}{2} M_2 \dot{y}_2^2$$
$$\mathcal{U} = \tfrac{1}{2} k_1 y_1^2 + \tfrac{1}{2} k_2 (y_2 - y_1)^2$$
$$\mathcal{W}_c = (-c\dot{y}_1) y_1 + [-c(\dot{y}_2 - \dot{y}_1)](y_2 - y_1)$$
$$\mathcal{W}_e = F_1(t) y_1 + F_2(t) y_2$$

Note that, in these expressions, it is important to include all the energy of the system. Note also that, as this system is defined, the damping force between the two masses is proportional to the relative velocity, and the distance through which that force moves while doing work is the relative displacement. The required derivatives are

$$\frac{\partial \mathcal{K}}{\partial \dot{y}_1} = M_1 \dot{y}_1 \qquad \frac{d}{dt}\left(\frac{\partial \mathcal{K}}{\partial \dot{y}_1}\right) = M_1 \ddot{y}_1$$

$$\frac{\partial \mathcal{K}}{\partial \dot{y}_2} = M_2 \dot{y}_2 \qquad \frac{d}{dt}\left(\frac{\partial \mathcal{K}}{\partial \dot{y}_2}\right) = M_2 \ddot{y}_2$$

$$\frac{\partial \mathcal{K}}{\partial y_1} = \frac{\partial \mathcal{K}}{\partial y_2} = 0$$

$$\frac{\partial \mathcal{U}}{\partial y_1} = k y_1 - k_2 y_2 + k_2 y_1$$

$$\frac{\partial \mathcal{U}}{\partial y_2} = k_2 y_2 - k_2 y_1$$

$$\frac{\partial \mathcal{W}_c}{\partial y_1} = -c\dot{y}_1 + c(\dot{y}_2 - \dot{y}_1) \qquad \frac{\partial \mathcal{W}_c}{\partial y_2} = -c(\dot{y}_2 - \dot{y}_1)$$

$$\frac{\partial \mathcal{W}_e}{\partial y_1} = F_1(t) \qquad \frac{\partial \mathcal{W}_e}{\partial y_2} = F_2(t)$$

116   Introduction to Structural Dynamics

Substitution of the derivatives with respect to y_1 (that is, $y_1 = q_i$) into Eq. (3.36), followed by the substitution of derivatives with respect to y_2, yields the two equations of motion

$$M_1\ddot{y}_1 + k_1 y_1 - k_2(y_2 - y_1) + c\dot{y}_1 - c(\dot{y}_2 - \dot{y}_1) = F_1(t)$$
$$M_2\ddot{y}_2 + k_2(y_2 - y_1) + c(\dot{y}_2 - \dot{y}_1) = F_2(t)$$

These equations are readily verified by the consideration of dynamic equilibrium as indicated in Fig. 3.15b. The above method is obviously an inefficient way to write the equations of motion. Furthermore, it should be recognized that the Lagrange equation is merely a device for writing the equation of motion, and is not a method of solution. However, as will be seen in the remainder of this chapter and in following chapters, the Lagrange equation has some very important uses.

3.7   Modal Analysis of Multidegree Systems*

Having developed the ideas and procedures presented in the foregoing sections, we are now in a position to determine the response of multidegree systems due to applied forces or initial conditions. This will be accomplished by the modal method, in which the responses in the normal modes are determined separately, and then superimposed to provide the total response. As will be proved below, the important point is that *each normal mode may be treated as an independent one-degree system*.

The applicability of the modal method of analysis is limited to linearly elastic systems and to cases in which all forces applied to the structure have the same time variation. These are rather severe limitations. When these conditions are not met, numerical analysis must be used, as demonstrated in Sec. 3.9.

a. Modal Equations

We shall now demonstrate that the normal modes are indeed independent and at the same time develop the governing modal equations of motion. This objective is conveniently accomplished by the use of the Lagrangian equation.

Consider a lumped-mass system having j masses, s springs, and N normal modes. The system may be close- or far-coupled. At any instant the total kinetic energy in the system is

$$\mathcal{K} = \sum_{r=1}^{j} \tfrac{1}{2} M_r \left(\sum_{n=1}^{N} \dot{a}_{rn} \right)^2 \tag{3.37}$$

---

* An alternative development of the modal method of analysis using matrix notation is given in the Appendix.

where $\dot{a}_{rn}$ is the velocity component of mass r associated with the nth mode. The total strain energy in the springs of the system is

$$\mathfrak{U} = \sum_{g=1}^{s} \tfrac{1}{2} k_g \left(\sum_{n=1}^{N} \Delta_{gn} \right)^2 \tag{3.38}$$

where Δ_{gn} is the distortion of spring g (i.e., the relative displacement of its ends) in the nth mode, and k_g is the stiffness of that spring. Both Eqs. (3.37) and (3.38) are based on the fact that any displacement or velocity is equal to the sum of the modal components.

The squared series in Eq. (3.37) is equivalent to the sum of the squares of all modal components of $\dot{a}_r$ plus twice the sum of all cross products of these components. When summed over all masses, the total of these cross products must be zero, according to the orthogonality condition [Eq. (3.13a)]. To demonstrate, consider a two-mass two-mode system for which

$$\sum_{r=1}^{2} \tfrac{1}{2} M_r \left(\sum_{n=1}^{2} \dot{a}_{rn} \right)^2 = \sum_{r=1}^{2} \tfrac{1}{2} M_r (\dot{a}_{r1}^2 + 2\dot{a}_{r1}\dot{a}_{r2} + \dot{a}_{r2}^2)$$

$$= \sum_{r=1}^{2} \tfrac{1}{2} M_r \sum_{n=1}^{2} \dot{a}_{rn}^2 + \sum_{r=1}^{2} M_r \dot{a}_{r1} \dot{a}_{r2}$$

The second series is identical with Eq. (3.13a), except that the modal terms are velocities rather than displacements. However, since characteristic shapes also apply to velocity vectors, this series must also be zero. Therefore Eq. (3.37) may be written as

$$\mathfrak{K} = \sum_{r=1}^{j} \tfrac{1}{2} M_r \sum_{n=1}^{N} \dot{a}_{rn}^2 \tag{3.39}$$

If exactly the same reasoning is applied to Eq. (3.38) and use is made of the second orthogonality condition [Eq. (3.13b)], it is clear that Eq. (3.38) can be written as

$$\mathfrak{U} = \sum_{g=1}^{s} \tfrac{1}{2} k_g \sum_{n=1}^{N} \Delta_{gn}^2 \tag{3.40}$$

Considering the external forces F_r acting at the r masses, we find that the work done in terms of the displacements is

$$\mathfrak{W}_e = \sum_{r=1}^{j} F_r \sum_{n=1}^{N} a_{rn} \tag{3.41}$$

For each mode it is convenient to select a modal displacement A_n so that all individual mass displacements may be expressed in terms of this

118   *Introduction to Structural Dynamics*

one variable. A_n is usually taken as the displacement of one arbitrarily selected mass. Thus

$$a_{rn} = A_n \left(\frac{a_{rn}}{A_n}\right) = A_n \phi_{rn}$$

$$\dot{a}_{rn} = \dot{A}_n \left(\frac{\dot{a}_{rn}}{\dot{A}_n}\right) = \dot{A}_n \phi_{rn}$$

$$\Delta_{gn} = A_n \left(\frac{\Delta_{gn}}{A_n}\right) = A_n \phi_{\Delta gn}$$

where ϕ_{rn} and $\phi_{\Delta gn}$ are constants for a given mode. A set of such constants defines the characteristic shape or may be determined therefrom. Equations (3.39) to (3.41) may therefore be written as

(a) $$\mathcal{K} = \sum_{r=1}^{j} \tfrac{1}{2} M_r \sum_{n=1}^{N} \dot{A}_n^2 \phi_{rn}^2$$

(b) $$\mathcal{U} = \sum_{g=1}^{s} \tfrac{1}{2} k_g \sum_{n=1}^{N} A_n^2 \phi_{\Delta gn}^2 \qquad (3.42)$$

(c) $$\mathcal{W}_e = \sum_{r=1}^{j} F_r \sum_{n=1}^{N} A_n \phi_{rn}$$

For use in Lagrange's equation, the following partial derivatives are obtained:

$$\frac{\partial \mathcal{K}}{\partial \dot{A}_n} = \sum_{r=1}^{j} M_r \dot{A}_n \phi_{rn}^2 \qquad \frac{d}{dt}\left(\frac{\partial \mathcal{K}}{\partial \dot{A}_n}\right) = \ddot{A}_n \sum_{r=1}^{j} M_r \phi_{rn}^2$$

$$\frac{\partial \mathcal{U}}{\partial A_n} = \sum_{g=1}^{s} k_g A_n \phi_{\Delta gn}^2 = A_n \sum_{g=1}^{s} k_g \phi_{\Delta gn}^2 \qquad (3.43)$$

$$\frac{\partial \mathcal{W}_e}{\partial A_n} = \sum_{r=1}^{j} F_r \phi_{rn}$$

where A_n is the modal displacement and also an arbitrary coordinate which will replace q_i in Eq. (3.36). Note that, in each case, only one term in the modal series $\sum_{n=1}^{N}$ has a derivative. Substitution of Eqs. (3.43) in Eq. (3.36) produces

$$\ddot{A}_n \sum_{r=1}^{j} M_r \phi_{rn}^2 + A_n \sum_{g=1}^{s} k_g \phi_{\Delta gn}^2 = \sum_{r=1}^{j} F_r \phi_{rn} \qquad (3.44)$$

Comparison of this equation with that for a one-degree system

$$M\ddot{y} + ky = F(t)$$

reveals the important fact that the *modal equation of motion* (3.44) is in exactly the same form as that for a one-degree system. Furthermore, the three series may be associated with an equivalent one-degree system as follows:

$$\sum_{r=1}^{j} M_r \phi_{rn}^2 = \text{equivalent mass}$$

$$\sum_{g=1}^{s} k_g \phi_{\Delta gn}^2 = \text{equivalent spring constant} \qquad (3.45)$$

$$\sum_{r=1}^{j} F_r \phi_{rn} = \text{equivalent force}$$

The modal equation may also be written in the form

$$\ddot{A}_n + \omega_n^2 A_n = \frac{\sum_{r=1}^{j} F_r \phi_{rn}}{\sum_{r=1}^{j} M_r \phi_{rn}^2}$$

or since F_r is a function of time and $F_r = F_{r1}[f(t)]$,

$$\ddot{A}_n + \omega_n^2 A_n = \frac{f(t) \sum_{r=1}^{j} F_{r1} \phi_{rn}}{\sum_{r=1}^{j} M_r \phi_{rn}^2} \qquad (3.46)$$

This is the most important equation in this chapter. The two summations in Eq. (3.46) are constants for a given mode and loading, and the modal response in terms of A_n may be determined by this equation without difficulty.* Since ω_n must have been previously determined (in order to obtain the modal shape), there is no need to compute the equivalent spring constant. Inspection of Eq. (3.46) makes it apparent that the modal method as formulated here can be used only if $f(t)$ is the same for all forces acting on the structure. However, this restriction can be removed if the modal equations are solved by numerical methods.

* In some treatments the arbitrary amplitudes of a mode are *normalized* so that $\sum_{r=1}^{j} M_r \phi_{rn}^2 = 1$. Thus the denominator on the right side of Eq. (3.46) would always be unity. In the present application there is no significant advantage in this procedure.

Carrying the analogy to a one-degree system further, we see that the *modal static deflection* must be given by

$$A_{nst} = \frac{\sum_{r=1}^{j} F_{r1}\phi_{rn}}{\omega_n^2 \sum_{r=1}^{j} M_r \phi_{rn}^2} \qquad (3.47)$$

since for a one-degree system $y_{st} = F_1/k = F_1/\omega^2 M$. The response is therefore given by

(a) $\qquad\qquad\qquad A_n(t) = A_{nst}(\text{DLF})_n \qquad (3.48)$
(b) $\qquad\qquad\qquad (A_n)_{\max} = A_{nst}(\text{DLF})_{n,\max}$

where $(\text{DLF})_n$ depends only on $f(t)$ and ω_n. All solutions and charts for DLF given in Chap. 2 may therefore be applied to the analysis of multidegree systems. This is an important conclusion.

The total deflection at any point r is obtained by superimposing the modes.

$$y_r(t) = \sum_{n=1}^{N} A_{nst}\phi_{rn}(\text{DLF})_n \qquad (3.49)$$

Given the modal frequencies and characteristic shapes, the analysis of multidegree systems using Eqs. (3.47) to (3.49) becomes relatively simple.

The analogy between a mode and an equivalent one-degree system may be stated quite simply as follows. The equivalent one-degree system is one for which the kinetic energy, internal strain energy, and work done by all external forces are at all times equal to the same quantities for the complete multidegree system when vibrating in this normal mode alone. Referring to Fig. 3.16, where the equivalent system is defined by k_e, M_e, and F_e, we may translate the foregoing statement as follows:

Equating kinetic energies,

$$\tfrac{1}{2}M_e \dot{A}_n^2 = \sum_{r=1}^{j} \tfrac{1}{2} M_r (\dot{A}_n \phi_{rn})^2$$

and therefore $\qquad\qquad M_e = \sum_{r=1}^{j} M_r \phi_{rn}^2$

Equating strain energies,

$$\tfrac{1}{2} k_e A_n^2 = \sum_{g=1}^{s} \tfrac{1}{2} k_g (A_n \phi_{\Delta gn})^2$$

and therefore $\qquad\qquad k_e = \sum_{g=1}^{s} k_g \phi_{\Delta gn}^2$

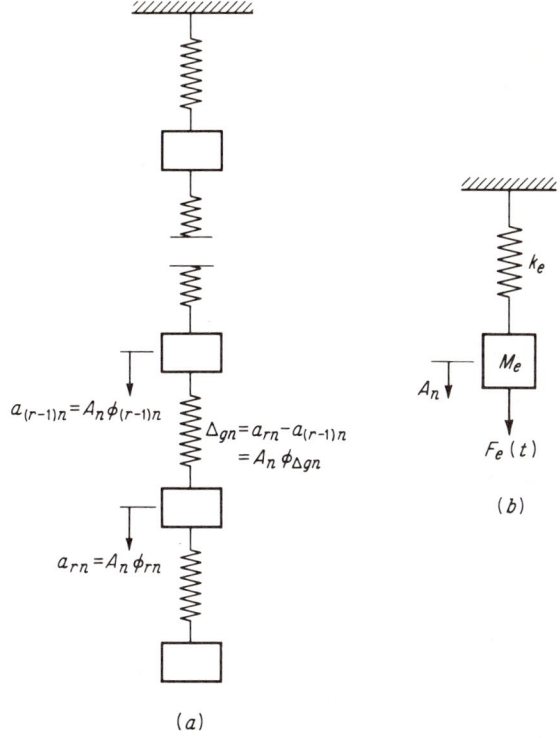

FIGURE 3.16 Modal analysis. (a) Multidegree system vibrating in mode n; (b) equivalent one-degree system for mode n.

Equating work by external forces,

$$F_e(t)A_n = \sum_{r=1}^{j} F_r(t)A_n\phi_{rn}$$

and therefore

$$F_e(t) = \sum_{r=1}^{j} F_r(t)\phi_{rn}$$

These expressions for equivalent-system parameters are identical with those of Eq. (3.45). There is, of course, a unique set of equivalent parameters for each normal mode. This way of visualizing the basic principle is sometimes useful.

b. Example

The response of the three-degree system shown in Fig. 3.17, resulting from the loads shown in the same figure, is to be determined. The

122   *Introduction to Structural Dynamics*

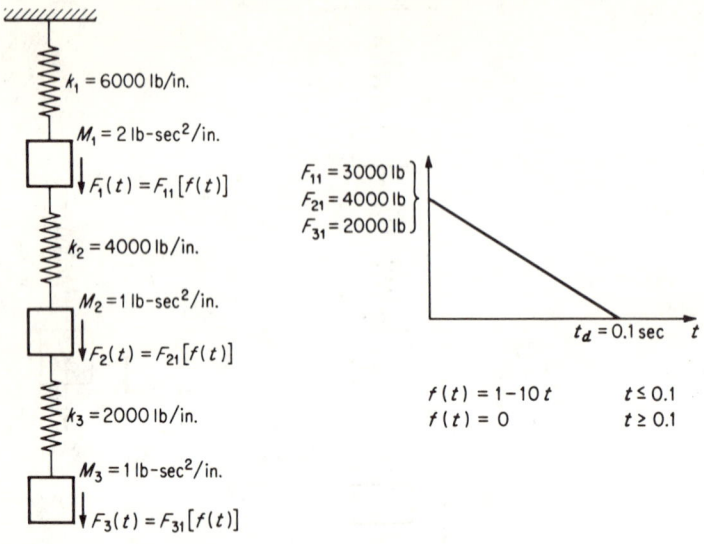

FIGURE 3.17   Example. Modal analysis.

normal modes were obtained in Sec. 3.4 for the same numerical values of the parameters, and the resulting natural frequencies and characteristic shapes are shown in Fig. 3.9. The first step is to evaluate the equivalent peak load and mass for use in Eq. (3.46). These rather simple computations are shown in Table 3.4. For each mode, A_n is

Table 3.4   Modal Analysis; Modal Static Deflections for System in Fig. 3.17

Mode	Mass point	ϕ_{rn}	F_{ri}	$F_{ri}\phi_{rn}$	M_r	$M_r\phi_{rn}^2$	Eq. (3.47) A_{nst}
1	1	1.00	3000	3000	2	2.0	$\omega_1^2 = 780$
	2	2.11	4000	8400	1	4.45	$A_{1st} = \dfrac{4440}{780 \times 18.70}$
	3	3.50	−2000	−7000	1	12.25	
	Σ			4440		18.70	$= 0.304$ in.
2	1	1.000	3000	3000	2	2.000	$\omega_2^2 = 3470$
	2	0.765	4000	3060	1	0.59	$A_{2st} = \dfrac{8126}{3470 \times 3.66}$
	3	−1.033	−2000	+2066	1	1.07	
	Σ			8126		3.66	$= 0.640$ in.
3	1	1.000	3000	3000	2	2.00	$\omega_3^2 = 8720$
	2	−1.866	4000	−7464	1	3.48	$A_{3st} = \dfrac{-5572}{8720 \times 5.79}$
	3	0.554	−2000	−1108	1	0.31	
	Σ			−5572		5.79	$= -0.111$ in.

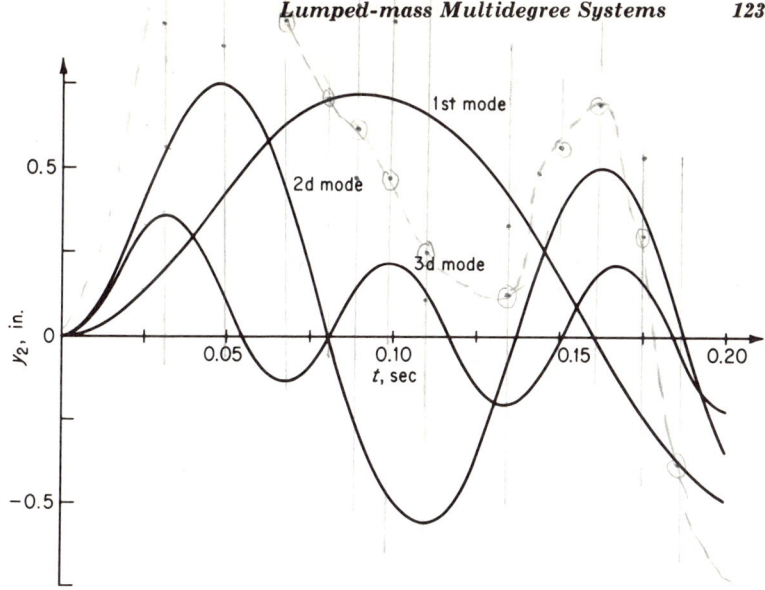

FIGURE 3.18 Example. Response of three-degree system in Fig. 3.17.

arbitrarily taken to be the deflection of mass 1. Equation (3.47) is then used to compute the modal static deflections A_{nst}.

For each mode, DLF is given by Eq. (2.17b) of Sec. 2.3c, which may be written as

$$(\text{DLF})_n = 1 - \cos \omega_n t + \frac{\sin \omega_n t}{\omega_n t_d} - \frac{t}{t_d} \quad \text{for } t < t_d$$

Equation (2.18b) would be used for $t > t_d$. In either case $(\text{DLF})_n$ is merely the value for a one-degree system of natural frequency ω_n and subjected to a triangular load pulse as shown in Fig. 3.17. The total deflection at any point is given by Eq. (3.49). For example, the deflection of mass 2 as a function of time is

$$\begin{aligned}
y_2(t) &= A_{1st}\phi_{21}(\text{DLF})_1 + A_{2st}\phi_{22}(\text{DLF})_2 + A_{3st}\phi_{23}(\text{DLF})_3 \\
&= (+0.304)(+2.11)(\text{DLF})_1 + (+0.640)(+0.765)(\text{DLF})_2 \\
&\quad + (-0.111)(-1.866)(\text{DLF})_3 \\
&= 0.64(\text{DLF})_1 + 0.49(\text{DLF})_2 + 0.21(\text{DLF})_3 \quad (3.50)
\end{aligned}$$

The modal components of $y_2(t)$ [i.e., the separate terms in Eq. (3.50)] are plotted in Fig. 3.18. These were obtained by evaluating Eqs. (2.17b) and (2.18b) for each mode. The total deflection at any time is of course the sum of the modal components.

It should be observed that the degree to which a mode participates in the total vibration is greatly affected by the distribution of load.

124   Introduction to Structural Dynamics

This fact is reflected in the quantity $\Sigma F_{r1}\phi_{rn}$. The more nearly the load values are similar to the corresponding amplitudes of the characteristic shape, the greater is the participation. In fact, if the loads at all points were proportional to the characteristic amplitudes of a certain mode at the same points, the response would be entirely in that mode and that mode alone.

The determination of maximum deflection at point 2 would involve differentiation of Eq. (3.50) with respect to time in order to obtain first the time of maximum response. This is obviously a very difficult process. In many cases the practical solution is to proceed graphically as in Fig. 3.18 and from this plot approximately deduce the time of maximum response.

An upper limit for the maximum response may be obtained by adding numerically the maximums of the modes taken separately. This can easily be done by use of Fig. 2.7, which gives $(\text{DLF})_{\text{max}}$. In this particular case we have

First mode:
$$T = 0.225 \text{ sec} \qquad \frac{t_d}{T} = 0.445 \qquad (\text{DLF})_{\text{max}} = 1.12$$

Second mode:
$$T = 0.107 \qquad \frac{t_d}{T} = 0.94 \qquad (\text{DLF})_{\text{max}} = 1.53$$

Third mode:
$$T = 0.067 \qquad \frac{t_d}{T} = 1.49 \qquad (\text{DLF})_{\text{max}} = 1.68$$

Therefore the upper bound of $y_{2,\text{max}}$ is

$$y_{2,\text{max}} < 0.64(1.12) + 0.49(1.53) + 0.21(1.68) = 1.82 \text{ in.}$$

Inspection of Fig. 3.18 indicates that, for this example, the value just computed is a rather conservative estimate of the maximum displacement. In fact, the true value, occurring at about 0.045 sec, is 1.31 in. However, if one mode had been more dominant and if the differences between the natural frequencies had been greater, both of which conditions usually occur, the upper bound would have been more acceptable.

If it is desired to compute a stress maximum in the structure, or in the above example, a maximum spring force, the procedure is essentially the same as that used above for maximum displacement. For example, the characteristic distortions of spring 2 are

$$\phi_{\Delta 21} = \phi_{21} - \phi_{11} = 1.11$$
$$\phi_{\Delta 22} = \phi_{22} - \phi_{12} = -0.235$$
$$\phi_{\Delta 23} = \phi_{23} - \phi_{13} = -2.866$$

The spring force at any time is given by

$$k_2\Delta_2(t) = \sum_{n=1}^{N} k_2 A_{nst}\phi_{\Delta 2n}(\text{DLF})_n$$
$$= k_2[(+0.304)(+1.11)(\text{DLF})_1 + (+0.640)(-0.235)(\text{DLF})_2$$
$$+ (-0.111)(-2.866)(\text{DLF})_3]$$
$$= k_2[0.34(\text{DLF})_1 - 0.15(\text{DLF})_2 + 0.32(\text{DLF})_3]$$

The upper bound of the maximum spring force is therefore

$$(k_2\Delta_2)_{\max} < 4000[0.34(1.12) + 0.15(1.53) + 0.32(1.68)]$$
$$< 4600 \text{ lb}$$

3.8 Multistory Rigid Frames Subjected to Lateral Loads

To illustrate further the use of the principles developed in the preceding sections, application is now made to actual structures. The examples selected are rigid building frames subjected to horizontal disturbances, e.g., dynamic loading due to blast or wind gust. The majority of such structures may be considered, without appreciable error, to be lumped-mass systems, with the masses concentrated at floor levels. Only horizontal motions are considered, and these are assumed to be independent of vertical motions. This assumption is permissible because vertical motion due to changes in column length or flexure of girders has relatively small amplitude and hence little effect on the horizontal response.

a. Frames with Rigid Girders

In many practical cases the girder stiffnesses relative to the columns are sufficiently large so that they may be assumed infinite. Since this simplifies the analysis somewhat, responses are often computed on this basis. Structures of this type are sometimes called *shear buildings.*

The building to be analyzed is the simple steel rigid frame shown in Fig. 3.19. The weights of the floors and walls are indicated and are assumed to include the structural weight. The building consists of a series of such frames spaced at 15 ft. It is assumed that both structural properties and loading are uniform along the length of the building, and therefore that the analysis to be made of an interior frame yields the response of the entire building. The loads which are concentrated at floor levels are an acceptable idealization of a distributed dynamic pressure applied to the walls. All loads have the time function indicated in Fig. 3.19.

Under the assumptions stated, the entire building may be represented by the close-coupled spring-mass system shown in Fig. 3.19. The con-

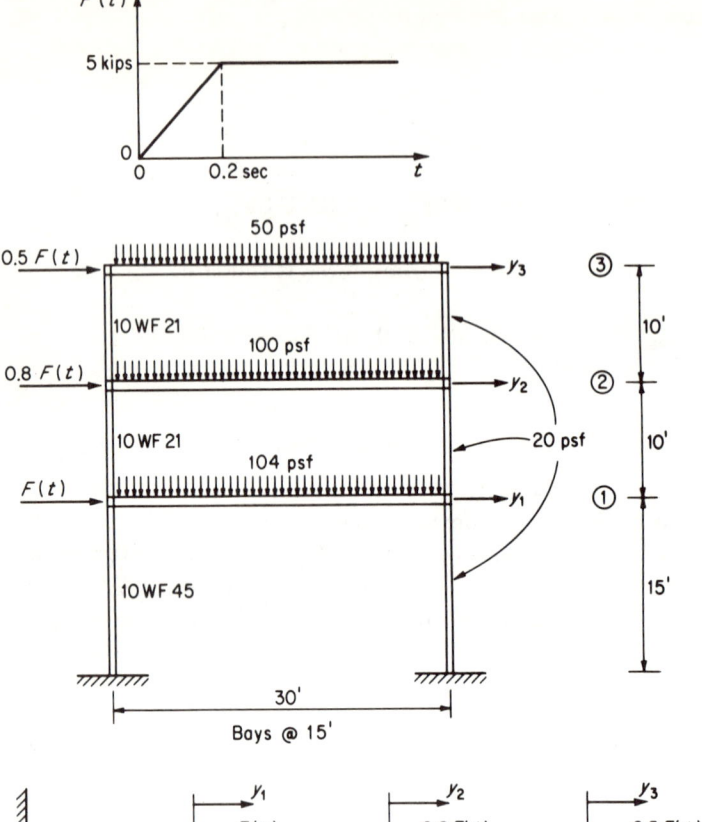

FIGURE 3.19 Example. Steel frame with rigid girders.

centrated weights which are each taken as the total floor weight plus that of the tributary wall area are computed as follows:

$$W_1 = 104(30)(15) + 20(12.5)(15)(2) = 54{,}300 \text{ lb}$$
$$M_1 = 141 \text{ lb-sec}^2/\text{in.}$$
$$W_2 = 100(30)(15) + 20(10)(15)(2) = 51{,}000 \text{ lb}$$
$$M_2 = 132 \text{ lb-sec}^2/\text{in.}$$
$$W_3 = 50(30)(15) + 20(5)(15)(2) = 25{,}500 \text{ lb}$$
$$M_3 = 66 \text{ lb-sec}^2/\text{in.}$$

It should be pointed out that the weights used in the dynamic analysis should be those expected to exist at the time of response and are not necessarily related to the design live floor loads. Since the girders are assumed rigid, as is the column base, each spring constant is given by

$$k = \frac{12EI(2)}{h^3}$$

and the individual values for the steel-column sections indicated are

$$k_1 = \frac{12(30 \times 10^6)(248.6)(2)}{(15 \times 12)^3} = 30{,}700 \text{ lb/in.}$$
$$k_2 = k_3 = \frac{12(30 \times 10^6)(106.3)(2)}{(10 \times 12)^3} = 44{,}400 \text{ lb/in.}$$

The equations of motion for the system, deduced by considering the dynamic equilibrium of each mass, are

$$M_1\ddot{y}_1 + k_1 y_1 - k_2(y_2 - y_1) = F(t)$$
$$M_2\ddot{y}_2 + k_2(y_2 - y_1) - k_3(y_3 - y_2) = 0.8F(t)$$
$$M_3\ddot{y}_3 + k_3(y_3 - y_2) = 0.5F(t)$$

The first step is to obtain the natural frequencies and characteristic shapes of the three modes. In the usual manner, the equations of motion are modified by taking the right sides equal to zero and substituting the modal components ($y = a \sin \omega t$) for the displacements and accelerations, to obtain

$$(-M_1\omega_n{}^2 + k_1 + k_2)a_{1n} + (-k_2)a_{2n} = 0$$
$$(-k_2)a_{1n} + (-M_2\omega_n{}^2 + k_2 + k_3)a_{2n} + (-k_3)a_{3n} = 0$$
$$(-k_3)a_{2n} + (-M_3\omega_n{}^2 + k_3)a_{3n} = 0$$

From the last set of equations the normal modes may be obtained either by direct determination (i.e., setting the determinant of the coefficients equal to zero as in Sec. 3.2) or by the Stodola-Vianello method (Sec. 3.4). In this particular case, the two methods are about equally convenient, because of the zero terms in the determinant. Otherwise the latter method is preferable. The natural periods and characteristic shapes resulting from these computations are given in Fig. 3.20.

Next, the modal static deflections are computed by the evaluation of Eq. (3.47). These computations are shown in Table 3.5. It is immediately apparent from the values of A_{nst} that the fundamental mode dominates this response.

In order to define completely the response (as used here response is taken to mean any displacement or stress), use is made of Eq. (3.49), in which (DLF)$_n$ is given by Eq. (2.20) for this particular load-time func-

Mode	ω^2	T, sec	a_1	a_2	a_3
1	69.3	0.755	+1.00	+1.471	+1.639
2	579	0.261	+1.00	−0.146	−1.041
3	1231	0.179	+1.00	−2.220	+2.680

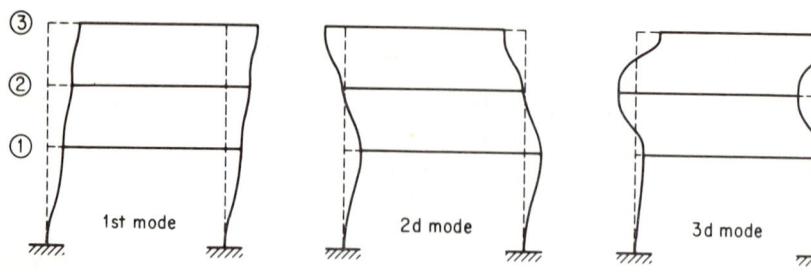

FIGURE 3.20 Characteristic shapes of frame in Fig. 3.19.

tion. The maximum dynamic load factor for any mode may be obtained from Fig. 2.9. The latter values for this example are:

Mode	t_r/T	$(DLF)_{max}$
1	0.265	1.89
2	0.77	1.28
3	1.12	1.11

The maximum roof deflections (for example) given by Eq. (3.49) for the modes separately are therefore

$$y_{31} = (+0.358)(+1.639)(1.89) = +1.11 \text{ in.}$$
$$y_{32} = (0.0146)(-1.041)(1.28) = -0.02 \text{ in.} \quad (3.51)$$
$$y_{33} = (+0.0018)(+2.680)(1.11) = +0.005 \text{ in.}$$

Since the first mode provides by far the major contribution, the upper bound (i.e., the numerical sum) of maximum deflection may be used without appreciable error. Therefore

$$y_{3,max} \cong 1.13 \text{ in.}$$

This overriding influence of the first mode on deflections is typical of building frames, provided only that all the loads act in the same direction.

The effect of the higher modes on stresses may, however, be more significant. For example, consider the column bending moments in the top story for which the associated characteristic amplitude is $\phi_{\Delta 3} = \phi_3 - \phi_2$. Thus the maximum relative story displacement in

Table 3.5 Modal Static Deflections for Frame in Figs. 3.19 and 3.20

Floor	F_{r1}	M_r	First mode			Second mode			Third mode		
			ϕ_{r1}	$F_{r1}\phi_{r1}$	$M_r\phi_{r1}^2$	ϕ_{r2}	$F_{r1}\phi_{r2}$	$M_r\phi_{r2}^2$	ϕ_{r3}	$F_{r1}\phi_{r3}$	$M_r\phi_{r3}^2$
1	5000	141	1.000	5000	141	1.000	5000	141	1.000	5000	141
2	4000	132	1.471	5884	286	−0.146	−548	3	−2.220	−8880	650
3	2500	66	1.639	4097	177	−1.041	−2602	72	2.680	6700	474
				14,981	604		1814	216		2820	1265

Eq. (3.47):

$$A_{1st} = \frac{14{,}981}{69.3 \times 604} = 0.358 \text{ in.}$$

$$A_{2st} = \frac{1814}{579 \times 216} = 0.0146 \text{ in.}$$

$$A_{3st} = \frac{2820}{1231 \times 1265} = 0.00181 \text{ in.}$$

each mode is

$$\Delta_{3n,\max} = A_{nst}(\phi_{\Delta 3n})(\text{DLF})_{n,\max}$$
$$\Delta_{31,\max} = (+0.358)(1.639 - 1.471)(1.89) = +0.114 \text{ in.}$$
$$\Delta_{32,\max} = (+0.0146)(-1.041 + 0.146)(1.28) = -0.017 \text{ in.}$$
$$\Delta_{33,\max} = (+0.00181)(2.680 + 2.220)(1.11) = +0.010 \text{ in.}$$

It may be observed that, although the first mode still dominates, the higher modes are relatively more important than in the previous computation. The time histories of the modal components of story displacement are plotted in Fig. 3.21. These are based on the time variation

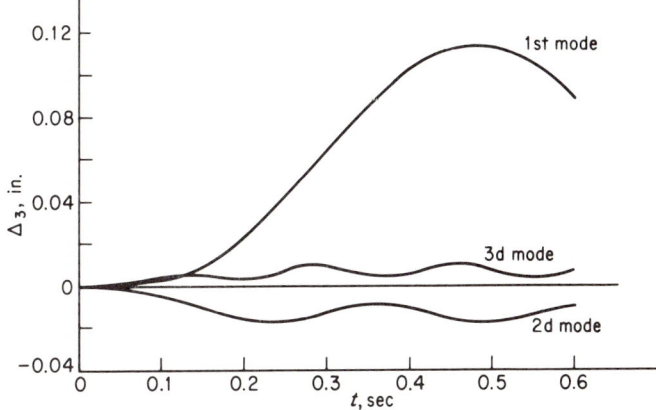

FIGURE 3.21 Modal components of top-story distortion for example defined in Fig. 3.19.

130 Introduction to Structural Dynamics

of DLF as given by Eqs. (2.20). Inspection of Fig. 3.21 reveals that, in this particular situation, little error would result if the maximum story displacement were assumed equal to the *algebraic* sum of maximum modal displacements, or 0.107 in. The maximum bending moment at any column end in the top story is therefore

$$\mathfrak{M} = \frac{6EI\Delta_{3,\max}}{h^2} = \tfrac{1}{4} k_3 \Delta_{3,\max} h$$
$$= \tfrac{1}{4}(44{,}400)(0.107)(10 \times 12)$$
$$= 142{,}500 \text{ lb-in.}$$

The use of the algebraic sum is proper in this example only because the loads continue to act on the structure. If the loads had diminished to zero, all DLFs would have been ± and the numerical sum would have been more appropriate. It may be noted that the algebraic sum could also have been used in computing the maximum roof deflection.

The above is intended to illustrate the fact that maximum responses can often be satisfactorily estimated without going through the tedious job of maximizing the sum of the modal responses mathematically. One must proceed with caution, however, and it is always conservative to use the upper bound, or numerical sum, of the maximum modal responses.

b. Frames with Flexible Girders

To illustrate the procedure for taking into account girder flexibility and also to indicate the magnitude of the effect, an analysis is now made of the same building frame as shown in Fig. 3.19, except that the girders are assumed to consist of the steel sections shown in Fig. 3.22. The loading and mass distributions are identical.

The stiffness coefficients, which are derived from a conventional elastic analysis of the frame, are shown in Fig. 3.22. The procedure is simply to impose a unit deflection at each floor in turn and to compute the resulting holding forces by moment distribution or any other appropriate method. A positive coefficient corresponds to a holding force in the positive y direction. Also shown in Fig. 3.22 are the column end moments corresponding to the unit distortions.

In writing the equations of motion, it must be recognized that the stiffness coefficients correspond to internal column shears, which resist the motion if the coefficient is positive. For example, referring to Fig. 3.22, we see that k_{11} resists the positive motion of the first floor if y_1 is positive, while k_{21}, being negative, increases the positive motion of the second floor when y_1 is positive. Thus the equations of motion for each mass have the form

$$M\ddot{y} = F(t) - \sum ky$$

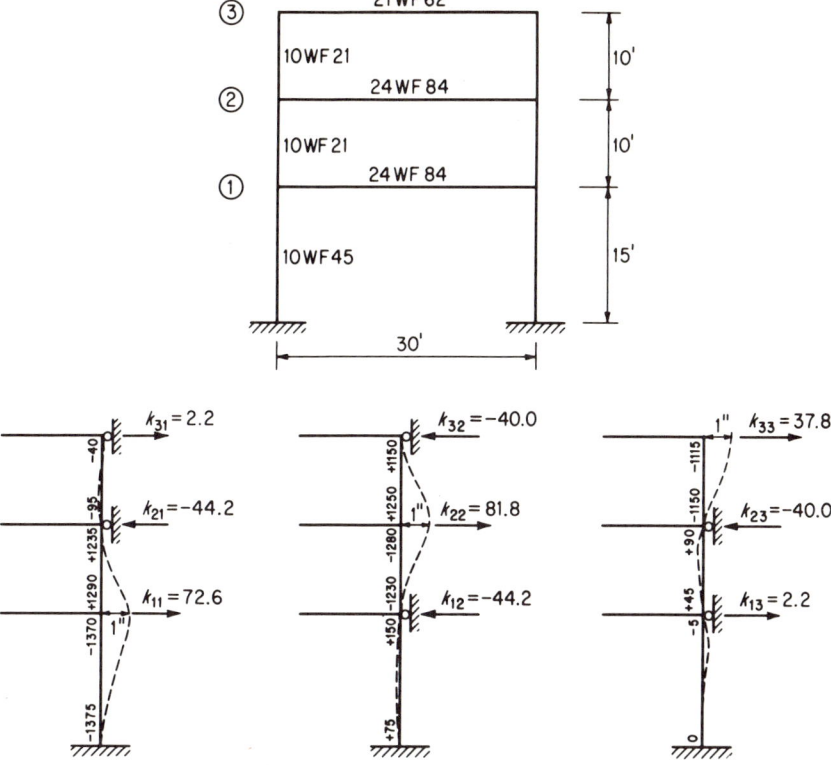

FIGURE 3.22 Example. Rigid frame with flexible girders. Stiffness coefficients. (Moments are given in kip-inches, and stiffness in kips per inch.)

The complete set of equations for this system is

$$M_1\ddot{y}_1 + k_{11}y_1 + k_{12}y_2 + k_{13}y_3 = F_1(t)$$
$$M_2\ddot{y}_2 + k_{21}y_1 + k_{22}y_2 + k_{23}y_3 = F_2(t)$$
$$M_3\ddot{y}_3 + k_{31}y_1 + k_{32}y_2 + k_{33}y_3 = F_3(t)$$

After we set the right sides equal to zero, replace the y's by the modal forms ($a \sin \omega t$), and substitute numerical values (Sec. 3.8a), these equations become

$$0.141\omega_n{}^2 a_{1n} = 72.6a_{1n} - 44.2a_{2n} + 2.2a_{3n}$$
$$0.132\omega_n{}^2 a_{2n} = -44.2a_{1n} + 81.8a_{2n} - 40.0a_{3n}$$
$$0.066\omega_n{}^2 a_{3n} = +2.2a_{1n} - 40.0a_{2n} + 37.8a_{3n}$$

where the units are kips and inches. By applying the Stodola-Vianello procedure as in previous examples, the natural frequencies and characteristic shapes shown in Table 3.6 are obtained. Comparison of these values with those in Fig. 3.20 reveals the effect of girder flexibility. As

Table 3.6 *Modal Shapes and Modal Static Deflections for Frame in Fig. 3.22*

Mode	ω^2	T	a_1	a_2	a_3
1	60	0.810	1.000	1.541	1.758
2	513	0.276	1.000	−0.0515	−1.123
3	1132	0.187	1.000	−1.873	2.075

			First mode			Second mode			Third mode		
Floor	F_{r1}	M_r	ϕ_{r1}	$F_{r1}\phi_{r1}$	$M_r\phi_{r1}^2$	ϕ_{r2}	$F_{r1}\phi_{r2}$	$M_r\phi_{r2}^2$	ϕ_{r3}	$F_{r1}\phi_{r3}$	$M_r\phi_{r3}^2$
1	5000	141	1.000	5000	141	1.000	5000	141	1.000	5000	141
2	4000	132	1.541	6164	313	−0.0515	−206	0	−1.873	−7492	463
3	2500	66	1.758	4395	204	−1.123	−2807	83	2.075	5187	284
Σ				15,559	658		1987	224		2695	888

Eq. (3.47): $\quad A_{1st} = \dfrac{15{,}559}{60.0 \times 658} = 0.394$ in. $\quad A_{2st} = \dfrac{1987}{513 \times 224} = 0.0173$ in. $\quad A_{3st} = \dfrac{2695}{1132 \times 888} = 0.00267$ in.

would be expected, the natural periods are increased, but only slightly. The characteristic shapes are also affected, but not radically changed.

The computations leading to the modal static deflections are tabulated in Table 3.6. Next, the maximum DLFs are obtained from Fig. 2.9 as follows (for the load function shown in Fig. 3.19):

Mode	t_r/T	$(DLF)_{max}$
1	0.25	1.90
2	0.73	1.33
3	1.07	1.05

Multiplication of the modal static deflection, the characteristic amplitude, and the DLF yields the following maximum modal floor deflections:

Mode	$A_{nst} \times \phi_{rn} \times (DLF)_{n,max}$	Max deflection of floor		
		3	2	1
1	$(+0.394) \begin{Bmatrix} +1.758 \\ +1.541 \\ +1.000 \end{Bmatrix} (1.90)$	+1.32	+1.15	+0.75
2	$(+0.0173) \begin{Bmatrix} -1.123 \\ -0.0515 \\ +1.000 \end{Bmatrix} (1.33)$	−0.026	0	+0.023
3	$(+0.00267) \begin{Bmatrix} +2.075 \\ -1.873 \\ +1.000 \end{Bmatrix} (1.05)$	+0.006	−0.005	+0.003

Comparison of the third-floor deflections with those obtained for rigid girders [Eqs. (3.51)] shows that, as expected, all have been slightly increased. This increase is primarily due to the smaller stiffness of the building, as would be true for static loading, although in addition the DLFs are slightly different because of the longer natural periods.

In order to compute a bending moment in the frame, the moments due to unit deflections given in Fig. 3.22 are multiplied by the actual deflections. For example, the bending moment at the bottom of the top-story column may be computed for each mode as follows:

First mode:

$$+1.32(-1150) + 1.15(+1250) + 0.75(-95) = -153 \text{ in.-kips}$$

134 *Introduction to Structural Dynamics*

Second mode:

$$-0.026(-1150) + 0.0(+1250) + 0.023(-95) = +28 \text{ in.-kips}$$

Third mode:

$$+0.006(-1150) - 0.005(+1250) + 0.003(-95) = -13 \text{ in.-kips}$$

The algebraic sum of these moments is 138 kip-in. compared with the sum of 142.5 kip-in. computed previously, when girder flexibility was not considered.

As illustrated by the two preceding examples, the only additional effort required to take into account girder flexibilities lies in the computation of stiffness coefficients. Although for most typical building frames the effect is not great, this effort may be worthwhile, to increase the accuracy of the analysis, particularly if the girders are unusually flexible. The basis for the decision is essentially the same as one would use for a static analysis of horizontal deflection.

c. Frames Subjected to Pulsating Forces

To illustrate further the response of multistory building frames, the effect of a sinusoidal horizontal force is now investigated. This condition might be caused by the operation of machinery on one of the floors. Although this effect seldom causes major structural damage, cases in which the resulting vibration is objectionable for one reason or another are fairly common.

Consider again the frame with flexible girders as in Sec. 3.8b (Fig. 3.22). Suppose that a horizontal force $F_1 \sin \Omega t$ were applied to the first floor as indicated in Fig. 3.23a. A general solution will be developed which will apply to all possible values of Ω. The analytical procedure is the same as for previous examples except that now there is only one load; i.e.,

$$y_{rn} = A_{nst}\phi_{rn}(\text{DLF})_n \qquad (3.52a)$$

where
$$A_{nst} = \frac{F_1\phi_{1n}}{\omega_n^2 \sum_{r=1}^{j} M_r\phi_{rn}^2} \qquad (3.52b)$$

and $(\text{DLF})_n$ is equal to that for a one-degree system subjected to sinusoidal force as developed in Sec. 2.5. More specifically, if damping is included, the DLF corresponding to maximum steady-state response is given by Eq. (2.41) or Fig. 2.18.

After substitution of the numerical values given in Table 3.6 for this structure, Eq. (3.52b) yields the following static deflections for the

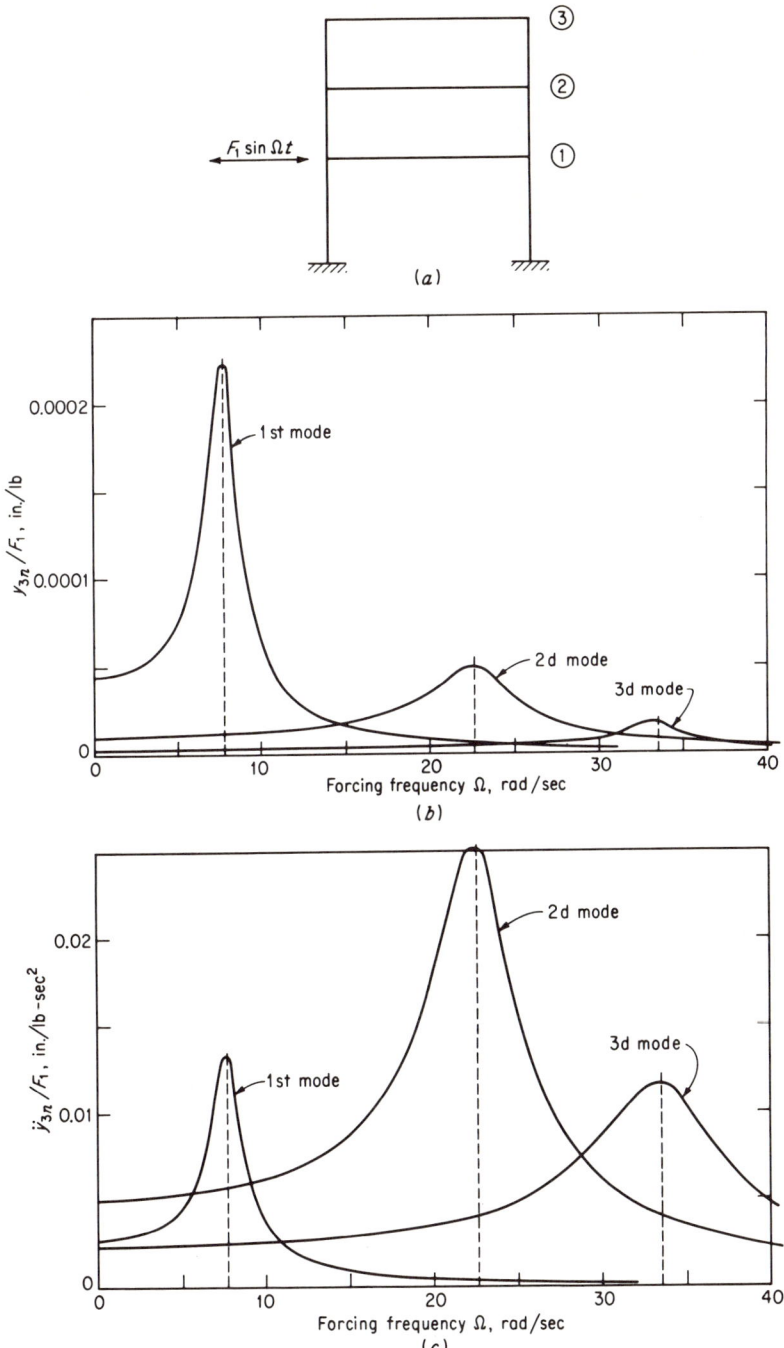

FIGURE 3.23 Rigid frame subjected to pulsating force. Response of top floor.

three modes:

$$A_{1st} = \frac{F_1}{39,500}$$

$$A_{2st} = \frac{F_1}{115,000}$$

$$A_{3st} = \frac{F_1}{1,010,000}$$

If the damping is taken to be 10 percent of critical in each mode, the maximum DLF at resonance is 5.0, according to Fig. 2.18. Thus the maximum possible modal components of deflection at any point may be computed by Eq. (3.52a). For example, these values for the roof are

$$y_{31,max} = A_{1st} \times 1.758 \times 5.0 = F_1(22.2 \times 10^{-5}) \quad \text{in.}$$
$$y_{32,max} = A_{2st} \times 1.123 \times 5.0 = F_1(4.89 \times 10^{-5}) \quad \text{in.}$$
$$y_{33,max} = A_{3st} \times 2.075 \times 5.0 = F_1(1.03 \times 10^{-5}) \quad \text{in.}$$

These maximums occur only when the forcing frequency is equal to the frequency of the particular mode. A complete description of possible responses is given by Fig. 3.23b, which was derived by the use of Eq. (3.52a), and Fig. 2.18, which provides DLF as a function of Ω/ω_n. It is apparent that the first mode produces by far the largest deflections, but these occur over a rather small range of forcing frequency. For higher frequencies of the applied force, the second and third modes become more important than the first. For a given Ω, the total maximum roof displacement could be conservatively taken as the numerical sum of the modal amplitudes.

Although the maximum first-mode amplitude is large, it should not be implied that the higher modes may be disregarded. In fact, the modal acceleration may be more significant than the amplitude since the inertia forces applied to the contents of the building are proportional to acceleration. Such forces are more likely to cause damage to the building or discomfort to persons within the building than is the mere occurrence of displacement. Therefore it is of interest to investigate maximum modal accelerations, and these are plotted in Fig. 3.23c. Since maximum acceleration equals maximum displacement times frequency squared, the ordinates of Fig. 3.23c are merely ω_n^2 times the ordinates of Fig. 3.23b. It is immediately apparent that response in the first mode is not the most serious with respect to acceleration.

d. Frames with Flexible Foundations

In all previous examples it has been assumed that the structure is supported on a rigid foundation. If the structure is founded on relatively

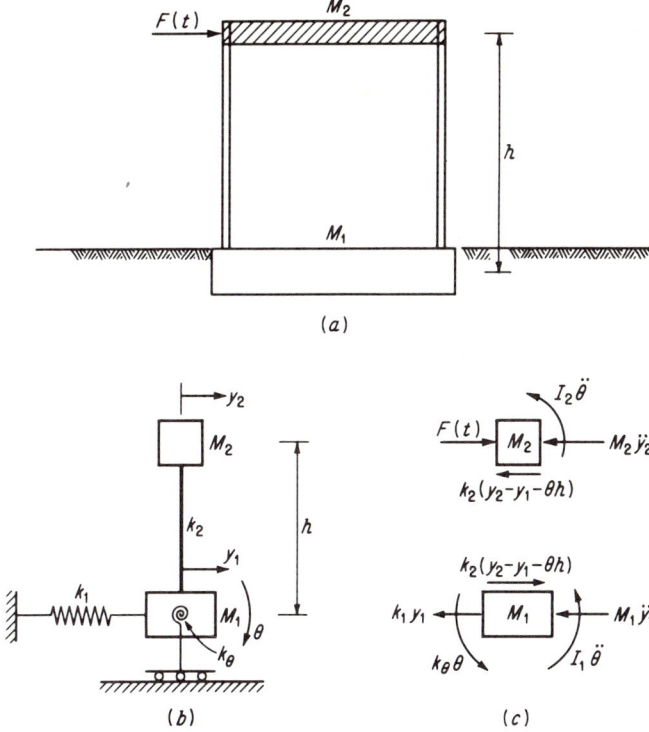

FIGURE 3.24 Frame on flexible foundation. Equivalent three-degree system.

soft soil, it may be necessary to include the effects of distortion within the earth adjacent to the foundation. For example, suppose a simple frame were supported by a foundation mat as shown in Fig. 3.24a. As forces are transmitted from the structure to the surrounding soil (or vice versa, in the case of ground motions), the stresses and resulting deformation in the latter permit the mass of the foundation to move. If attention is restricted to horizontal motion (vertical motion would usually be an uncoupled phenomenon) and the base mat is assumed rigid, the system may be represented as shown in Fig. 3.24b. The spring constants k_1 and k_θ (rotational spring) may be determined at least approximately, using basic procedures of soil mechanics.

The system has three degrees of freedom associated with the coordinates y_1, y_2, and θ. If changes in column length are negligible, both masses rotate by the same amount. As indicated by the equilibrium diagrams in Fig. 3.24c, the three equations of motion are as follows. The first states the horizontal equilibrium of mass 1, and the second of mass 2, and the third states the overall rotational equilibrium about the base:

$$M_1\ddot{y}_1 + k_1 y_1 - k_2(y_2 - y_1 - \theta h) = 0$$
$$M_2\ddot{y}_2 + k_2(y_2 - y_1 - \theta h) = F(t)$$
$$I_1\ddot{\theta} + I_2\ddot{\theta} + M_2\ddot{y}_2 h + k_\theta \theta = F(t)h$$

where I is the mass moment of inertia. Note that, since the y's are absolute displacements, the distortion of spring 2 is the relative motion of the two masses minus θh, which is that portion of the relative motion that does not cause a spring force. Based on these equations, analysis may proceed as for any three-degree system.

The type of analysis indicated above could of course also be made for structures on spread footings or pile foundations. For most rigid frame structures on typical foundations, the effect of soil distortion is not significant and may be safely ignored. Investigations of earthquake response have indicated that foundation flexibility has little effect on stresses in the building frame, but does affect the acceleration input; i.e., the acceleration of the frame foundation is not the same as the free-field acceleration away from the structure.

The support flexibility may have a significant effect on the natural frequencies, and if these are in themselves important, e.g., when the structure is part of a machine, it should be taken into account. This has been found to be true of the supporting structures of certain tracking radars controlled by frequency-sensitive servomechanisms.

3.9 Elasto-plastic Analysis of Multidegree Systems

Rigorous analysis of inelastic multidegree systems is in most cases not practical, and therefore numerical analysis is usually employed. This can be done in a straightforward manner, using the procedures presented in Chap. 1.

When dealing with building frames which deflect into the plastic range, it is prudent to consider the structure to be a shear building; i.e., the girder flexibilities are ignored. This is true because the system is changed each time a plastic hinge is formed at any point in the structure. For example, the stiffness coefficients shown in Fig. 3.22 are correct only as long as the complete structure is elastic. If a single hinge is formed, the structural system is altered and all coefficients change. These will change again when a second hinge appears or when one is eliminated. A completely rigorous analysis would therefore require the evaluation of many coefficients. Furthermore, the process of keeping track of the changing pattern of hinges is exceedingly tedious. The effort required for this type of analysis, while not impossible,[18] is seldom worth-

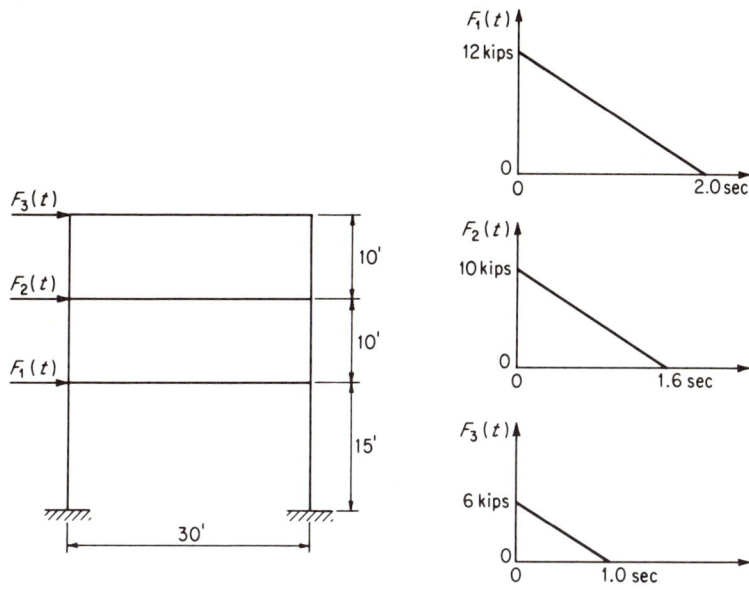

FIGURE 3.25 Example. Elasto-plastic frame analysis. Structure same as in Fig. 3.19.

while. The error introduced by adopting a shear building or close-coupled model is, except in rare cases, tolerable. This statement follows from the fact that, if the response is elastic, the girder flexibilities are of secondary importance (Sec. 3.8b). These flexibilities are even less significant after plastic hinges have formed, because the inelastic behavior is usually restricted to the columns, and hence the plastic shear resistance of a story is not affected by the girder stiffnesses.

An elasto-plastic analysis of the frame in Fig. 3.19 will be made to illustrate the detailed procedure. The applied load functions are shown in Fig. 3.25. For generality, the load-time functions, i.e., the durations, have been made different for the three loads. The computations are shown in Table 3.7, as are the equations of motion. The maximum resistance in each story is given by $4\mathfrak{M}_P/h$, where $\mathfrak{M}_P$ is the plastic bending strength of the column section. The resistance functions are assumed to be bilinear, as discussed in Sec. 1.5a. The numerical procedure represented by Table 3.7 is identical with that used in Chap. 1 and employs the lumped-impulse method as indicated by the recurrence formula given. Since the smallest natural period is 0.179 sec (Fig. 3.20), a time interval of 0.02 sec, or approximately one-tenth of that period, is used. If a larger interval had been used, the computations would not properly have reflected the participation of the third mode.

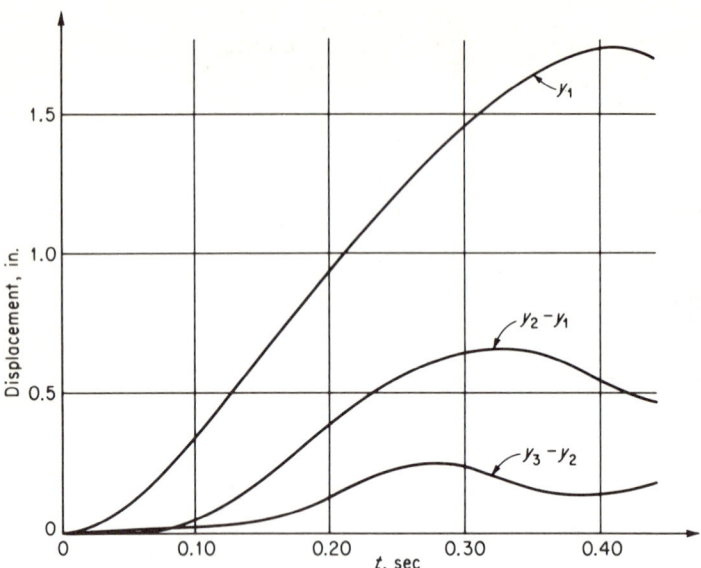

FIGURE 3.26 Response computed in Table 3.7 for frame shown in Fig. 3.25.

The result of the analysis is plotted in Fig. 3.26, where the relative story displacements are shown. In the first story, only the first-mode response is apparent, but the effect of the higher modes is evident in the upper-story responses. Note that only the top story remains elastic. The maximum deflection in the bottom story is about 1.3 times the elastic limit. During the time covered by these computations, the effect of damping is small, but if a longer period of response were to be investigated, damping should be included.

3.10 Damping in Multidegree Systems

The inclusion of damping in multidegree analysis involves some rather troublesome problems. This is true because there is little theoretical means for determining the nature of the damping. In terms of the idealized system, one cannot be sure what arrangement of dampers to assume and, if assumed, what coefficients to assign. Experimental investigations have thrown little light on the subject.

If a modal analysis as in Secs. 3.7 and 3.8 is being made, one may simply assume a reasonable percentage of critical damping in each mode. With this assumption, the analysis merely involves taking that amount of damping into account when computing the DLFs for the modes, which may then be superimposed as before. This is a satisfactory procedure since the analyst can usually estimate the percentage of critical damping better than he can estimate the individual damping coefficients.

If a numerical analysis as in Sec. 3.9 is to be made, including damping, actual coefficients must in some way be determined. In order to be sure that the coefficients are reasonable, it is advisable to relate these to the percentage of critical damping in each mode. This relation may be established by the following procedure. The general equation of free motion for the rth mass of a lumped-parameter system may be written as

$$M_r \ddot{y}_r + \sum^i k_i y_i + \sum^i c_{ri} \dot{y}_i = 0 \qquad (3.53)$$

where $\sum^i$ indicates a series in which there is one term for each of the i displacements. c_{ri} is the damping coefficient, which applies to the ith velocity in the rth equation of motion. Equation (3.53) may be converted into a modal equation for the nth mode by the substitutions

$$y_i = y_r \frac{\phi_{in}}{\phi_{rn}} \qquad \text{and} \qquad \dot{y}_i = \dot{y}_r \frac{\phi_{in}}{\phi_{rn}}$$

Thus we obtain

$$M_r \ddot{y}_r + \left(\sum^i k_i \frac{\phi_{in}}{\phi_{rn}} \right) y_r + \left(\sum^i c_{ri} \frac{\phi_{in}}{\phi_{rn}} \right) \dot{y}_r = 0$$

Since the last equation is in the same form as the equation of motion for a one-degree system, it is apparent that critical damping in the nth mode is defined by [Eq. (2.25)]

$$\left(\sum^i c_{ri} \frac{\phi_{in}}{\phi_{rn}} \right)_{cr,n} = 2 M_r \omega_n$$

and that the equation which may be used to compute the c's is

$$\left(\sum^i c_{ri} \frac{\phi_{in}}{\phi_{rn}} \right)_n = 2 M_r \omega_n C_n \qquad (3.54)$$

where C_n is the ratio of actual to critical damping in the nth mode.

In a system having N degrees of freedom, there are N^2 damping coefficients to be computed. These are provided by Eq. (3.54), which represents N^2 equations; i.e., for each of the N modes, there is one equation for each of the N masses. It is not necessary to solve the entire set simultaneously, since the N equations written for a particular mass are independent and may be solved for the N coefficients associated with that mass. Furthermore, the matrix of coefficients is symmetric; that is, $c_{12} = c_{21}$, etc.

Table 3.7 Numerical Analysis of Elasto-plastic Three-degree System (Fig. 3.25)

$$M_1 \ddot{y}_1 = F_1(t) - \left\{ \begin{array}{c} k_1 y_1 \\ R_{m1} \end{array} \right\} + \left\{ \begin{array}{c} k_2(y_2 - y_1) \\ R_{m2} \end{array} \right\}$$

$$M_2 \ddot{y}_2 = F_2(t) - \left\{ \begin{array}{c} k_2(y_2 - y_1) \\ R_{m2} \end{array} \right\} + \left\{ \begin{array}{c} k_3(y_3 - y_2) \\ R_{m3} \end{array} \right\}$$

$$M_3 \ddot{y}_3 = F_3(t) - \left\{ \begin{array}{c} k_3(y_3 - y_2) \\ R_{m3} \end{array} \right\}$$

$$y^{(s+1)} = 2y^{(s)} - y^{(s-1)} + \ddot{y}^{(s)} (\Delta t)^2$$

$(\Delta t)^2 = 0.0004$

$k_1 = 30.7$ kips/in.; $R_{m1} = 39.8$ kips; $M_1 = 0.141$ kip-sec²/in.
$k_2 = 44.4$ kips/in.; $R_{m2} = 26.2$ kips; $M_2 = 0.132$ kip-sec²/in.
$k_3 = 44.4$ kips/in.; $R_{m3} = 26.2$ kips; $M_3 = 0.066$ kip-sec²/in.

t	F_1	F_2	F_3	R_1	R_2	R_3	$\ddot{y}_1 (\Delta t)^2$	y_1	$\ddot{y}_2 (\Delta t)^2$	y_2	$\ddot{y}_3 (\Delta t)^2$	y_3	$y_2 - y_1$	$y_3 - y_2$
0	12.00	10.00	6.00	0	0	0	0.0340	0	0.0303	0	0.0364	0		
0.02	11.88	9.87	5.88	0.52	-0.08	0.14	0.0320	0.0170	0.0305	0.0151	0.0348	0.0182	-0.0019	0.0031
0.04	11.76	9.75	5.76	2.03	-0.24	0.47	0.0269	0.0660	0.0317	0.0607	0.0320	0.0712	-0.0053	0.0105
0.06	11.64	9.62	5.64	4.35	-0.17	0.81	0.1202	0.1419	0.0321	0.1380	0.0292	0.1562	-0.0039	0.0182
0.08	11.52	9.50	5.52	7.30	0.42	1.02	0.0131	0.2380	0.0306	0.2474	0.0272	0.2704	0.0094	0.0230
0.10	11.40	9.37	5.40	10.66	1.78	1.08	0.0072	0.3472	0.0263	0.3874	0.0262	0.4118	0.0402	0.0244
0.12	11.28	9.25	5.28	14.22	4.00	1.14	0.0030	0.4636	0.0193	0.5537	0.0250	0.5794	0.0901	0.0257
0.14	11.16	9.12	5.16	17.91	6.94	1.45	0.0005	0.5830	0.0110	0.7393	0.0225	0.7720	0.1563	0.0327
0.16	11.04	9.00	5.04	21.58	10.34	2.27	-0.0006	0.7029	0.0028	0.9359	0.0168	0.9871	0.2330	0.0512
0.18	10.92	8.87	4.92	25.24	13.90	3.72	-0.0012	0.8222	-0.0040	1.1353	0.0073	1.2190	0.3131	0.0837

Table 3.7 Numerical Analysis of Elasto-plastic Three-degree System (Fig. 3.25) (Continued)

t	F_1	F_2	F_3	R_1	R_2	R_3	$\ddot{y}_1 (\Delta t)^2$	y_1	$\ddot{y}_2 (\Delta t)^2$	y_2	$\ddot{y}_3 (\Delta t)^2$	y_3	$y_2 - y_1$	$y_3 - y_2$
0.20	10.80	8.75	4.80	28.86	17.33	5.66	−0.0021	0.9403	−0.0088	1.3307	−0.0052	1.4582	0.3904	0.1275
0.22	10.68	8.62	4.68	32.43	20.47	7.76	−0.0036	1.0563	−0.0124	1.5173	−0.0187	1.6922	0.4610	0.1749
0.24	10.56	8.50	4.56	35.88	23.21	9.59	−0.0060	1.1687	−0.0155	1.6915	−0.0305	1.9075	0.5228	0.2160
0.26	10.44	8.37	4.44	39.14	25.53	10.75	−0.0090	1.2751	−0.0194	1.8502	−0.0382	2.0923	0.5751	0.2421
0.28	10.32	8.25	4.32	39.80	26.20	11.07	−0.0093	1.3725	−0.0208	1.9895	−0.0409	2.2389	0.6170	0.2494
0.30	10.20	8.12	4.20	39.80	26.20	10.50	−0.0096	1.4606	−0.0230	2.1080	−0.0382	2.3446	0.6474	0.2366
0.32	10.08	8.00	4.08	39.80	26.20	9.26	−0.0100	1.5391	−0.0271	2.2035	−0.0314	2.4121	0.6644	0.2086
0.34	9.96	7.87	3.96	39.80	26.20	7.83	−0.0103	1.6076	−0.0318	2.2719	−0.0234	2.4482	0.6643	0.1763
0.36	9.84	7.75	3.84	39.80	25.24	6.77	−0.0134	1.6658	−0.0325	2.3085	−0.0177	2.4609	0.6427	0.1524
0.38	9.72	7.65	3.72	39.80	23.43	6.36	−0.0189	1.7106	−0.0285	2.3126	−0.0160	2.4559	0.6020	0.1433
0.40	9.60	7.50	3.60	39.80	21.20	6.51	−0.0256	1.7365	−0.0218	2.2882	−0.0176	2.4349	0.5517	0.1467
0.42	9.48	7.37	3.48	39.80	19.13	6.85	−0.0318	1.7368	−0.0149	2.2420	−0.0204	2.3963	0.5052	0.1543
0.44								1.7053		2.1809		2.3373	0.4756	0.1564

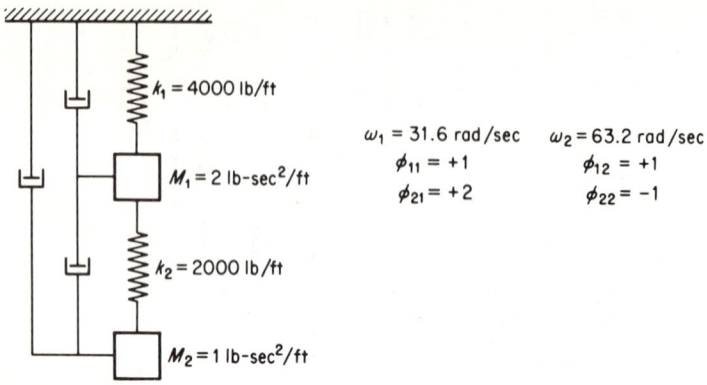

FIGURE 3.27 Example. Normal modes of damped two-degree system.

To illustrate the above procedure, consider the two-degree system shown in Fig. 3.27, where the natural frequencies and characteristic shapes are given. There are four coefficients to be determined from the following four equations, all derived from Eq. (3.54):

First mode ($n = 1$):

$$M_1(r = 1): c_{11} \times 1 + c_{12} \times 2/1 = 2 \times 2 \times 31.6 \times C_1$$
$$M_2(r = 2): c_{21} \times 1/2 + c_{22} \times 1 = 2 \times 1 \times 31.6 \times C_1$$

Second mode ($n = 2$):

$$M_1(r = 1): c_{11} \times 1 + c_{12} \times (-1/1) = 2 \times 2 \times 63.2 \times C_2$$
$$M_2(r = 2): c_{21} \times (-1/1) + c_{22} \times 1 = 2 \times 1 \times 63.2 \times C_2$$

Suppose we desire to have 5 percent of critical damping in the first mode ($C_1 = 0.05$) and 10 percent in the second ($C_2 = 0.1$). The last equations then become

$$c_{11} + 2c_{12} = 6.32$$
$$\tfrac{1}{2}c_{21} + c_{22} = 3.16$$
$$c_{11} - c_{12} = 25.28$$
$$-c_{21} + c_{22} = 12.64$$

The first and third are solved independently, as are the second and fourth, and thus we obtain

$$c_{11} = +18.96 \qquad c_{21} = -6.32$$
$$c_{12} = -6.32 \qquad c_{22} = +6.32$$

all in the units pound-seconds per inch. As expected, $c_{21} = c_{12}$.

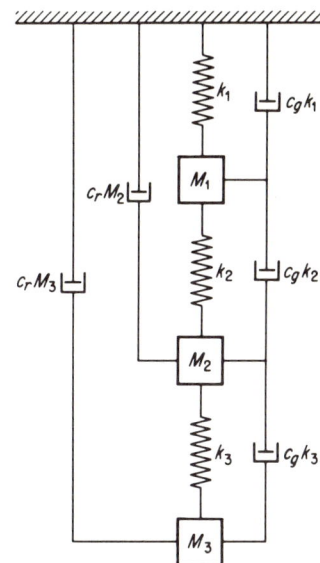

FIGURE 3.28 Three-degree system with absolute and relative damping.

The two equations of motion for the damped system may now be written by substituting into Eq. (3.53) the damping coefficients computed above and the other numerical parameters given in Fig. 3.27.

$$2\ddot{y}_1 + 6000y_1 - 2000y_2 + 18.96\dot{y}_1 - 6.32\dot{y}_2 = 0$$
$$\ddot{y}_2 - 2000y_1 + 2000y_2 - 6.32\dot{y}_1 + 6.32\dot{y}_2 = 0$$

These provide the desired amount of damping.

When dealing with a system having many degrees of freedom, the above procedure is rather cumbersome. Furthermore, if only a few of the lower modes are being considered, it may not be desirable to compute all natural frequencies as required by this procedure.

It is possible to approximate the damping coefficients and thereby avoid the difficulties of the more rigorous approach given above. One way of doing this is to assume two sets of dampers, one associated with the springs and the other with the masses. The former are given coefficients, each proportional to the corresponding spring stiffness, and the latter are given coefficients proportional to the masses. Such a system of dampers is shown in Fig. 3.28, where c_g and c_r are constants common to the coefficients in each set. Note that the damping forces are proportional to relative velocity in the first case and absolute velocity in the other. Values of c_g and c_r may be determined so as to give reasonable damping in each mode. This is accomplished as follows.

The general equation of free motion of a mass r in a system damped in this manner may be written

$$M_r \ddot{y}_r + \sum^i k_i y_i + \sum^i c_g k_i \dot{y}_i + c_r M_r \dot{y}_r = 0 \qquad (3.55)$$

Operating on this equation as was done on Eq. (3.53), we obtain the modal equation

$$M_r \ddot{y}_r + \left(\sum^i k_i \frac{\phi_{in}}{\phi_{rn}} \right) y_r + \left(c_g \sum^i k_i \frac{\phi_{in}}{\phi_{rn}} \right) \dot{y}_r + c_r M_r \dot{y}_r = 0$$

which may also be written

$$M_r \ddot{y}_r + k'_{rn} y_r + (c_g k'_{rn} + c_r M_r) \dot{y}_r = 0$$

Since this equation of motion is of the same form as that for a one-degree system, it is apparent that critical damping in the nth mode is defined by [Eq. (2.25)]

$$(c_g k'_{rn} + c_r M_r)_{cr,n} = 2 M_r \omega_n$$

Dividing both sides by M_r and noting that M_r / k'_{rn} must be ω_n^2,

$$(c_g \omega_n^2 + c_r)_{cr,n} = 2\omega_n$$

or $\qquad (c_g \omega_n^2 + c_r)_n = C_n (2\omega_n) \qquad (3.56)$

where C_n is the ratio of actual to critical damping. The advantage in assuming this particular damping arrangement is now apparent. Equation (3.56) involves only the natural frequency, and hence relates directly the coefficients and the percent of critical damping for each of the modes.

Since there are two coefficients to be determined, namely, c_g and c_r, we may control the percentage of critical in two modes, but no more. The usual procedure is to adjust the two coefficients until a reasonable result is obtained. Note that, if only one form of damping had been included (e.g., relative damping), only one mode could have been controlled. In general, c_g is more effective in the higher modes, and c_r in the lower.

To illustrate application of the above, suppose that, for the three-degree system of Sec. 3.8b (Fig. 3.22), it was desired to have 10 percent damping in the first mode and 5 percent in the third. The natural frequencies are $\omega_1 = 7.75$, $\omega_2 = 22.7$, and $\omega_3 = 33.6$ rad/sec. Writing Eq. (3.56) for the first and third modes,

$$c_g (60) + c_r = 0.1 (2 \times 7.75)$$
$$c_g (1132) + c_r = 0.05 (2 \times 33.6)$$

When solved, these simultaneous equations yield

$$c_g = 0.00169 \text{ sec} \qquad c_r = 1.45 \text{ sec}^{-1}$$

Substitution of these values into Eq. (3.56) yields the damping ratio for the second mode.

$$C_2 = \frac{0.00169(513) + 1.45}{2 \times 22.7} = 0.051, \text{ or } 5.1\%$$

This may well be considered a proper representation of damping in the system. If not, the values c_g and c_r can be adjusted until a satisfactory result, considering all three modes, is obtained. The equations of motion to be used in a numerical analysis of the damped system are given directly by Eqs. (3.55). For example, the first of these is

$$M_1 \ddot{y}_1 + k_{11} y_1 + k_{12} y_2 + k_{13} y_3 + c_g(k_{11} \dot{y}_1 + k_{12} \dot{y}_2 + k_{13} \dot{y}_3) + c_r M_1 \dot{y}_1 = 0$$

which, together with the other two equations, can be solved in the usual manner.

Problems

3.1 A two-degree system (Fig. 3.2) has the following parameters: $M_1 = 4$ lb-sec^2/ft, $M_2 = 2$ lb-sec^2/ft, $k_1 = 4000$ lb/ft, $k_2 = 2000$ lb/ft. Using a direct determination, obtain the natural frequencies and characteristic shapes of both modes. Demonstrate orthogonality of the modes.

Answer
$\omega_1 = 22.3$ rad/sec
$\omega_2 = 44.6$ rad/sec
$a_{11} = +1;\ a_{21} = +2$
$a_{12} = +1;\ a_{22} = -1$

3.2 For the system shown in Fig. 3.29, write the frequency equation in terms of M_1, M_2, EI, l, and k_s. Neglect the mass of the beams.

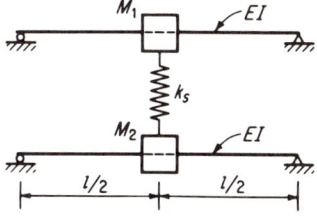

FIGURE 3.29 Problem 3.2.

3.3 Referring to the massless beam supporting two concentrated weights as shown in Fig. 3.5, suppose that the masses are at the $\frac{1}{3}$ points of the span and that $l = 200$ in., $EI = 0.5 \times 10^{10}$ lb-in.^{2}, $M_1 = M_2 = 10{,}000$ lb. Determine the natural frequencies and characteristic shapes.

148   Introduction to Structural Dynamics

Answer

$\omega_1 = 28.0$ rad/sec
$\omega_2 = 108.3$ rad/sec
$a_{11} = +1; a_{21} = +1$
$a_{12} = +1; a_{22} = -1$

3.4 For a two-degree system as in Fig. 3.2, $M_1 = M_2 = 1$ lb-sec²/in., and it is known that, for the first mode, $\omega_1 = 100$ rad/sec, $a_{11} = +1$, $a_{21} = +3$. Determine the characteristic shape and natural frequency of the second mode.

Answer

$a_{12} = +1; a_{22} = -\frac{1}{3}$
$\omega_2 = 245$ rad/sec

3.5 A three-story building frame as in Fig. 3.4a is to be considered as a shear building, i.e., a close-coupled system as indicated in Fig. 3.4b. The following data are given: $M_1 = M_2 = 3$ kip-sec²/ft, $M_3 = 2$ kip-sec²/ft, $k_1 = 1000$ kip/ft, $k_2 = 800$ kip/ft, $k_3 = 600$ kip/ft. Using the Stodola-Vianello procedure based on stiffness coefficients, determine the natural frequencies and characteristic shapes of all modes.

Answer

$\omega_1 = 8.7$ rad/sec
$\omega_2 = 21.0$ rad/sec
$\omega_3 = 29.2$ rad/sec

3.6 Referring to Prob. 3.5, determine the flexibility coefficients for the system and write the equations of motion in these terms. Demonstrate that the Stodola-Vianello procedure now converges first on the fundamental mode.

3.7 For the system in Prob. 3.5, obtain the fundamental-mode frequency, using the Rayleigh method based on the dead-load shape.

3.8 Continuing Prob. 3.7, refine the fundamental-mode shape by additional cycles and obtain the higher-mode frequencies and shapes, using the Schmidt orthogonalization procedure.

3.9 Demonstrate the validity of the Lagrange equation by using it to write the equations of motion for the two-degree system in Fig. 3.6.

3.10 Write the modal equations of motion for the system of Prob. 3.1.

3.11 Write the modal equations of motion for the system of Prob. 3.3.

3.12 Consider the two-degree system and the modal parameters shown in Fig. 3.27. Using modal analysis, derive expressions for the displacements of the two masses as functions of time for the following cases: (a) a suddenly applied constant force of 1 kip applied to mass 2; (b) mass 2 is given a downward displacement of 1 in. (while mass 1 is held in place), and both masses are then suddenly released. Evaluate all numerical terms. *Hint:* In (b) the initial displacements may be broken down into modal components.

3.13 The three-story building frame of Prob. 3.5 is subjected to a suddenly applied constant horizontal force of 50 kips at the second floor. Make a rigorous modal analysis, and plot the deflection of the top floor up to the first peak of response.

Answer

$y_{3,\text{max}} = 0.236$ ft

3.14 Referring to Prob. 3.13, determine the total shear in the top-story columns at the time of maximum roof deflection.

Answer

$V = 17.7$ kips at $t = 0.38$ sec

3.15 The two-degree system shown in Fig. 3.30 is subjected to the force $F_1 = 100 \sin \Omega t$ lb. Assume no damping.

  a. For $\Omega = 50$ rad/sec, estimate the maximum steady-state displacement of M_2 by adding numerically the modal components.

  b. For $\Omega = \omega_2$, the natural frequency of the second mode, estimate the displacement of M_2 after one cycle of load. *Hint:* See Sec. 2.5a.

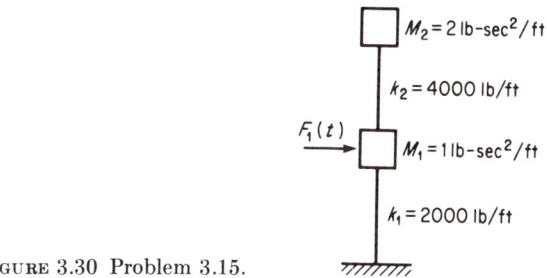

FIGURE 3.30 Problem 3.15.

3.16 A single-story frame on a flexible foundation may be represented by the system shown in Fig. 3.31. It is assumed that only rotation (no vertical or horizontal translation) of the foundation is possible. k_θ is the rotational spring constant at the base, and I_1 and I_2 are the mass moments of inertia. Determine the natural frequencies, and write the modal equations for the case of a horizontal force as shown.

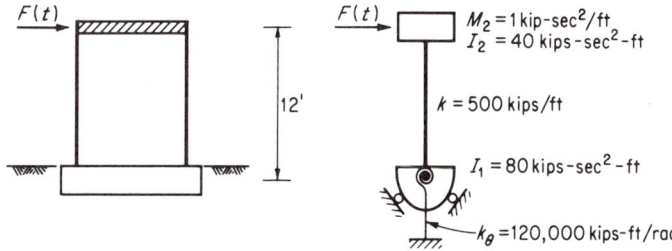

FIGURE 3.31 Problem 3.16.

3.17 Rewrite the equations of motion for the three-degree system of Prob. 3.5 so as to include damping. The amount of damping is to be 10 percent of critical in the first mode and 5 percent in the second and third modes.

4
Structures with Distributed Mass and Load

4.1 Introduction

All structures are in reality distributed mass systems since massless springs as assumed in previous chapters are physically impossible. However, some structures may be closely approximated by lumped-mass systems if, as in the case of the building frames discussed in Chap. 3, the mass of the springs is small compared with the mass concentrations at points between springs. For other classes of structures, such as beams with mass distributed along the span, it is often easier to solve the equations based on the continuous-mass system than to convert the member into an equivalent lumped-parameter system. It is the latter class which is the subject of this chapter.

If the mass is continuously distributed, there are an infinite number of degrees of freedom, since any small element could be considered as a discrete particle connected by springs to all other elements. However, only a few of the lower modes have responses of any significance for practical purposes, and in some cases only the fundamental mode is of importance. Therefore analysis usually begins with the isolation of the lower modes and determination of the natural frequencies and characteristic shapes of these modes. The modal responses may then be computed and superimposed in much the same manner as for a lumped-mass system.

With the exception of Sec. 4.8, the treatment in this chapter is restricted to linearly elastic systems. Approximate methods for handling inelastic response are given in Chap. 5.

4.2 Single-span Beams—Normal Modes of Vibration

Consider the beam shown in Fig. 4.1, where m is the mass intensity in lb-sec^2/in.^{2} and p is the load intensity in lb/in. The mass is, for the present, assumed uniform along the span, but p may be a function of both t and x; that is, p varies both with time and position along the span. The dynamic equilibrium of an element of length is also depicted in Fig. 4.1, where $\mathfrak{M}$ is bending moment and V is shear, both positive in the conventional sense as shown. The net load intensity w on the element is

$$w = p(t, x) - m\ddot{y}$$

positive load being in the same direction as positive y. Since moment, load, and deflection are related by

$$\mathfrak{M} = -EI \frac{\partial^2 y}{\partial x^2}$$

$$\frac{\partial^2 \mathfrak{M}}{\partial x^2} = -EI \frac{\partial^4 y}{\partial x^4} = -w$$

we may write

$$EI \frac{\partial^4 y}{\partial x^4} + m\ddot{y} = p(t, x) \tag{4.1}$$

where EI is the rigidity of the beam, which is assumed constant along the span. Partial derivatives are indicated since y is a function of t as well as x. Equation (4.1) is the equation of motion governing transverse vibration of the beam. If this equation can be solved, the result will be the beam deflection as a function of both time and position along the span.

Several noteworthy approximations have been made in the derivation of Eq. (4.1). First, shear deformation of the member has been ignored, and second, rotation of the element in Fig. 4.1 has not been considered.

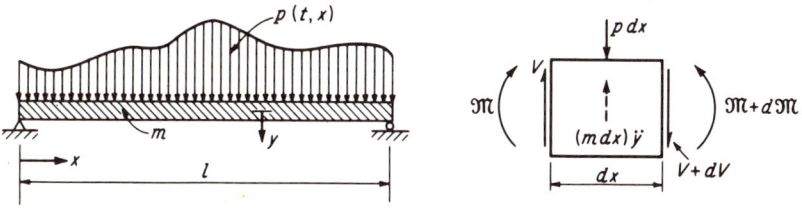

FIGURE 4.1 Simple beam with distributed mass and load.

As the beam deflects, such rotation must occur, and this results in rotational inertia moments which affect the equations of dynamic equilibrium. For slender beams, a classification which includes a great majority of actual cases, neither shear nor rotational effects are important and no further attention is given to them herein. Many investigations of this subject have been made and reported in the literature.[6,8,19]

In any normal mode, by definition,

$$p(t, x) = 0 \quad \text{and} \quad y_n(t, x) = f_n(t)\Phi_n(x)$$

where $f_n(t)$ is a time function, and $\Phi_n(x)$ is the characteristic shape with some undetermined amplitude. We may also write

$$\ddot{y}_n = \ddot{f}_n(t)\Phi_n(x) \quad \text{and} \quad \frac{\partial^4 y}{\partial x^4} = f_n(t)\frac{d^4}{dx^4}\Phi_n(x)$$

where $\ddot{y}_n$ is the second partial derivative of y_n with respect to t. Substitution in Eq. (4.1) provides

$$EIf_n(t)\frac{d^4}{dx^4}\Phi_n(x) + m\Phi_n(x)\ddot{f}_n(t) = 0$$

or

$$\frac{EI}{m\Phi_n(x)}\frac{d^4}{dx^4}\Phi_n(x) = -\frac{\ddot{f}_n(t)}{f_n(t)} \quad (4.2)$$

Since the left side of Eq. (4.2) varies only with x and the right side only with t, each must be equal to a constant, which, as will be seen below, is equal to ω_n^2. Thus, by setting each side equal to ω_n^2, we may write the two equations

(a) $$\ddot{f}_n(t) + \omega_n^2 f_n(t) = 0$$
(b) $$\frac{d^4}{dx^4}\Phi_n(x) - \frac{m\omega_n^2}{EI}\Phi_n(x) = 0 \quad (4.3)$$

The solution for the first of these is

$$f_n(t) = C_1 \sin \omega_n t + C_2 \cos \omega_n t \quad (4.4)$$

which merely indicates that the time function is harmonic with natural frequency ω_n, and hence that Eq. (4.2) is valid for normal modes. The solution of Eq. (4.3b) is

$$\Phi_n(x) = \mathcal{A}_n \sin a_n x + \mathcal{B}_n \cos a_n x + \mathcal{C}_n \sinh a_n x + \mathcal{D}_n \cosh a_n x \quad (4.5)$$

where

$$a_n = \sqrt[4]{\frac{m\omega_n^2}{EI}}$$

Equation (4.5) is general in that it may be applied to spans with any type of end restraints. The constants may be determined by consideration of the boundary conditions of the particular problem.

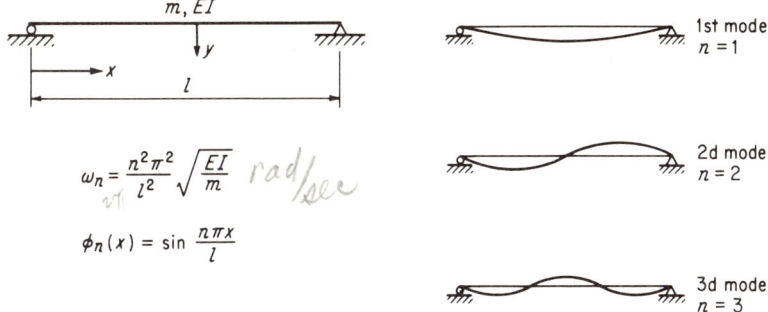

FIGURE 4.2 Normal modes of simple beam with uniform mass.

a. Hinged Supports

At both ends of a beam with simple hinged supports (Fig. 4.2), the boundary conditions are $y = 0$ and $\mathfrak{M} = 0$ or $\partial^2 y/\partial x^2 = 0$. Since $y_n = f_n(t)\Phi_n(x)$ and $f_n(t)$ cannot be zero at all times, it follows from these boundary conditions that

$$\Phi_n(x) = 0 \quad \text{at } x = 0 \text{ and } x = l$$
$$\frac{d^2}{dx^2}\Phi_n(x) = 0 \quad \text{at } x = 0 \text{ and } x = l$$

Substituting $x = 0$ into Eq. (4.5) and into the second derivative of that equation, we find

$$0 = \mathfrak{B}_n + \mathfrak{D}_n$$
$$0 = -\mathfrak{B}_n a_n{}^2 + \mathfrak{D}_n a_n{}^2$$

from which it follows that both $\mathfrak{B}_n$ and $\mathfrak{D}_n$ must be zero. Making use of this conclusion and substituting $x = l$ into Eq. (4.5) and its second derivative, we obtain

$$\Phi_n(l) = 0 = \mathfrak{A}_n \sin a_n l + \mathfrak{C}_n \sinh a_n l$$
$$\frac{d^2}{dx^2}\Phi_n(l) = 0 = -\mathfrak{A}_n a_n{}^2 \sin a_n l + \mathfrak{C}_n a_n{}^2 \sinh a_n l$$

Adding and subtracting these two expressions after canceling $a_n{}^2$ from the second,

$$2\mathfrak{C}_n \sinh a_n l = 0$$
$$2\mathfrak{A}_n \sin a_n l = 0$$

Since $\sinh a_n l$ cannot be zero, $\mathfrak{C}_n$ must be zero. Furthermore, $\mathfrak{A}_n = 0$ is a trivial solution; i.e., it represents no vibration, and therefore the frequency equation must be

$$\sin a_n l = 0$$

or

$$a_n = \frac{n\pi}{l} \tag{4.6}$$

where n is any integer between 1 and infinity. Based on this result and the fact that $\mathcal{B}_n = \mathcal{C}_n = \mathcal{D}_n = 0$, the natural frequencies and characteristic shapes are determined. From Eq. (4.5),

$$a_n^4 = \frac{m\omega_n^2}{EI}$$

Therefore
$$\omega_n = \frac{n^2\pi^2}{l^2}\sqrt{\frac{EI}{m}} \tag{4.7}$$

and
$$\Phi_n(x) = \mathcal{A}_n \sin\frac{n\pi x}{l} \tag{4.8a}$$

Since $\mathcal{A}_n$ is arbitrary, we may let it be unity and define the characteristic shape as

$$\phi_n(x) = \sin\frac{n\pi x}{l} \tag{4.8b}$$

The fundamental mode is given by $n = 1$, the second mode by $n = 2$, etc. The natural frequencies vary with n^2 and therefore are in proportion to 1, 4, 9, 16, etc. The characteristic shapes are all sine waves, that for the fundamental mode being a one-half cycle, for the second mode a full cycle, etc. The first three characteristic shapes are shown in Fig. 4.2. It is of significance that odd modes are symmetrical and even modes antisymmetrical. The total deflection, for any given situation, obtained by superimposing modes is simply

$$y(t, x) = \sum_{n=1}^{\infty} A_n(t) \sin\frac{n\pi x}{l} \tag{4.9}$$

where the A_n's are determined by the loading conditions as discussed in Sec. 4.3.

b. Fixed Supports

If the beam is fixed against translation and rotation at both ends, the boundary conditions are

$$y = 0 \quad \text{and} \quad \frac{dy}{dx} = 0 \quad \text{at } x = 0, l$$

or
$$\Phi_n(x) = 0 \quad \text{and} \quad \frac{d}{dx}\Phi_n(x) = 0 \quad \text{at } x = 0, l$$

Substitution of the latter and $x = 0$ into Eq. (4.5) and its derivative yields

$$\Phi_n(0) = 0 = \mathcal{B}_n + \mathcal{D}_n \qquad \mathcal{D}_n = -\mathcal{B}_n$$

$$\frac{d}{dx}\Phi_n(0) = 0 = \mathcal{A}_n a_n + \mathcal{C}_n a_n \qquad \mathcal{C}_n = -\mathcal{A}_n$$

Therefore Eq. (4.5) becomes

$$\Phi_n(x) = \mathcal{A}_n(\sin a_n x - \sinh a_n x) + \mathcal{B}_n(\cos a_n x - \cosh a_n x)$$

and, for $x = l$,

$$\Phi_n(l) = 0 = \mathcal{A}_n(\sin a_n l - \sinh a_n l) + \mathcal{B}_n(\cos a_n l - \cosh a_n l)$$

$$\frac{d}{dx}\Phi_n(l) = 0 = \mathcal{A}_n a_n(\cos a_n l - \cosh a_n l) + \mathcal{B}_n a_n(-\sin al - \sinh a_n l)$$

For $\mathcal{A}_n$ and/or $\mathcal{B}_n$ to be other than zero, which is a necessary condition for vibration, the determinant of the coefficients must be zero.

$$\begin{vmatrix} (\sin a_n l - \sinh a_n l) & (\cos a_n l - \cosh a_n l) \\ (\cos a_n l - \cosh a_n l) & (-\sin a_n l - \sinh a_n l) \end{vmatrix} = 0$$

When the determinant is expanded, this equation reduces to

$$\cos a_n l \cosh a_n l - 1 = 0 \qquad (4.10)$$

which is the frequency equation for a fixed-ended beam. The roots of this equation are closely approximated by

$$a_n l = (n + \tfrac{1}{2})\pi \qquad n = 1, 2, 3, \ldots$$

or

$$a_n = (n + \tfrac{1}{2})\frac{\pi}{l}$$

and the natural frequencies are given by

$$a_n^4 = \frac{m\omega_n^2}{EI} = (n + \tfrac{1}{2})^4 \frac{\pi^4}{l^4}$$

Therefore

$$\omega_n = \frac{(n + \tfrac{1}{2})^2 \pi^2}{l^2}\sqrt{\frac{EI}{m}} \qquad (4.11)$$

and hence the modal frequencies are in proportion to $(1.5)^2$, $(2.5)^2$, $(3.5)^2$, etc. It may be noted that the first-mode frequency is 2.25 times that for a simply supported beam [Eq. (4.7)].

The characteristic shapes are more complex in this case. It is apparent from the expression for $\Phi_n(l)$ above that

$$\left(\frac{\mathcal{A}}{\mathcal{B}}\right)_n = \frac{\cos a_n l - \cosh a_n l}{\sinh a_n l - \sin a_n l}$$

Therefore the characteristic shapes (after reversing signs for convenience) may be expressed by

$$\Phi_n(x) = A_n\left[\left(\frac{\mathcal{A}}{\mathcal{B}}\right)_n (\sinh a_n x - \sin a_n x) + \cosh a_n x - \cos a_n x\right] \qquad (4.12)$$

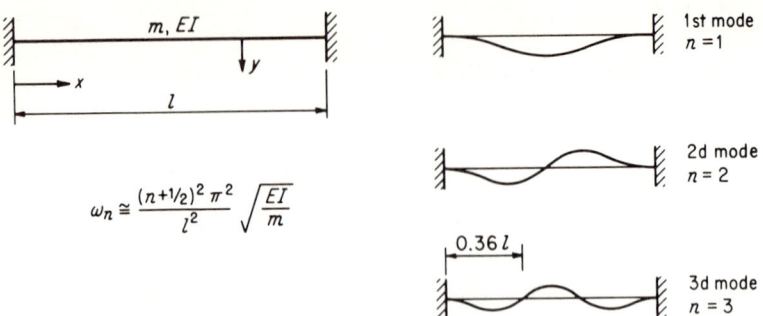

FIGURE 4.3 Normal modes of fixed-ended beam with uniform mass.

where A_n is an arbitrary amplitude, but $(\alpha/\mathbb{B})_n$ is a definite constant for each mode. Values of the latter are tabulated in Table 4.1. The shapes of the first three modes are indicated in Fig. 4.3.

c. Fixed-hinged Beam (Fig. 4.4)

Proceeding in exactly the same manner as before, we introduce the boundary conditions

$$y = 0 \quad \text{at } x = 0, l$$

$$\frac{dy}{dx} = 0 \quad \text{at } x = 0 \quad \text{and} \quad \frac{d^2y}{dx^2} = 0 \quad \text{at } x = l$$

This leads to the frequency equation

$$\tan a_n l = \tanh a_n l$$

The roots are given with sufficient accuracy by

$$a_n l = (n + \tfrac{1}{4})\pi \qquad n = 1, 2, 3, \ldots$$

and the natural frequencies are

$$\omega_n = \frac{(n + \tfrac{1}{4})^2 \pi^2}{l^2} \sqrt{\frac{EI}{m}} \tag{4.13}$$

The first mode has a frequency 1.56 times that of a simply supported beam. The characteristic shapes may be expressed as

$$\Phi_n(x) = A_n \left[\left(\frac{\alpha}{\mathbb{B}}\right)_n (\sinh a_n x - \sin a_n x) + \cosh a_n x - \cos a_n x \right] \tag{4.14}$$

where

$$\left(\frac{\alpha}{\mathbb{B}}\right)_n = \frac{\cos a_n l - \cosh a_n l}{\sinh a_n l - \sin a_n l}$$

The first three modal shapes are indicated in Fig. 4.4.

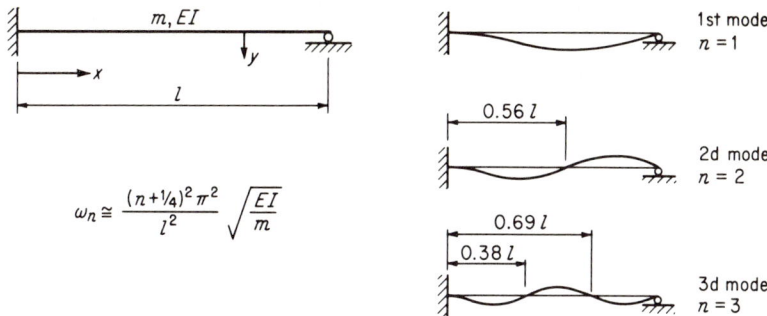

FIGURE 4.4 Normal modes of fixed-simply supported beam with uniform mass.

d. Cantilever Beam

The boundary conditions for this case are

$$y = \frac{dy}{dx} = 0 \quad \text{at } x = 0$$

$$\frac{d^2y}{dx^2} = \frac{d^3y}{dx^3} = 0 \quad \text{at } x = l$$

The third derivative must be zero at the free end since, in a normal mode (free vibration without external forces), there can be no shear at this point. By introducing these boundary conditions into Eq. (4.5) and proceeding as in previous cases, the frequency equation is found to be

$$\cos a_n l \cosh a_n l + 1 = 0$$

The first root of this equation is $a_n l = 1.875$, and the higher roots may be closely approximated by

$$a_n l = (n - \tfrac{1}{2})\pi \quad n = 2, 3, 4, \ldots$$

Thus the natural frequencies are

$$\omega_1 = \frac{(0.597\pi)^2}{l^2}\sqrt{\frac{EI}{m}}$$

and
$$\omega_n = \frac{(n - \tfrac{1}{2})^2 \pi^2}{l^2}\sqrt{\frac{EI}{m}} \quad n > 1 \quad (4.15)$$

It may be noted that the frequencies for a cantilever are the same as the next lower mode of a fixed beam; i.e., the second mode of the cantilever has the same frequency as the fundamental mode of a beam fixed at both ends. The natural frequency of the first mode is 0.356 times that for a simply supported beam.

158 Introduction to Structural Dynamics

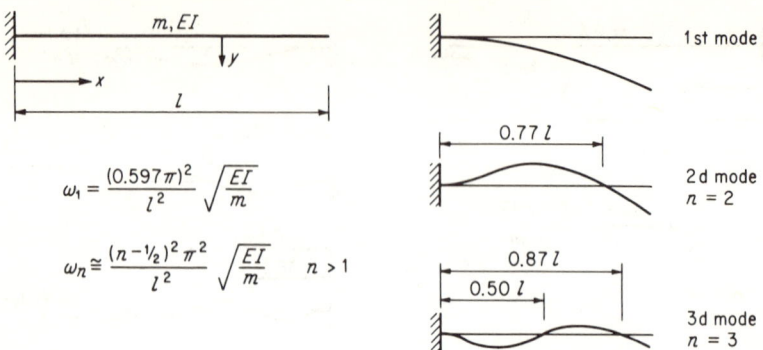

FIGURE 4.5 Normal modes of cantilever beam with uniform mass.

The characteristic shapes for a cantilever beam are given by

$$\Phi_n(x) = A_n \left[\left(\frac{\alpha}{\beta}\right)_n (\sinh a_n x - \sin a_n x) + \cosh a_n x - \cos a_n x \right] \quad (4.16)$$

where

$$\left(\frac{\alpha}{\beta}\right)_n = -\frac{\cos a_n l + \cosh a_n l}{\sin a_n l + \sinh a_n l}$$

These are indicated in Fig. 4.5.

4.3 Forced Vibration of Beams

To determine the response of beams due to applied dynamic forces, use will be made of Lagrange's equation in the same manner as previously for lumped-mass systems. The dynamic deflection may be represented by the summation of the modal components:

$$y(t, x) = \sum_{}^{n} A_n \phi_n(x) \quad (4.17)$$

where A_n is the modal amplitude (which varies with time), and $\phi_n(x)$ is the characteristic shape. The velocity is given by

$$\dot{y}(t, x) = \sum_{}^{n} \dot{A}_n \phi_n(x) \quad (4.18)$$

For use in Lagrange's equation the kinetic energy of the complete system is expressed as

$$\mathcal{K} = \tfrac{1}{2} m \int_0^l \dot{y}^2 \, dx = \tfrac{1}{2} m \int_0^l \left[\sum_{}^{n} \dot{A}_n \phi_n(x) \right]^2 dx$$

where m is the uniformly distributed mass, and the integration provides the summation of all kinetic energy along the length of the beam.

Expanding the series, the last equation may be written as

$$\mathcal{K} = \tfrac{1}{2}m \int_o^l \left[\sum_{}^{n} \dot{A}_n{}^2 \phi_n{}^2(x) \right] dx + m \int_o^l \left\{ \sum [\dot{A}_n \phi_n(x)][\dot{A}_m \phi_m(x)] \right\} dx$$

where the series in the second term indicates the sum of all the modal cross products. The contribution of one of these cross products may be written as

$$m \int_o^l \dot{A}_n \phi_n(x) \dot{A}_m \phi_m(x) \, dx$$

or replacing the integral by a summation,

$$\sum_{r=1}^{j} [\dot{A}_n \phi_n(x_r)][\dot{A}_m \phi_m(x_r)] m \, \Delta x$$

where j is the number of discrete elements into which the beam is divided for the purpose of summation. The expression may also be written as

$$\dot{A}_n \dot{A}_m \sum_{r=1}^{j} M_r \phi_n(x_r) \phi_m(x_r) \tag{4.19}$$

The orthogonality condition as expressed by Eq. (3.13a),

$$\sum_{r=1}^{j} M_r a_{rn} a_{rm} = 0$$

indicates that expression (4.19) must also be zero since $\phi(x_r)$ and a_r are identical in meaning. Thus the entire second term of the expression for $\mathcal{K}$ is zero and

$$\mathcal{K} = \tfrac{1}{2} m \int_o^l \left[\sum_{}^{n} \dot{A}_n{}^2 \phi_n{}^2(x) \right] dx$$

$$= \tfrac{1}{2} m \sum_{}^{n} \dot{A}_n{}^2 \int_o^l \phi_n{}^2(x) \, dx \tag{4.20}$$

Furthermore,
$$\frac{\partial \mathcal{K}}{\partial \dot{A}_n} = m \dot{A}_n \int_o^l \phi_n{}^2(x) \, dx$$

and
$$\frac{d}{dt} \frac{\partial \mathcal{K}}{\partial \dot{A}_n} = m \ddot{A}_n \int_o^l \phi_n{}^2(x) \, dx$$

The work done by external dynamic forces during an arbitrary distortion is

$$\mathcal{W}_e = \int_o^l p(t, x) \left[\sum_{}^{n} A_n \phi_n(x) \right] dx$$

$$= \int_o^l f(t) p_1(x) \left[\sum_{}^{n} A_n \phi_n(x) \right] dx \tag{4.21}$$

where $f(t)$ is the load-time function, and $p_1(x)$ is the load distribution along the span, which may be any function of x. The rate of change of external work with respect to A_n is therefore

$$\frac{\partial W_e}{\partial A_n} = f(t) \int_0^l p_1(x)\phi_n(x)\, dx$$

Writing the Lagrange equation (3.36) with damping omitted and substituting from the above, we obtain

$$\frac{d}{dt}\frac{\partial \mathcal{K}}{\partial \dot{A}_n} + \frac{\partial \mathcal{U}}{\partial A_n} = \frac{\partial W_e}{\partial A_n}$$

$$m\ddot{A}_n \int_0^l \phi_n^2(x)\, dx + \frac{\partial \mathcal{U}}{\partial A_n} = f(t)\int_0^l p_1(x)\phi_n(x)\, dx \qquad (4.22)$$

It is unnecessary to evaluate the strain energy $\mathcal{U}$ since we know by previous experience that, if the last equation is divided by the coefficient of $\ddot{A}_n$, the coefficient of A_n becomes ω_n^2. Thus

$$\ddot{A}_n + \omega_n^2 A_n = \frac{f(t)\int_0^l p_1(x)\phi_n(x)\, dx}{m\int_0^l \phi_n^2(x)\, dx} \qquad (4.23)$$

which is the equation of motion for the nth mode and completely analogous to Eq. (3.46) for lumped-mass systems. The modal static deflection is defined by

$$A_{nst} = \frac{\int_0^l p_1(x)\phi_n(x)\, dx}{\omega_n^2 m \int_0^l \phi_n^2(x)\, dx} \qquad (4.24)$$

The modal response is given by

$$A_n(t) = A_{nst}(\text{DLF})_n \qquad (4.25a)$$

and the total response by

$$y(x, t) = \sum^n A_n(t)\phi_n(x) \qquad (4.25b)$$

where $(\text{DLF})_n$ is the dynamic load factor for the equivalent one-degree system of the nth mode. These equations are analogous to Eqs. (3.47) to (3.49) for lumped-mass systems and could, in fact, have been deduced directly therefrom.

Equation (4.23) is completely general and applies to beams with any support conditions and with any type of load distribution. If the loads are concentrated rather than distributed, the integral in the numerator of the right-hand side merely becomes a summation having one term

Table 4.1 Integrals of Characteristic Functions for Beams with Various Support Conditions

$$\phi(x) = \left(\frac{\alpha}{\mathcal{B}}\right)_n (\sinh a_n x - \sin a_n x) + \cosh a_n x - \cos a_n x$$

Beam type	Mode	$(\alpha/\mathcal{B})_n$	$\int_0^l \phi(x)\,dx$
(fixed-fixed)	1	-0.9825	$0.8308l$
	2	-1.0007	0
	3	-1.0000	$0.3640l$
(fixed-pinned)	1	-1.0007	$0.8604l$
	2	-1.0000	$0.0829l$
	3	-1.0000	$0.3343l$
(fixed-free)	1	-0.7341	$0.7830l$
	2	-1.0184	$0.4340l$
	3	-0.9992	$0.2544l$

For all beam types shown and all modes $\int_0^l [\phi(x)]^2\,dx = l$

for each load. The execution of the integration becomes tedious for other than simply supported beams because the characteristic shapes are rather complicated functions. Values of the integrals are given in Table 4.1 for the first three modes of some commonly encountered beam types.

a. Concentrated Loads

Consider the simply supported beam shown in Fig. 4.6a, which is subjected to a concentrated dynamic load at midspan. The modal characteristic shapes [Eq. (4.8b)] are

$$\phi_n(x) = \sin \frac{n\pi x}{l}$$

and the numerator integral in the right side of (4.23) is replaced by

$$\int_0^l p_1(x)\phi_n(x)\,dx = \sum^F F\phi_n(c_F)$$

where c_F is the location of the load F. In this particular case, there is only one load, and $c_F = l/2$. Therefore

$$\sum^F F\phi_n(c_F) = F_1 \sin \frac{n\pi}{2}$$

162 Introduction to Structural Dynamics

and

$$\int_0^l \phi_n^2(x)\, dx = \int_0^l \sin^2 \frac{n\pi x}{l}\, dx = \frac{l}{2}$$

The modal equation of motion obtained by substitution into Eq. (4.23) is

$$\ddot{A}_n + \omega_n^2 A_n = \frac{f(t)F_1 \sin(n\pi/2)}{ml/2} \quad (4.26)$$

the solution of which may be obtained as for any one-degree system. By Eq. (4.24), the modal static deflection is

$$A_{nst} = \frac{F_1 \sin(n\pi/2)}{\omega_n^2 ml/2}$$

From the latter it is apparent that all even modes contribute nothing to the deflection at any point since $\sin(n\pi/2) = 0$. This is true because such modes are antisymmetrical (Fig. 4.2) and are not excited by a symmetrical load. It may also be stated that there is no response in a mode when the load is applied at a node in the characteristic shape of that mode.

To define the example further, let us suppose that the force F_1 were suddenly applied and constant. Then, by Eq. (2.12),

$$(\text{DLF})_n = 1 - \cos \omega_n t$$

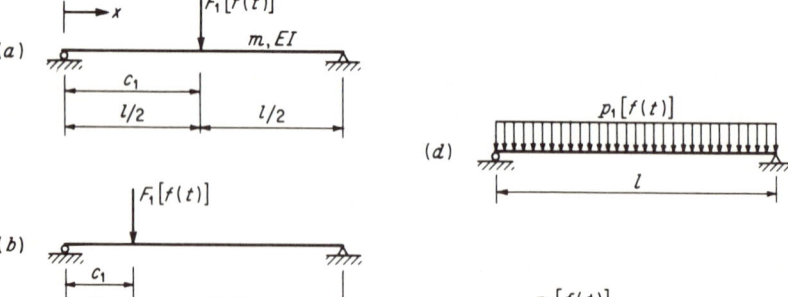

FIGURE 4.6 Simple beam subjected to various load distributions.

Structures with Distributed Mass and Load

and the modal response [Eq. (4.25a)] is

$$A_n = \frac{F_1 \sin(n\pi/2)}{\omega_n^2 ml/2}(1 - \cos \omega_n t)$$

The modal deflection at any point along the span is

$$y_n = A_n \phi_n(x) = \frac{F_1 \sin(n\pi/2)}{\omega_n^2 ml/2}(1 - \cos \omega_n t) \sin \frac{n\pi x}{l}$$

The total deflection at any point and at any time is obtained merely by superimposing modes.

$$y = \frac{2F_1}{ml}\sum^n \frac{1}{\omega_n^2} \sin \frac{n\pi}{2}(1 - \cos \omega_n t) \sin \frac{n\pi x}{l} \qquad (4.27)$$

To illustrate further, the dynamic deflection at midspan ($x = l/2$) equals

$$y\left(x = \frac{l}{2}\right) = \frac{2F_1}{ml}\sum^n \frac{1}{\omega_n^2} \sin^2 \frac{n\pi}{2}(1 - \cos \omega_n t) \qquad (4.28)$$

It is also of interest to compare the contributions of the various modes to the midspan deflection. This will be done on the basis of maximum modal amplitudes without regard to the manner in which the modal displacements actually combine. The amplitudes will indicate, in general, the relative importance of the modes.

In Eq. (4.28) the DLFs all have maximum values of 2, and may be eliminated from consideration. Furthermore, since the sines are all unity for the odd modes, the modal contributions are simply in proportion to $1/\omega_n^2$. Therefore the maximum modal deflections are in proportion to 1, $\frac{1}{81}$, and $\frac{1}{625}$ for the first, third, and fifth modes, respectively. It is apparent that, in this example, the higher modes contribute very little to the midspan deflection.

Suppose now that the concentrated load is at the left quarter point of the beam, as in Fig. 4.6b. The load integral in Eq. (4.23) is, in this case,

$$\sum^F F\phi_n(c_F) = F_1 \phi_n\left(\frac{l}{4}\right) = F_1 \sin \frac{n\pi}{4}$$

and the total deflection would be given by

$$y(x) = \frac{2F_1}{ml}\sum^n \frac{1}{\omega_n^2} \sin \frac{n\pi}{4}(\text{DLF})_n \sin \frac{n\pi x}{l} \qquad (4.29)$$

which corresponds to Eq. (4.27). In contrast to the previous case, which involved a load at midspan, the second mode now makes a contribution

164   Introduction to Structural Dynamics

to the dynamic deflection. If the maximum DLFs are the same in all modes, Eq. (4.29) indicates that the ratio of the first- and second-mode amplitudes is

$$\frac{A_1}{A_2} = \frac{\omega_2{}^2}{\omega_1{}^2} \frac{\sin \pi/4}{\sin \pi/2} = 16(0.707) = 11.3$$

Thus, even in this case, the major contribution is made by the first mode.

If the load distribution were antisymmetrical, only the even modes would be excited. For example, if equal but opposite loads were applied as in Fig. 4.6c,

$$\sum^F F\phi_n(c_F) = F_1\phi_n(c_1) - F_1\phi_n(l - c_1) = F_1\left[\sin\frac{n\pi c_1}{l} - \sin\frac{n\pi(l - c_1)}{l}\right]$$

where the negative sign indicates that the load is upward. This expression is zero for all odd modes because these are symmetrical and the two sine terms are equal. The total dynamic deflection could be expressed by

$$y(x) = \frac{2F_1}{ml} \sum^n \frac{1}{\omega_n{}^2}\left[\sin\frac{n\pi c_1}{l} - \sin\frac{n\pi(l - c_1)}{l}\right](\text{DLF})_n \sin\frac{n\pi x}{l} \quad (4.30)$$

b. Distributed Loads

If the loads are distributed, the integration indicated in the right side of Eq. (4.23) must be executed. For example, consider a simply supported beam with uniform load distribution as in Fig. 4.6d. The load integral is

$$\int_0^l p_1(x)\phi_n(x)\,dx = p_1 \int_0^l \sin\frac{n\pi x}{l}\,dx$$

$$= -\frac{p_1 l}{n\pi}(\cos n\pi - 1)$$

$$= \frac{2p_1 l}{n\pi} \quad n \text{ odd only}$$

Thus, as would be expected with a symmetrical loading, only the odd modes contribute. The total response as given by Eq. (4.25) is therefore

$$y(x) = \frac{4p_1}{m\pi} \sum^n \frac{1}{n\omega_n{}^2}(\text{DLF})_n \sin\frac{n\pi x}{l} \quad n = 1, 3, 5, \ldots \quad (4.31)$$

where $(\text{DLF})_n$ is simply the DLF for a one-degree system subjected to the same load-time function as that for p_1. Since n appears in the denominator, it is apparent that higher modes are even less important here than in the case of concentrated loads. This is true because the load distribution is similar to the characteristic shape of the first mode. In fact, if the load distribution were the same as that shape, i.e., if

$p(x) = p_1 \sin(\pi x/l)$, only the first mode would be excited and the contributions of all other modes would be exactly zero.

If the load had been uniformly distributed over one-half the beam as in Fig. 4.6e,

$$\int_0^l p_1(x)\phi_n(x)\,dx = p_1 \int_0^{l/2} \sin\frac{n\pi x}{l}\,dx$$

$$= \frac{p_1 l}{n\pi}\left(1 - \cos\frac{n\pi}{2}\right)$$

The total dynamic beam deflection would then be given by

$$y(x) = \frac{2p_1}{m\pi}\sum^n \frac{1}{n\omega_n^2}\left(1 - \cos\frac{n\pi}{2}\right)(\text{DLF})_n \sin\frac{n\pi x}{l} \qquad (4.32)$$

c. Dynamic Stresses

In order to determine dynamic stresses we need only apply the well-known relationships

$$\mathfrak{M} = -EI\frac{\partial^2 y}{\partial x^2}$$

$$V = \frac{\partial \mathfrak{M}}{\partial x}$$

for bending moment and shear. Thus the computation involves only the differentiation with respect to x of the expressions given above for dynamic deflection. For example, in the case of a simply supported beam with uniformly distributed load, Eq. (4.31) may be operated on to obtain the following:

(a) $\quad \mathfrak{M} = \dfrac{4\pi EI p_1}{ml^2} \sum^n \dfrac{n}{\omega_n^2}(\text{DLF})_n \sin\dfrac{n\pi x}{l}$

(b) $\quad V = \dfrac{4\pi^2 EI p_1}{ml^3} \sum^n \dfrac{n^2}{\omega_n^2}(\text{DLF})_n \cos\dfrac{n\pi x}{l} \qquad n = 1, 3, 5, \ldots \quad (4.33)$

Note that, in going from deflection to moment to shear, the higher modes become increasingly important, as indicated by the increasing power of n. To illustrate, the amplitudes of the first and third modes, neglecting possible differences in DLF, are in the following ratios (note that ω_n^2 is proportional to n^4):

$$\frac{y_1}{y_3} = 3^5 = 243$$

$$\frac{\mathfrak{M}_1}{\mathfrak{M}_3} = 3^3 = 27$$

$$\frac{V_1}{V_3} = 3^2 = 9$$

This tendency is generally true of beam response.

166   Introduction to Structural Dynamics

Equation (4.33b) may be used to obtain the dynamic beam reaction by substituting $x = 0$ or $x = l$:

$$V_{x=0} = \frac{4EIp_1\pi^2}{ml^3} \sum^{n} \frac{n^2}{\omega_n^2} (\text{DLF})_n \qquad n = 1, 3, 5, \ldots \qquad (4.34)$$

When the first mode dominates the response, it is possible to obtain approximate deflections or stresses directly from the static values of these quantities. For example, the maximum dynamic bending moment at $x = l/2$ for the uniformly loaded beam may be closely approximated by

$$\mathfrak{M}_{x=l/2} = \frac{p_1 l^2}{8} (\text{DLF})_1$$

The corresponding value given by Eq. (4.33a), neglecting higher modes, is

$$\mathfrak{M}_{x=l/2} = \frac{4\pi EIp_1}{ml^2\omega_1^2} (\text{DLF})_1$$

or since $\omega_1^2 = \pi^4 EI/ml^4$,

$$\mathfrak{M}_{x=l/2} = \frac{4}{\pi^3} p_1 l^2 (\text{DLF})_1 = 0.129 p_1 l^2 (\text{DLF})_1$$

The close agreement between these two computations is due to the fact that static deflections can also be expressed in terms of modal components, and for a uniformly loaded beam the first mode dominates both static and dynamic response.

d. Examples

To illustrate application of the foregoing developments, suppose we were interested in the maximum deflection and stresses produced in a

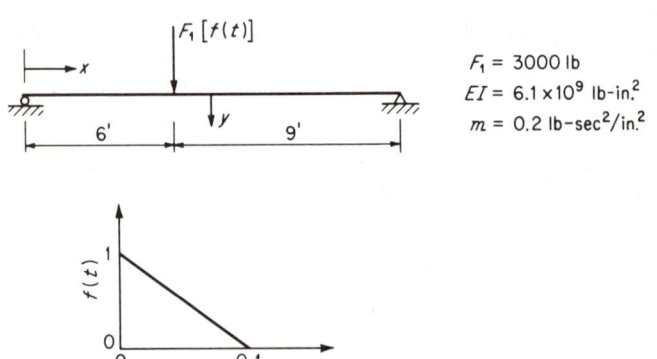

FIGURE 4.7 Example. Simple beam with concentrated dynamic force.

simply supported beam by a concentrated force such as shown in Fig. 4.7. The mass, which includes the weight of the beam, is assumed to be uniformly distributed. The natural frequencies, by Eq. (4.7), are

$$\omega_n = \frac{n^2\pi^2}{l^2}\sqrt{\frac{EI}{m}} = \frac{n^2\pi^2}{(180)^2}\sqrt{\frac{6.1 \times 10^9}{0.2}} = 53n^2$$

$$\omega_1 = 53 \qquad \omega_2 = 212 \qquad \omega_3 = 477 \qquad \text{rad/sec}$$

and
$$T_1 = 0.118 \qquad T_2 = 0.0296 \qquad T_3 = 0.0132 \qquad \text{sec}$$

Higher modes will be neglected.

It is apparent from the discussion of Sec. 4.3a that the dynamic deflection is given by

$$y(x) = \frac{2F_1}{ml}\sum_{}^{n}\frac{1}{\omega_n^2}\left(\sin\frac{n\pi c_1}{l}\right)(\text{DLF})_n \sin\frac{n\pi x}{l}$$

where $c_1 = 6$ ft. Substituting the numerical values given, we obtain the modal deflections

$$y_1(x) = +0.0565(\text{DLF})_1 \sin\frac{\pi x}{l} \qquad \text{in.}$$

$$y_2(x) = +0.00218(\text{DLF})_2 \sin\frac{2\pi x}{l} \qquad \text{in.} \qquad (4.35)$$

$$y_3(x) = -0.00043(\text{DLF})_3 \sin\frac{3\pi x}{l} \qquad \text{in.}$$

Restricting attention to maximum modal deflections, the DLFs may be obtained from Fig. 2.7 as follows:

$$\frac{t_d}{T_1} = \frac{0.1}{0.118} = 0.85 \qquad (\text{DLF})_{1,\max} = 1.48$$

$$\frac{t_d}{T_2} = \frac{0.1}{0.0296} = 3.4 \qquad (\text{DLF})_{2,\max} = 1.85$$

$$\frac{t_d}{T_3} = \frac{0.1}{0.0132} = 7.6 \qquad (\text{DLF})_{3,\max} = 1.94$$

Inserting these into Eqs. (4.35), we obtain the modal deflections (at the time of maximum modal response).

$$y_1 = +0.084 \sin\frac{\pi x}{l} \qquad \text{in.}$$

$$y_2 = +0.0040 \sin\frac{2\pi x}{l} \qquad \text{in.} \qquad (4.36)$$

$$y_3 = -0.00084 \sin\frac{3\pi x}{l} \qquad \text{in.}$$

168   Introduction to Structural Dynamics

It is apparent that the second mode contributes little and the third mode could well be ignored. These maximums could be added numerically to obtain an upper bound of deflection. A more exact determination would require consideration of the time variation of modal responses and could, if necessary, be accomplished by the use of Eqs. (2.17b) and (2.18b), which give DLF as a function of time.

The maximum (timewise) modal bending moments may be obtained by double differentiation of Eqs. (4.36).

$$\mathfrak{M} = -EI \frac{\partial^2 y}{\partial x^2}$$

$$\mathfrak{M}_1 = +0.084 \frac{\pi^2 EI}{l^2} \sin \frac{\pi x}{l} = (15.6 \times 10^4) \sin \frac{\pi x}{l} \quad \text{in.-lb}$$

$$\mathfrak{M}_2 = +0.0040 \frac{4\pi^2 EI}{l^2} \sin \frac{2\pi x}{l} = (3.0 \times 10^4) \sin \frac{2\pi x}{l} \quad \text{in.-lb}$$

$$\mathfrak{M}_3 = -0.00084 \frac{9\pi^2 EI}{l^2} \sin \frac{3\pi x}{l} = (-1.4 \times 10^4) \sin \frac{3\pi x}{l} \quad \text{in.-lb}$$

At the point of loading ($x = 6$ ft), these moments become

$$\mathfrak{M}_1 = 14.8 \times 10^4 \quad \text{in.-lb}$$
$$\mathfrak{M}_2 = 1.7 \times 10^4 \quad \text{in.-lb}$$
$$\mathfrak{M}_3 = 0.8 \times 10^4 \quad \text{in.-lb}$$

The maximum (timewise) modal shears are obtained by

$$V = \frac{\partial \mathfrak{M}}{\partial x}$$

$$V_1 = 2720 \cos \frac{\pi x}{l} \quad \text{lb}$$

$$V_2 = 1050 \cos \frac{2\pi x}{l} \quad \text{lb}$$

$$V_3 = -740 \cos \frac{3\pi x}{l} \quad \text{lb}$$

The maximum (spanwise) shears which occur at $x = 0$ are also the left beam reactions and are equal to the coefficients of the cosines just given. The second and third modes are in this respect more important, and the numerical addition of modal shears or reactions may be too conservative. If so, one would have to consider the actual time variation of DLF in each mode.

As a second example, let us consider a prismatic, fixed-ended beam subjected to a uniformly distributed dynamic load. Only the first mode will be considered. The characteristic shapes are given by Eq. (4.12). The general equation of motion is Eq. (4.13). The values of the integrals on the right side of the latter equation are given in Table 4.1, and therefore the equation becomes

$$\ddot{A}_1 + \omega_1^2 A_1 = \frac{p_1 f(t) \times 0.8308 l}{m \times l} = 0.831 \frac{p_1 f(t)}{m}$$

where $p_1 f(t)$ is the time-varying load and m the mass, both per unit length. The modal static deflection as given by Eq. (4.24) is

$$A_{1st} = 0.831 \frac{p_1}{\omega_1^2 m}$$

and the dynamic response is

$$A_1(t) = 0.831 \frac{p_1}{\omega_1^2 m} (\text{DLF})_1$$

where $(\text{DLF})_1$ depends upon $f(t)$.

The dynamic deflection at any point along the span is given by

$$y(x, t) = A_1(t) \phi_1(x)$$
$$= A_1(t)[-0.9825(\sinh a_1 x - \sin a_1 x) + \cosh a_1 x - \cos a_1 x]$$

where $a_1 = 3\pi/2l$, and the quantity -0.9825 is $\alpha/\mathcal{B}$ as defined in Sec. 4.2b and given in Table 4.1. The deflection at midspan, obtained by substituting $x = l/2$ and the expression given above for $A_1(t)$, is equal to

$$[y(t)]_{x=l/2} = 1.32 \frac{p_1}{\omega_1^2 m} (\text{DLF})_1$$

$$= \frac{p_1 l^4}{3.84 \pi^4 EI} (\text{DLF})_1$$

The dynamic bending moment is found by

$$\mathfrak{M}_1(x, t) = -EI \frac{\partial^2 y(x, t)}{\partial x^2} = -EI A_1(t) \frac{\partial^2 \phi_1(x)}{\partial x^2}$$

After substituting for $A_1(t)$ and $\partial^2 \phi_1(x)/\partial x^2$, this expression may be reduced to

$$\mathfrak{M}_1(x, t) = 0.0376 p_1 l^2 (\text{DLF})_1 [-0.9825(\sinh a_1 x - \sin a_1 x) + \cosh a_1 x - \cos a_1 x]$$

from which the maximum moment in the span is found to be

$$[\mathfrak{M}_1(t)]_{x=0} = -0.075 p_1 l^2 (\text{DLF})_1$$

The responses in higher modes could be obtained in the same manner, making use of the numerical values in Table 4.1.

4.4 Beams with Variable Cross Section and Mass

In the preceding sections we have dealt only with prismatic beams. If the cross-section properties and/or the mass vary along the span, it is usually necessary, or at least advisable, to adopt numerical methods. Perhaps the best procedure is to use the modified Rayleigh method (Sec. 3.5) to obtain natural frequencies and characteristic shapes. With this accomplished, modal analysis of the response may be executed in the usual manner. This procedure is illustrated below by the analysis of a simply supported, nonprismatic beam. The method is, however, perfectly general. It could be applied to single spans with any end conditions and also to continuous beams.

a. Example

The simple-span beam of Fig. 4.8 has in its central half a mass intensity of m_2 and a stiffness EI_2, while in the outer quarters of the span the corresponding quantities are m_1 and EI_1. We desire to obtain the response (deflection and bending moment) due to a uniformly distributed dynamic load $p(t)$ which has the time function shown.

In order to obtain the natural frequency and characteristic shape, the beam will be converted into a lumped-parameter system. For this purpose the span is divided into 20 equal segments, but because of symmetry, only one-half the span need be considered. Each section will have a mass $M_r = m\,(\Delta x)$ and an applied load $F_r(t) = p(t)\,(\Delta x)$. All quantities (deflection, bending moment, etc.) will be computed at

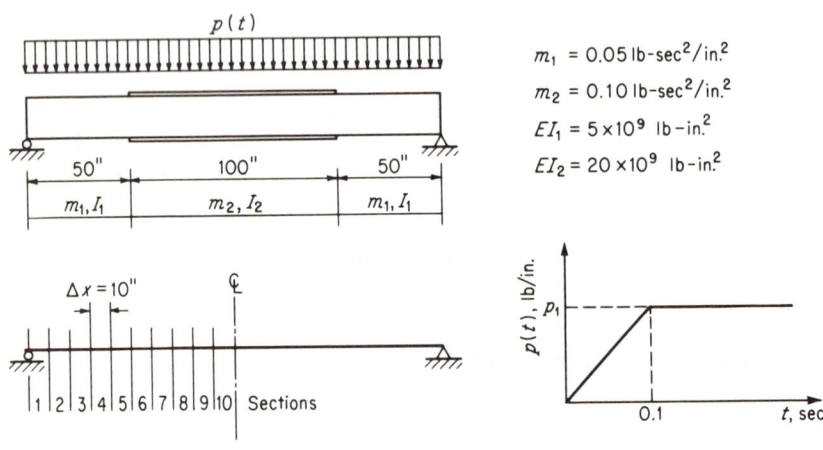

FIGURE 4.8 Example. Analysis of simple beam with variable mass and stiffness.

the centers of the segments. The precision of the solution obviously increases with the number of segments.

In Sec. 3.5 the natural frequency is given by Eq. (3.25), or

$$\omega^2 = \frac{\sum_{r=1}^{j} F_{ri}\phi_r''}{A'' \sum_{r=1}^{j} M_r(\phi_r'')^2}$$

where $F_{ri} = M_r\phi_r'$, ϕ_r' is an assumed shape, and ϕ_r'' is the shape caused by the forces F_{ri} (normalized so that $A''\phi_r''$ is the true deflection at r). In this example only the fundamental mode is considered. Higher modes could be included by the method given in Sec. 3.5b, but it is obvious in this case that these would not be important.

The computations leading to natural frequency are shown in Table 4.2. The assumed shape in the first cycle is taken as a sine curve because it is expected that the true shape will not be radically different from that for a prismatic beam. The deflections due to F_{ri}, that is, $A''\phi_r''$, are calculated by the conjugate-beam method, in which the bending moment due to the elastic load is equal to the deflection. Any other procedure for computing deflections could have been used. The deflection at section 10 is arbitrarily taken to be A'', and when all deflections are divided by this value, the result is the computed shape ϕ_r''. Comparison of ϕ_r'' and the assumed shape ϕ_r' is indicative of the accuracy. For the second cycle the shape assumed is that computed in the first cycle. The final shape in the second cycle is very close to that obtained in the first, and therefore convergence is satisfactory. Based on the summations at the end of the second cycle, the natural frequency of the first mode is

$$\omega_1^2 = \frac{5.017}{112{,}760 \times 10^{-9} \times 5.030} = 8850$$
$$\omega_1 = 94.1 \text{ rad/sec} \qquad T_1 = 0.0667 \text{ sec}$$

The equation of motion is obtained by Eq. (3.46),

$$\ddot{A}_1 + \omega_1^2 A_1 = \frac{f(t) \sum_{r=1}^{j} F_{r1}\phi_r''}{\sum_{r=1}^{j} M_r(\phi_r'')^2}$$

where F_{r1} is the maximum load at segment r and is for all segments equal to $p_1(\Delta x)$. The summations are evaluated in Table 4.2. The modal

Table 4.2 Modified Rayleigh Method for Natural Frequency; Nonprismatic Beam; Computations for Beam in Fig. 4.8 (Half-span)

		Sec. 1	Sec. 2	Sec. 3	Sec. 4	Sec. 5	Sec. 6	Sec. 7	Sec. 8	Sec. 9	Sec. 10	Σ
Cycle 1	M_r											
	Assumed shape, ϕ_r'	0.5	0.5	0.5	0.5	0.5	1.0	1.0	1.0	1.0	1.0	
	$F_{ri} = M_r \phi_r'$	0.078	0.234	0.382	0.522	0.650	0.760	0.852	0.924	0.973	0.996	
	Inertia shear, V_r (at left of load)	0.039	0.117	0.191	0.261	0.325	0.760	0.852	0.924	0.973	0.996	
	Inertia moment, $\mathfrak{M}$ (at midsection)	5.438	5.399	5.282	5.091	4.830	4.505	3.745	2.893	1.969	0.996	
	Elastic load, $(\mathfrak{M}/EI)\,\Delta x \times 10^9$	27.19	81.18	134.00	184.91	233.21	279.26	315.71	344.64	364.33	374.29	
	Elastic shear $\times 10^9$	54.38	162.36	268.00	369.82	466.42	139.13	157.85	172.32	182.16	187.14	
	Elastic moment, $A''\phi_r'' \times 10^9$	2159	2105	1943	1675	1305	839	699	542	369	187	
	ϕ_r'' ($A'' = 107{,}440 \times 10^{-9}$)	10,800	31,850	51,280	68,030	81,080	89,470	96,460	101,880	105,570	107,440	
	$F_{ri}\phi_r''$	0.100	0.296	0.477	0.633	0.755	0.831	0.896	0.946	0.982	1.000	$\Sigma F_{ri}\phi_r'' = 4.763$
	$M_r(\phi_r'')^2$	0.004	0.035	0.091	0.165	0.245	0.631	0.764	0.875	0.956	0.996	
		0.005	0.044	0.114	0.200	0.285	0.691	0.804	0.895	0.965	1.000	$\Sigma M_r(\phi_r')^2 = 5.003$
Cycle 2	Assumed shape, ϕ_r'	0.100	0.296	0.477	0.633	0.755	0.831	0.896	0.946	0.982	1.000	
	$F_{ri} = M_r \phi_r'$	0.050	0.148	0.238	0.316	0.378	0.831	0.896	0.946	0.982	1.000	
	Inertia shear, V_r (at left of load)	5.785	5.735	5.587	5.349	5.033	4.655	3.824	2.928	1.982	1.000	
	Inertia moment $\mathfrak{M}$ (at mid-section)	28.92	86.27	142.14	195.63	245.96	292.51	330.75	360.03	379.85	389.85	
	Elastic load, $(\mathfrak{M}/EI)\,\Delta x \times 10^9$	57.84	172.54	284.38	391.26	491.92	146.25	165.37	180.01	189.92	194.92	
	Elastic shear $\times 10^9$	2274	2216	2044	1760	1368	876	730	565	385	195	
	Elastic moment, $A''\phi_r'' \times 10^9$	11,370	33,530	53,970	71,570	85,250	94,010	101,310	106,960	110,810	112,760	
	ϕ_r'' ($A'' = 112{,}760 \times 10^{-9}$)	0.101	0.298	0.479	0.636	0.757	0.835	0.900	0.950	0.984	1.000	$\Sigma F_{ri}\phi_r'' = 5.017$
	$F_{ri}\phi_r''$	0.005	0.044	0.114	0.201	0.286	0.695	0.807	0.899	0.966	1.000	
	$M_r(\phi_r'')^2$	0.005	0.044	0.115	0.202	0.287	0.697	0.810	0.902	0.968	1.000	$\Sigma M_r(\phi_r')^2 = 5.030$
	$F_{ri}\phi_r''$	1.01p₁	2.98p₁	4.79p₁	6.36p₁	7.57p₁	8.35p₁	9.00p₁	9.50p₁	9.84p₁	10.00p₁	$\Sigma F_{ri}\phi_r'' = 69.40$p₁

172

static deflection is, by Eq. (3.47),

$$A_{1st} = \frac{\sum_{r=1}^{j} F_{r1}\phi_r''}{\omega_1^2 \sum_{r=1}^{j} M_r(\phi_r'')^2}$$

$$= \frac{69.40 \times 2p_1}{8850 \times 5.030 \times 2} = (1.56p_1)10^{-3} \quad \text{in.}$$

The maximum DLF for this load-time function (Fig. 4.8) may be obtained from Fig. 2.9. Entering this chart with

$$t_r/T = 0.1/0.0667 = 1.5$$

we obtain

$$(\text{DLF})_{\max} = 1.20$$

Thus the maximum dynamic deflection (at section 10 since A'' is based on this point) is

$$(y_{10})_{\max} = A_{1st}(\text{DLF})_{\max} = (1.87p_1)10^{-3} \quad \text{in.}$$

and the deflection at midspan would be only slightly larger. The maximum dynamic bending moment at midspan may be closely approximated by the static moment times the maximum DLF.*

$$\mathfrak{M}_{\max} = \tfrac{1}{8}p_1 l^2 \times 1.20 = 0.15p_1 l^2$$

Higher modes do not contribute appreciably to either deflection or bending moment. As discussed in Sec. 4.3, it might be desirable to include the third mode (the second does not contribute in this symmetrical case) if shears or reactions were to be computed, although the contribution of this mode would not be great.

The above method is a powerful tool because of its generality. It may be used for any variation in stiffness and mass (including concentrated masses), any load distribution or time function, and any combination of end conditions. The basic computation is the determination of deflected shape, which may be done by any convenient method. If important, the effect of shear distortion, elastic supports, or any other special conditions of the particular problem could be included in the deflections, and hence in the dynamic analysis.

* The bending moment based upon the first mode could be computed from the data in Table 4.2. The bending moment due to F_{ri} is the modal moment, and 389.85 is the moment at point 10, corresponding to a deflection of $112{,}760 \times 10^{-9}$ in. Since moment is proportional to deflection, the maximum dynamic moment is easily computed for the actual deflection of $1.87p_1 \times 10^{-3}$ in. The value given above, based on the static moment, is somewhat more accurate since the higher modes are approximately accounted for.

174 Introduction to Structural Dynamics

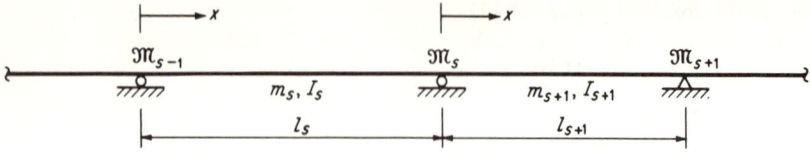

FIGURE 4.9 Continuous beams—notation.

4.5 Continuous Beams

The dynamic analysis of continuous beams by any rigorous method is a very cumbersome procedure. This is primarily due to the fact that the characteristic shapes are rather complicated mathematically and not convenient for manipulation. The difficulty is compounded because there are usually several natural modes with frequencies that are not radically different and all of which might contribute significantly to the response. Although the presentation which follows is rigorous, the engineer must in many practical cases resort to approximate solutions for response such as discussed in Chap. 5.

a. General Frequency Equation

Consider the general case of a continuous beam as shown in Fig. 4.9, where it is assumed that each span has uniform mass distribution and stiffness. We begin by noting that Eq. (4.5) applies to spans with any end conditions, and hence the characteristic shape for the nth mode and the s span is defined by

$$\Phi_{ns}(x) = \mathcal{A}_{ns} \sin a_{ns}x + \mathcal{B}_{ns} \cos a_{ns}x + \mathcal{C}_{ns} \sinh a_{ns}x + \mathcal{D}_{ns} \cosh a_{ns}x$$

$$\Phi'_{ns}(x) = \mathcal{A}_{ns}a_{ns} \cos a_{ns}x - \mathcal{B}_{ns}a_{ns} \sin a_{ns}x + \mathcal{C}_{ns}a_{ns} \cosh a_{ns}x + \mathcal{D}_{ns}a_{ns} \sinh a_{ns}x$$

$$\Phi''_{ns}(x) = -\mathcal{A}_{ns}a_{ns}^2 \sin a_{ns}x - \mathcal{B}_{ns}a_{ns}^2 \cos a_{ns}x + \mathcal{C}_{ns}a_{ns}^2 \sinh a_{ns}x + \mathcal{D}_{ns}a_{ns}^2 \cosh a_{ns}x$$

where Φ' and Φ'' are the first and second derivatives with respect to x and

$$a_{ns} = \sqrt[4]{\frac{m_s \omega_n^2}{EI_s}}$$

It is now convenient to introduce the following boundary conditions for two adjacent spans (Fig. 4.9):

$$
\begin{align}
(a) & \quad \Phi_{ns}(0) = 0 \\
(b) & \quad \Phi_{ns}(l_s) = 0 \\
(c) & \quad \Phi'_{ns}(l_s) = \Phi'_{n(s+1)}(0) \\
(d) & \quad EI_s\Phi''_{ns}(l_s) = EI_{s+1}\Phi''_{n(s+1)}(0) = -\mathfrak{M}_{ns} \\
(e) & \quad \Phi_{n(s+1)}(0) = 0
\end{align}
\qquad (4.37)
$$

Structures with Distributed Mass and Load

which state that, in the modal shape, the deflections at the supports are zero and that the slopes and bending moments of two adjacent spans at the common support must be equal. Substituting the expressions for modal shape and derivatives thereof into (a) to (e), we obtain the following:

(f) $$\mathcal{B}_{ns} + \mathcal{D}_{ns} = 0$$
(g) $$\mathcal{A}_{ns} \sin a_{ns}l_s + \mathcal{B}_{ns} \cos a_{ns}l_s + \mathcal{C}_{ns} \sinh a_{ns}l_s + \mathcal{D}_{ns} \cosh a_{ns}l_s = 0$$
(h) $$\mathcal{A}_{ns} \cos a_{ns}l_s - \mathcal{B}_{ns} \sin a_{ns}l_s + \mathcal{C}_{ns} \cosh a_{ns}l_s + \mathcal{D}_{ns} \sinh a_{ns}l_s = \frac{a_{n(s+1)}}{a_{ns}} (\mathcal{A}_{n(s+1)} + \mathcal{C}_{n(s+1)})$$
(i) $$-\mathcal{A}_{ns} \sin a_{ns}l_s - \mathcal{B}_{ns} \cos a_{ns}l_s + \mathcal{C}_{ns} \sinh a_{ns}l_s + \mathcal{D}_{ns} \cosh a_{ns}l_s = \frac{a_{n(s+1)}^2}{a_{ns}^2} \frac{I_{s+1}}{I_s} [-\mathcal{B}_{n(s+1)} + \mathcal{D}_{n(s+1)}]$$
(j) $$\mathcal{B}_{n(s+1)} + \mathcal{D}_{n(s+1)} = 0$$

Adding and subtracting (g) and (i) and substituting from (j),

$$\mathcal{D}_{n(s+1)} = -\mathcal{B}_{n(s+1)}$$

(k) $$\mathcal{C}_{ns} \sinh a_{ns}l_s - \mathcal{B}_{ns} \cosh a_{ns}l_s = -\mathcal{B}_{n(s+1)} \frac{a_{n(s+1)}^2}{a_{ns}^2} \frac{I_{s+1}}{I_s}$$

(l) $$\mathcal{A}_{ns} \sin a_{ns}l_s + \mathcal{B}_{ns} \cos a_{ns}l_s = \mathcal{B}_{n(s+1)} \frac{a_{n(s+1)}^2}{a_{ns}^2} \frac{I_{s+1}}{I_s}$$

from which

(m) $$\mathcal{C}_{ns} = \frac{+\mathcal{B}_{ns} \cosh a_{ns}l_s - \mathcal{B}_{n(s+1)}(a_{n(s+1)}^2/a_{ns}^2)(I_{s+1}/I_s)}{\sinh a_{ns}l_s}$$

(n) $$\mathcal{A}_{ns} = \frac{-\mathcal{B}_{ns} \cos a_{ns}l_s + \mathcal{B}_{n(s+1)}(a_{n(s+1)}^2/a_{ns}^2)(I_{s+1}/I_s)}{\sin a_{ns}l_s}$$

Adding (m) and (n), we may write

(o) $$\mathcal{A}_{ns} + \mathcal{C}_{ns} = \mathcal{B}_{ns} G_{ns} - \mathcal{B}_{n(s+1)} \frac{a_{n(s+1)}^2}{a_{ns}^2} \frac{I_{s+1}}{I_s} H_{ns}$$

where

(p) $$G_{ns} = \coth a_{ns}l_s - \cot a_{ns}l_s$$
(q) $$H_{ns} = \operatorname{cosech} a_{ns}l_s - \operatorname{cosec} a_{ns}l_s$$

If all subscripts in Eq. (o) are raised by 1, and the resulting expression for $(\mathcal{A}_{n(s+1)} + \mathcal{C}_{n(s+1)})$ is substituted into the right side of (h), and if (m) and (n) along with $\mathcal{D}_{ns} = -\mathcal{B}_{ns}$ are substituted into the left side, Eq. (h) may be written in the form

(r) $$\mathcal{B}_{ns} H_{ns} - \mathcal{B}_{n(s+1)} \left(\frac{a_{n(s+1)}^2 I_{s+1}}{a_{ns}^2 I_s} G_{ns} + \frac{a_{n(s+1)}}{a_{ns}} G_{n(s+1)} \right) + \mathcal{B}_{n(s+2)} \frac{a_{n(s+2)}^2 I_{s+2}}{a_{ns} a_{n(s+1)} I_{s+1}} H_{n(s+1)} = 0$$

From (d) we may write

$$-\mathfrak{M}_{ns} = EI_{s+1}(-\mathfrak{B}_{n(s+1)} + \mathfrak{D}_{n(s+1)})a_{n^2(s+1)}$$

or since $\mathfrak{D}_{n(s+1)} = -\mathfrak{B}_{n(s+1)}$,

(s) $$\mathfrak{B}_{n(s+1)} = \frac{\mathfrak{M}_{ns}}{2EI_{s+1}a_{n^2(s+1)}}$$

Finally, substituting (s) and its equivalents with the subscripts raised and lowered by 1 into (r) and canceling common terms,

$$\mathfrak{M}_{n(s-1)}\frac{H_{ns}l_s}{(a_{ns}l_s)I_s} - \mathfrak{M}_{ns}\left[\frac{G_{ns}l_s}{(a_{ns}l_s)I_s} + \frac{G_{n(s+1)}l_{s+1}}{(a_{n(s+1)}l_{s+1})I_{s+1}}\right]$$
$$+ \mathfrak{M}_{n(s+1)}\frac{H_{n(s+1)}l_{s+1}}{(a_{n(s+1)}l_{s+1})I_{s+1}} = 0 \quad (4.38)$$

Equation (4.38) is the *three-moment equation* which may be used to obtain the natural frequencies of normal modes. It is equivalent to the three-moment equation used in static analysis and also that used in determining buckling loads. As in those cases, Eq. (4.38) is applied to each pair of adjacent spans. If the end of the exterior span is hinged, the moment at that point is taken as zero. If the end is fixed, the equation is applied in such a way that $\mathfrak{M}_{ns}$ is the moment at the fixed end, and in so doing the I of the fictitious span outside the exterior support is taken to be infinite.

By the procedure outlined above, one equation is written for each support moment and the result is a set of simultaneous equations. The $\mathfrak{M}$'s are moments which occur during free vibration in that mode and of course cannot be determined. However, in order for any vibration to be possible, the determinant of the coefficients of the $\mathfrak{M}$'s must be zero. Expanding this determinant leads to the frequency equation, the roots of which are values of $a_n l$, which are directly related to ω_n. This procedure is illustrated by examples given below. Having obtained the frequencies, the characteristic shapes are determined by substituting each root in turn into boundary equations such as Eqs. (4.37f) to (4.37j), the number of equations required being one less than the number of coefficients ($\mathfrak{A}_{ns}$, etc.) to be computed.

b. Natural Frequencies for Special Cases

First we consider a *two-span* beam as shown in Fig. 4.10. Equation (4.38) need only be written once, with $\mathfrak{M}_{ns}$ taken to be the moment at the interior support and $\mathfrak{M}_{n(s-1)} = \mathfrak{M}_{n(s+1)} = 0$. Thus

$$-\mathfrak{M}_{ns}\left[\frac{G_{ns}l_s}{(a_{ns}l_s)I_s} + \frac{G_{n(s+1)}l_{s+1}}{(a_{n(s+1)}l_{s+1})I_{s+1}}\right] = 0 \quad (4.39a)$$

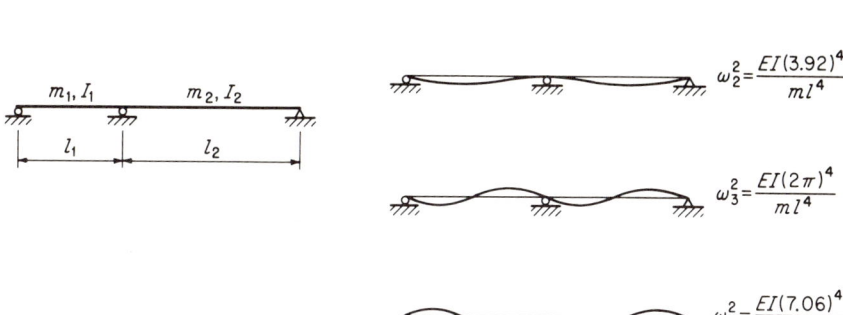

FIGURE 4.10 Normal modes of two-span beam.

and the frequency equation is

$$\frac{G_{n1}l_1}{(a_{n1}l_1)I_1} + \frac{G_{n2}l_2}{(a_{n2}l_2)I_2} = 0 \qquad (4.39b)$$

where $G_{n1} = \coth a_{n1}l_1 - \cot a_{n1}l_1$
$G_{n2} = \coth a_{n2}l_2 - \cot a_{n2}l_2$

$$a_{n1} = \sqrt[4]{\frac{m_1\omega_n^2}{EI_1}}$$

The problem is now to determine values of $a_n l$ which satisfy this equation. Note that $a_{n1}l_1$ and $a_{n2}l_2$ have a constant relationship for given beam properties, and hence one may be replaced by a constant times the other. In general, such frequency equations are not easily solved and a trial-and-error procedure must be employed. This process may be accelerated by the use of published tables giving values of G and H.[21]

To illustrate by a specific result, let us take the case of *two spans identical in stiffness, length, and mass*. The frequency equation now becomes

$$G_{n1} + G_{n2} = 0$$

This equation has two sets of roots, the first corresponding to

$$G_{n1} = G_{n2} = \pm \infty$$

and the second to $G_{n1} = G_{n2} = 0$. The roots are

$$a_{n1}l_1 = a_{n2}l_2 = \pi, 2\pi, 3\pi, \ldots$$
and $\qquad a_{n1}l_1 = a_{n2}l_2 = 3.97, 7.06, 10.2, \ldots$

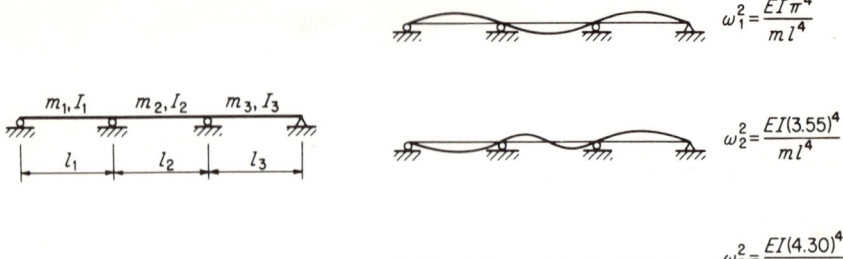

FIGURE 4.11 Normal modes of three-span beam.

The natural frequencies are therefore given by

$$\omega_n^2 = \frac{EI a_n^4}{m} = \frac{EI}{ml^4}(\pi^4, \overline{3.92}^4, \overline{2\pi}^4, \overline{7.06}^4, \ldots) \quad (4.40)$$

The characteristic shapes for the first four modes are shown in Fig. 4.10. Note that the first set of roots (the odd modes) are for antisymmetrical modes and are the same as the natural modes of a single simply supported span.* The second set of roots are for symmetrical modes which correspond to a single span fixed at one end and simply supported at the other. This similarity to a single span is obviously correct, since in an antisymmetrical mode there is a node at the center support and hence one span does not affect the other. On the other hand, in a symmetrical mode there can be no rotation at the center support, and the situation is the same as if that were a fixed support.

Second, consider a *three-span beam* such as in Fig. 4.11. Writing Eq. (4.38) twice,

$$-\mathfrak{M}_{n1}\left[\frac{G_{n1}l_1}{(a_{n1}l_1)I_1} + \frac{G_{n2}l_2}{(a_{n2}l_2)I_2}\right] + \mathfrak{M}_{n2}\left[\frac{H_{n2}l_2}{(a_{n2}l_2)I_2}\right] = 0$$

$$\mathfrak{M}_{n1}\left[\frac{H_{n2}l_2}{(a_{n2}l_2)I_2}\right] - \mathfrak{M}_{n2}\left[\frac{G_{n2}l_2}{(a_{n2}l_2)I_2} + \frac{G_{n3}l_3}{(a_{n3}l_3)I_3}\right] = 0$$

Expanding the determinant of the coefficients of $\mathfrak{M}$, we obtain the frequency equation

$$\left[\frac{G_{n1}l_1}{(a_{n1}l_1)I_1} + \frac{G_{n2}l_2}{(a_{n2}l_2)I_2}\right]\left[\frac{G_{n2}l_2}{(a_{n2}l_2)I_2} + \frac{G_{n3}l_3}{(a_{n3}l_3)I_3}\right] - \left[\frac{H_{n2}l_2}{(a_{n2}l_2)I_2}\right]^2 = 0 \quad (4.41)$$

* Note that Eq. (4.39a) is also satisfied by $\mathfrak{M}_{ns} = 0$, which indicates that these modes must be the same as for two independent spans.

Structures with Distributed Mass and Load   179

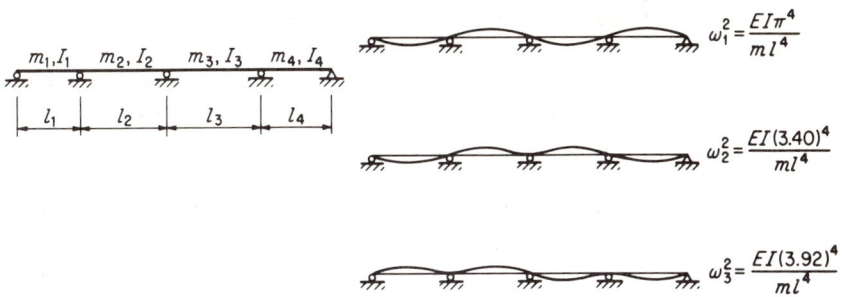

FIGURE 4.12 Normal modes of four-span beam.

For *three identical spans* this becomes

$$4G_n{}^2 - H_n{}^2 = 0$$
or
$$2G_n = \pm H_n$$

where G_n and H_n apply to any of the spans. The three sets of roots to this equation are (1) $G_n = H_n = \pm \infty$: the first root of the set is $a_1 l = \pi$ and the complete set is the same as for the modes of a single, simply supported span; (2) $2G_n = -H_n$: the first root of the set which is the second mode of the beam is a $a_2 l = 3.55$ and all modes of this set have a node at the center of the middle span; and (3) $2G_n = +H_n$: the first root of this set, or the third mode of the beam, is $a_3 l = 4.30$. These first three modes are shown in Fig. 4.11.

We could proceed in the same manner to investigate a *four-span beam*, but of course the frequency equation is much more complicated. However, for *identical spans* it reduces to[21]

$$\sqrt{2}\, G_n = \pm H_n$$

Again, there are three sets of roots, the first of which is the same as that for a simply supported, single span. The lowest three modes are shown in Fig. 4.12.

It should be apparent from the above discussion that, for any number of identical spans with hinges at exterior supports, the fundamental mode is the same as for a single, simply supported span. The higher modes of the single-span case are also higher modes of the multispan case, but these are interspersed with other modes. There is another group of symmetrical modes with only small rotations at interior supports which correspond roughly to a single fixed-ended span. In addition, there are various combinations of these two types, the number of possible combinations increasing with the number of spans.

Several investigators have devised ingenious ways of obtaining natural frequencies for continuous beams.[20,22] Although these are useful, they do not generally provide characteristic shapes, and it is an unfortunate fact that the latter are required for complete dynamic analysis. It is for this reason that one is usually forced to make approximate analyses.

c. Example of Response Calculation

To illustrate a complete dynamic analysis, let us consider a manageable problem such as the determination of response for the two-span beam shown in Fig. 4.13. Only the first two modes will be considered, since for this type of loading (i.e., a concentrated load near one midspan) higher modes would not make significant contributions to the response. In Sec. 4.5b it was shown that these modes are identified by

$$a_1 l = \pi \qquad \omega_1^2 = \frac{EI\pi^4}{ml^4}$$

$$a_2 l = 3.92 \qquad \omega_2^2 = \frac{EI(3.92)^4}{ml^4}$$

It is also known by deduction that the characteristic shapes are given by Eq. (4.8) (single, simply supported span) for the first mode and by Eq. (4.14) (single, fixed-hinged span) for the second mode. However, to illustrate the procedure that would be used for unequal spans, we shall follow a more general approach.

The characteristic shapes may be derived from the boundary conditions represented by Eqs. (4.37a) to (4.37e). Equations (a), (d), and (e) provide

$$\Phi_{n1}(0) = 0 = \mathcal{B}_{n1} + \mathcal{D}_{n1}$$
$$\Phi_{n1}''(0) = 0 = -\mathcal{B}_{n1} + \mathcal{D}_{n1}$$
$$\Phi_{n2}(0) = 0 = \mathcal{B}_{n2} + \mathcal{D}_{n2}$$

from which we obtain

$$\mathcal{B}_{n1} = \mathcal{D}_{n1} = 0 \quad \text{and} \quad \mathcal{D}_{n2} = -\mathcal{B}_{n2}$$

Making use of these relations between $\mathcal{B}$ and $\mathcal{D}$ and noting that, for identical spans, $a_{n1} = a_{n2} = a_n$, we may write

$$
\begin{aligned}
&(a) \quad \Phi_{n1}(l) = 0 && \mathcal{A}_{n1} \sin a_n l + \mathcal{C}_{n1} \sinh a_n l = 0 \\
&(b) \quad \Phi_{n1}'(l) = \Phi_{n2}'(0) && \mathcal{A}_{n1} \cos a_n l + \mathcal{C}_{n1} \cosh a_n l = \mathcal{A}_{n2} + \mathcal{C}_{n2} \\
&(c) \quad \Phi_{n1}''(l) = \Phi_{n2}''(0) && -\mathcal{A}_{n1} \sin a_n l + \mathcal{C}_{n1} \sinh a_n l = -2\mathcal{B}_{n2} \quad (4.42) \\
&(d) \quad \Phi_{n2}(l) = 0 && \mathcal{A}_{n2} \sin a_n l + \mathcal{B}_{n2}(\cos a_n l - \cosh a_n l) \\
& && \qquad + \mathcal{C}_{n2} \sinh a_n l = 0
\end{aligned}
$$

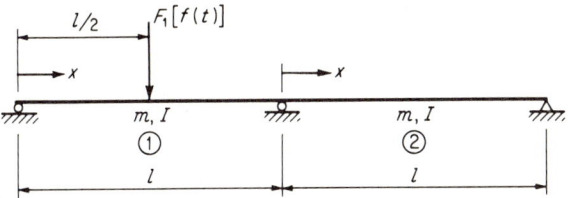

FIGURE 4.13 Example. Response of two-span beam.

For any value of $a_n l$, these four equations may be solved simultaneously to obtain four of the unknown coefficients in terms of the fifth. This will define the characteristic shape.

For the first mode we substitute $a_n l = \pi$ and obtain, from Eqs. (4.42),

$$\mathcal{C}_{11} = \mathcal{B}_{12} = \mathcal{C}_{12} = 0$$
$$\mathcal{A}_{11} = -\mathcal{A}_{12}$$

Letting $\mathcal{A}_{11} = A_1$, the arbitrary modal amplitude, and substituting into the general expression for $\Phi_{ns}(x)$, we obtain the characteristic shape

$$\Phi_{11}(x) = A_1 \sin a_1 x = A_1 \sin \frac{\pi x}{l}$$

$$\Phi_{12}(x) = -A_1 \sin \frac{\pi x}{l} \qquad (4.43)$$

or $\qquad \phi_{11}(x) = \sin \frac{\pi x}{l} \qquad \phi_{12}(x) = -\sin \frac{\pi x}{l}$

These are, as expected, the same shape as that for the first mode of a single, simply supported span.

For the second mode we substitute $a_n l = 3.92$ and solve the four equations (4.42) simultaneously. This produces the result

$$\mathcal{C}_{22} = -25.30 \mathcal{C}_{21} \qquad \mathcal{A}_{21} = +35.60 \mathcal{C}_{21}$$
$$\mathcal{B}_{22} = -25.19 \mathcal{C}_{21} \qquad \mathcal{D}_{22} = +25.19 \mathcal{C}_{21}$$
$$\mathcal{A}_{22} = +25.32 \mathcal{C}_{21}$$

where $\mathcal{C}_{21}$ has arbitrarily been taken as the reference coefficient. The second-mode characteristic shapes for the two spans are therefore

$$\Phi_{21}(x) = \mathcal{C}_{21}\left(35.60 \sin \frac{3.92x}{l} + \sinh \frac{3.92x}{l}\right)$$

$$\Phi_{22}(x) = \mathcal{C}_{21}\left(25.32 \sin \frac{3.92x}{l} - 25.19 \cos \frac{3.92x}{l} \right.$$
$$\left. - 25.30 \sinh \frac{3.92x}{l} + 25.19 \cosh \frac{3.92x}{l}\right)$$

182   Introduction to Structural Dynamics

Since it is known that this mode is symmetrical, the first of these may be used for the second span if for that span the direction of x is reversed and the origin taken at the right support. Furthermore, it is a convenience to refer Φ to the midspan deflection. If this is done,

$$\Phi_{21}(x) = A_2\left(0.975 \sin \frac{3.92x}{l} + 0.0274 \sinh \frac{3.92x}{l}\right)$$

$$\phi_{21}(x) = \frac{\Phi_{21}(x)}{A_2} \tag{4.44}$$

where A_2 is the modal displacement at either midspan.

The modal equations of motion may now be obtained by the direct application of Eq. (4.23), with the load integral replaced by one term, since we are dealing with a single concentrated load.

$$\ddot{A}_n + \omega_n^2 A_n = \frac{[f(t)]F_1[\phi_{n1}(l/2)]}{m \sum \int_0^l \phi_n^2(x)\,dx} \tag{4.45}$$

The summation in the right-side denominator indicates that the integration must include both spans. For the first mode, using Eq. (4.43),

$$F_1\left[\phi_{11}\left(\frac{l}{2}\right)\right] = F_1\left(\sin \frac{\pi}{2}\right) = F_1$$

$$m \sum \int_0^l \phi_1^2(x)\,dx = 2m \int_0^l \sin^2 \frac{\pi x}{l}\,dx = ml$$

In the last computation the sum for the two spans is twice the value of the integral for one span since the shapes differ only in sign. Thus the equation of motion for the first mode is

$$\ddot{A}_1 + \omega_1^2 A_1 = \frac{F_1[f(t)]}{ml} \tag{4.46}$$

and the modal static deflection by Eq. (4.24) is

$$A_{1st} = \frac{F_1}{\omega_1^2 ml} \tag{4.47}$$

for the second mode, and using Eq. (4.44),

$$F_1[\phi_{21}]_{x=l/2} = F_1(0.975 \sin 1.96 + 0.0274 \sinh 1.96) = F_1$$

$$m \sum \int_0^l \phi_2^2(x)\,dx = 2m \int_0^l \left(0.975 \sin \frac{3.92x}{l} + 0.0274 \sinh \frac{3.92x}{l}\right)^2 dx$$

$$= 0.95\,ml$$

In the last computation the symmetry of the mode has been recognized and advantage has been taken of the fact that the value of the integral must be the same for both spans.

The equation of motion for the second mode is therefore

$$\ddot{A}_2 + \omega_2{}^2 A_2 = \frac{F_1[f(t)]}{0.95ml} \tag{4.48}$$

and the modal static deflection is

$$A_{2st} = \frac{F_1}{(0.95ml)\omega_2{}^2} \tag{4.49}$$

Combining the two modes and using Eq. (4.25), we finally obtain the total midspan dynamic deflection.

$$\begin{aligned} y(t) &= A_{1st}(\text{DLF})_1 + A_{2st}(\text{DLF})_2 \\ &= \frac{F_1}{\omega_1{}^2 ml}(\text{DLF})_1 + \frac{F_1}{\omega_2{}^2(0.95ml)}(\text{DLF})_2 \\ &= \frac{F_1}{ml}\left[\frac{(\text{DLF})_1}{\omega_1{}^2} + \frac{(\text{DLF})_2}{0.95\omega_2{}^2}\right] \end{aligned} \tag{4.50}$$

where the DLF's depend on the load-time function $f(t)$ and are evaluated in the usual way as for a one-degree system. Equation (4.50) gives the deflection at the center of the left span since the A's were taken as the characteristic amplitude at that point. The dynamic deflection at the center of the right span would be

$$y(t) = \frac{F_1}{ml}\left[-\frac{(\text{DLF})_1}{\omega_1{}^2} + \frac{(\text{DLF})_2}{0.95\omega_2{}^2}\right]$$

It may be noted that, in this case, the contributions of the modes are roughly (neglecting possible differences in DLF) in proportion to $1/\omega^2$. Therefore the second-mode contribution is $(\pi/3.92)^4$, or 0.41 of the first. It may also be observed that, if a symmetric (downward) load had also been applied to the right span, the first mode would have contributed nothing and the second-mode displacement would have been twice that indicated above.

4.6 Beam-girder Systems

The analysis of floor systems or other structures consisting of combinations of beams or girders is obviously rather involved. Not only is each element a complex system, as we have seen in previous sections, but the interaction of elements creates the possibility of many significant

184   *Introduction to Structural Dynamics*

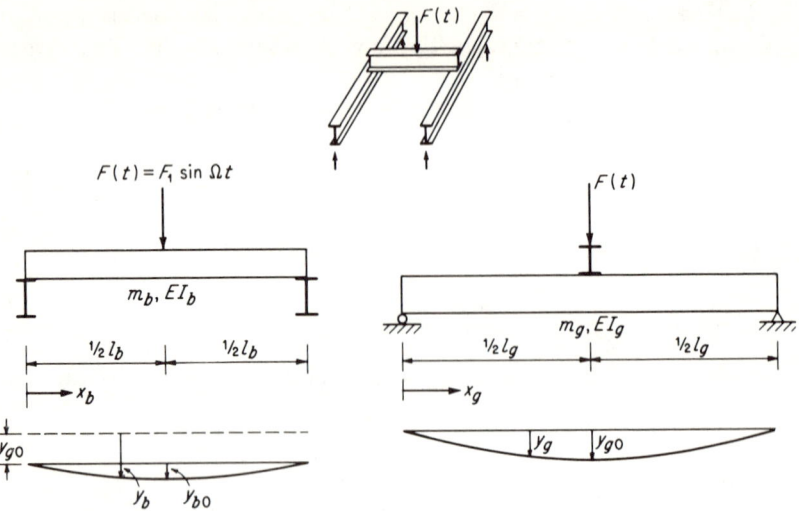

FIGURE 4.14 Beam-girder system—notation.

modal shapes. What follows is approximate in that only the most important of these shapes are included. The procedure is sufficiently accurate, however, for structures of the type discussed, namely, those which are essentially symmetrical with respect to both structural properties and loading.

We shall consider the simple system shown in Fig. 4.14, which consists of one floor beam supported by two identical girders. All elements are considered to be simply supported, and the dynamic load is applied at midspan of the beam. For this analysis it will be assumed that there is only one possible deflected shape for each member and that this is the same as the fundamental mode of a simply supported, independently acting beam. By this assumption the system has three degrees of freedom. However, since the structure and loading are both symmetrical, one of these will not be excited and need not be considered in the analysis. The deflections of the two girders will always be identical. The assumed shapes are given by

(a) $\qquad y_g = y_{go} \sin \dfrac{\pi x_g}{l_g}$

(b) $\qquad y_b = y_{go} + y_{bo} \sin \dfrac{\pi x_b}{l_b}$ (4.51)

where y_{bo} and y_{go} are the midspan ordinates. Since the two latter ordinates are unrelated, they completely define the two-degree dynamic model.

Structures with Distributed Mass and Load   185

The equations of motion may be derived by use of the Lagrangian equation (3.36). The total kinetic energy is given by

$$\mathcal{K} = \tfrac{1}{2}m_b \int_0^{l_b} \left(\dot{y}_{bo} \sin \frac{\pi x_b}{l_b} + \dot{y}_{go}\right)^2 dx_b + 2 \times \tfrac{1}{2}m_g \int_0^{l_g} \left(\dot{y}_{go} \sin \frac{\pi x_g}{l_g}\right)^2 dx_g$$

$$= \tfrac{1}{2}m_b \left(\dot{y}_{bo}^2 \frac{l_b}{2} + 2\dot{y}_{bo}\dot{y}_{go}\frac{2l_b}{\pi} + \dot{y}_{go}^2 l_b\right) + m_g \dot{y}_{go}^2 \frac{l_g}{2} \qquad (4.52)$$

The strain energy may be obtained by the following:

$$\mathcal{U} = \frac{EI_b}{2}\int_0^{l_b}\left(\frac{d^2y_b}{dx_b^2}\right)^2 dx_b + 2 \times \frac{EI_g}{2}\int_0^{l_g}\left(\frac{d^2y_g}{dx_g^2}\right)^2 dx_g$$

After substitution of the second derivatives of Eqs. (4.51), this becomes

$$\mathcal{U} = \frac{EI_b}{2}\int_0^{l_b}\left(-y_{bo}\frac{\pi^2}{l_b^2}\sin\frac{\pi x_b}{l_b}\right)^2 dx_b + EI_g\int_0^{l_g}\left(-y_{go}\frac{\pi^2}{l_g^2}\sin\frac{\pi x_g}{l_g}\right)^2 dx_g$$

$$= \frac{EI_b y_{bo}^2 \pi^4}{4l_b^3} + \frac{EI_g y_{go}^2 \pi^4}{2l_g^3} \qquad (4.53)$$

The general expression for external work by the load at midspan of the beam is

$$\mathcal{W}_e = F(t)(y_{bo} + y_{go}) \qquad (4.54)$$

Into Lagrange's equation

$$\frac{d}{dt}\left(\frac{\partial \mathcal{K}}{\partial \dot{q}_i}\right) + \frac{\partial \mathcal{U}}{\partial q_i} = \frac{\partial \mathcal{W}_e}{\partial q_i}$$

we substitute the necessary derivatives, first taking $q_i = y_{go}$ and then $q_i = y_{bo}$, to obtain

$$m_b l_b \left(\frac{2}{\pi}\ddot{y}_{bo} + \ddot{y}_{go}\right) + m_g l_g \ddot{y}_{go} + \frac{EI_g \pi^4}{l_g^3} y_{go} = F(t)$$
$$m_b l_b \left(\tfrac{1}{2}\ddot{y}_{bo} + \frac{2}{\pi}\ddot{y}_{go}\right) + \frac{EI_b \pi^4}{2l_b^3} y_{bo} = F(t) \qquad (4.55)$$

These are the equations of motion for the two-degree system.

For convenience, we now assign the following numerical values:

$$m_b = m_g = 0.1 \text{ lb-sec}^2/\text{in.}^2$$
$$EI_b = EI_g = 2 \times 10^{10} \text{ lb-in.}^2$$
$$l_b = l_g = 200 \text{ in.}$$

Substituting these values into Eqs. (4.55) and rearranging,

$$0.636\ddot{y}_{bo} + 2\ddot{y}_{go} + 12{,}180 y_{go} = 0.05 F(t)$$
$$0.500\ddot{y}_{bo} + 0.636\ddot{y}_{go} + 6090 y_{bo} = 0.05 F(t) \qquad (4.56)$$

To determine the natural frequencies, we substitute $y = a_n(\sin \omega_n t)$, $\ddot{y} = -a_n\omega_n^2 \sin \omega_n t$, and $F(t) = 0$, to obtain

$$(-0.636\omega_n^2)a_{bon} + (-2\omega_n^2 + 12{,}180)a_{gon} = 0 \\ (-0.500\omega_n^2 + 6090)a_{bon} + (-0.636\omega_n^2)a_{gon} = 0 \quad (4.57)$$

Setting the determinant of the coefficients equal to zero, expanding the determinant, and solving the resulting equation leads to

$$\omega_1^2 = 4810 \quad \omega_2^2 = 25{,}890$$

Substituting these into either one of Eqs. (4.57) gives the characteristic shapes

$$a_{bo1} = +0.835 a_{go1} \\ a_{bo2} = -2.40 a_{go2}$$

Thus, in the first mode, all beams vibrate in phase, while in the second, they are 180° out of phase, the beam distortion being considerably greater. It must be remembered that a_{bon} is the beam amplitude *relative* to the girder, and thus the total deflection is the algebraic sum of a_{bon} and a_{gon}.

To obtain the modal equations, we first rewrite the energy expressions in modal terms. We shall arbitrarily take the girder deflection y_{go} as the modal amplitude. Therefore, for the first mode, we substitute $y_{go} = A_1$ and $y_{bo} = 0.835 A_1$ into Eqs. (4.52) to (4.54) and insert numerical values for the parameters to obtain

$$\mathcal{K} = 34.10 \dot{A}_1^2 \\ \mathcal{U} = 164{,}000 A_1^2 \\ \mathcal{W}_e = 1.835 A_1 F(t)$$

For the second mode we substitute $y_{go} = A_2$ and $y_{bo} = -2.40 A_2$. Thus

$$\mathcal{K} = 18.22 \dot{A}_2^2 \\ \mathcal{U} = 472{,}000 A_2^2 \\ \mathcal{W}_e = -1.40 A_2 F(t)$$

We now write the Lagrangian equation for each mode separately to obtain the modal equations of motion:

$$\ddot{A}_1 + 4810 A_1 = 0.0269 F(t) \\ \ddot{A}_2 + 25{,}890 A_2 = -0.0384 F(t) \quad (4.58)$$

Denoting the applied load by $F(t) = F_1[f(t)]$ and referring back to Eqs. (4.23) and (4.24) (the equation of motion in general form), we see that the modal static deflections are merely the right sides of Eqs. (4.58), with the time function removed and divided by ω_n^2. Therefore

$$A_{1st} = \frac{0.0269 F_1}{4810} = +5.58 \times 10^{-6} F_1$$

$$A_{2st} = \frac{-0.0384 F_1}{25{,}890} = -1.48 \times 10^{-6} F_1$$

Furthermore, since A is the modal value of y_{go}, the dynamic deflections are given by

$$\begin{aligned}
y_{go}(t) &= A_{1st}(\text{DLF})_1 + A_{2st}(\text{DLF})_2 \\
&= +5.58 \times 10^{-6} F_1 (\text{DLF})_1 - 1.48 \times 10^{-6} F_1 (\text{DLF})_2 \\
y_{bo}(t) &= 0.835 A_{1st}(\text{DLF})_1 + (-2.40) A_{2st}(\text{DLF})_2 \\
&= +4.65 \times 10^{-6} F_1 (\text{DLF})_1 + 3.55 \times 10^{-6} F_1 (\text{DLF})_2
\end{aligned} \quad (4.59)$$

Returning now to the actual pulsating force, let us say that $F_1 = 5000$ lb and $\Omega = 45$ rad/sec. If we are interested in the steady-state response, assuming that the free part has been removed by damping, the maximum DLF is given by Eq. (2.36):

$$(\text{DLF})_{n,\max} = \frac{1}{1 - \Omega^2/\omega_n^2}$$

which, for the first and second modes, is $+1.72$ and $+1.08$, respectively. Insertion of these values into (4.59) gives the maximum deflections to be expected.

$$(y_{go})_{\max} = +0.040 \text{ in.} \qquad (y_{bo})_{\max} = +0.059 \text{ in.}$$

These are the amplitudes of harmonic motion, which, because the free vibration has been removed by damping, has the same frequency as the applied force. The positive signs indicate that the motion is in phase with that force. A negative sign would indicate a motion 180° out of phase. The total amplitude at the beam midspan is the sum, or $+0.099$ in.

The bending moments in the members are of course directly related to the midspan deflections, in accordance with the assumed shapes. Using Eqs. (4.51), we find

$$\mathfrak{M}_g = -EI_g \frac{\partial^2 y_g}{\partial x_g^2} = \frac{EI_g y_{go} \pi^2}{l_g^2} \sin \frac{\pi x_g}{l_g}$$

and

$$\mathfrak{M}_b = \frac{EI_g y_{bo} \pi^2}{l_b^2} \sin \frac{\pi x_b}{l_b}$$

The procedure given above could be extended to include systems with more than one floor beam. If the loading distribution were not symmetrical, it would be necessary to treat the two girders individually.

Introduction to Structural Dynamics

Furthermore, the analysis could be refined by including additional sine terms in the assumed shapes [Eq. (4.51)]. All this could be accomplished without changing the basic procedure. However, each such modification increases the number of degrees of freedom, and hence the difficulty of solution.[11]

4.7 Plates or Slabs Subjected to Normal Loads

In this section we shall investigate the dynamic behavior of rectangular plates or slabs. This is a three-dimensional problem and obviously somewhat more complex than analysis of the two-dimensional elements previously considered. Although the analysis which follows is for a homogeneous and isotropic material, it is commonly used for reinforced concrete slabs.

Consider the slab shown in Fig. 4.15, which is rectangular ($a \times b$) in plan, uniform in thickness h and mass m, simply supported on all edges, and subjected to a uniformly distributed dynamic load $p(t)$. The deflected shape may be taken as

$$y = \sum_{j=1}^{\infty} \sum_{i=1}^{\infty} A_{ji} \sin \frac{j\pi x}{a} \sin \frac{i\pi z}{b} \qquad (4.60)$$

where A_{ji} is the modal ordinate at the center of the plate. Equation (4.60) obviously satisfies the boundary conditions $y = 0$, $\partial^2 y/\partial x^2 = 0$, and $\partial^2 y/\partial z^2 = 0$ at all four edges. Each possible combination of integer values of j and i defines a modal shape. In the following, only the first mode will be considered, that is, $j = i = 1$, but the method of extension

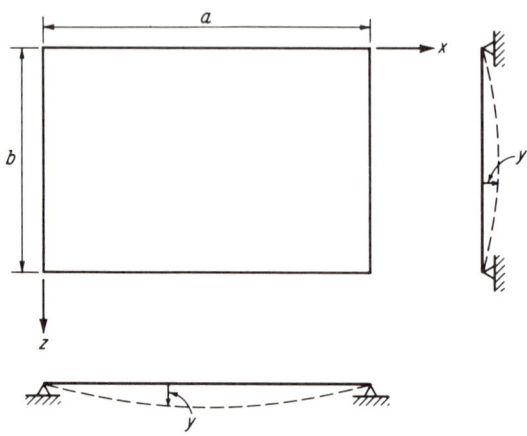

FIGURE 4.15 Simply supported rectangular slab.

to higher modes will be apparent. Thus we shall deal with the shape

$$y = A_1 \sin \frac{\pi x}{a} \sin \frac{\pi z}{b} \tag{4.61}$$

and use the Lagrangian equation

$$\frac{d}{dt}\left(\frac{\partial \mathcal{K}}{\partial \dot{A}_1}\right) + \frac{\partial \mathcal{U}}{\partial A_1} = \frac{\partial \mathcal{W}_e}{\partial A_1}$$

to obtain the modal equation of motion.

Since the kinetic energy of any element is given by

$$d\mathcal{K} = \tfrac{1}{2} m \dot{y}^2 \, dx \, dz$$

where m is the mass per unit area, the total kinetic energy obtained by integrating over the slab area is

$$\mathcal{K} = \tfrac{1}{2} m \int_o^b \int_o^a \left(\dot{A}_1 \sin \frac{\pi x}{a} \sin \frac{\pi z}{b} \right)^2 dx \, dz$$

$$= \tfrac{1}{8} m \dot{A}_1^2 ab$$

Therefore
$$\frac{d}{dt}\left(\frac{\partial \mathcal{K}}{\partial \dot{A}_1}\right) = \tfrac{1}{4} mab \ddot{A}_1 \tag{4.62}$$

The total strain energy by conventional plate theory is[24]

$$\mathcal{U} = \frac{Eh^3}{24(1-\nu^2)} \int_o^b \int_o^a \left[\left(\frac{\partial^2 y}{\partial x^2}\right)^2 + \left(\frac{\partial^2 y}{\partial z^2}\right)^2 + 2\nu \frac{\partial^2 y}{\partial x^2} \frac{\partial^2 y}{\partial z^2} \right.$$
$$\left. + 2(1-\nu)\left(\frac{\partial^2 y}{\partial x \, \partial z}\right)^2 \right] dx \, dz$$

where E = modulus of elasticity
h = plate thickness
ν = Poisson's ratio for the homogeneous material

Operating on Eq. (4.61), we obtain

$$\frac{\partial^2 y}{\partial x^2} = -A_1 \frac{\pi^2}{a^2} \sin \frac{\pi x}{a} \sin \frac{\pi z}{b}$$

$$\frac{\partial^2 y}{\partial z^2} = -A_1 \frac{\pi^2}{b^2} \sin \frac{\pi x}{a} \sin \frac{\pi z}{b}$$

$$\frac{\partial^2 y}{\partial x \, \partial z} = +A_1 \frac{\pi^2}{ab} \cos \frac{\pi x}{a} \cos \frac{\pi z}{b}$$

Substituting these expressions and integrating as indicated above, we find that the strain energy is given by

$$\mathcal{U} = \frac{Eh^3 \pi^4 ab}{96(1-\nu^2)} A_1^2 \left(\frac{1}{a^2} + \frac{1}{b^2} \right)^2 \tag{4.63}$$

The external work by a uniformly distributed load is

$$W_e = p(t) \int_0^b \int_0^a A_1 \sin\frac{\pi x}{a} \sin\frac{\pi z}{b} \, dx \, dz$$

$$= p(t) \left(\frac{4ab}{\pi^2}\right) A_1 \tag{4.64}$$

Finally, substitution of Eq. (4.62) and the derivatives of Eqs. (4.63) and (4.64) into the Lagrange equation yields the modal equation of motion

$$\tfrac{1}{4} mab \, \ddot{A}_1 + \frac{Eh^3\pi^4 ab}{48(1-\nu^2)}\left(\frac{1}{a^2}+\frac{1}{b^2}\right)^2 A_1 = p(t)\left(\frac{4ab}{\pi^2}\right)$$

or

$$\ddot{A}_1 + \frac{Eh^3\pi^4}{12(1-\nu^2)m}\left(\frac{1}{a^2}+\frac{1}{b^2}\right)^2 A_1 = p(t)\left(\frac{16}{m\pi^2}\right) \tag{4.65}$$

From the last, it is apparent that the natural frequency of the first mode is given by

$$\omega_1 = \pi^2 \left[\frac{Eh^3}{12(1-\nu^2)m}\right]^{1/2}\left(\frac{1}{a^2}+\frac{1}{b^2}\right) \tag{4.66}$$

If the general expression for deflected shape [Eq. (4.60)] had been used throughout the above development, it would have been found that the natural frequencies of all modes were given by[6]

$$\omega_n = \pi^2 \left[\frac{Eh^3}{12(1-\nu^2)m}\right]^{1/2}\left(\frac{j^2}{a^2}+\frac{i^2}{b^2}\right) \tag{4.67}$$

The higher modes are obtained by taking various integer combinations of j and i. The correspondence between these integers and the modes in the order of increasing frequency depends upon the ratio of a to b. In the case of a square slab, taking j or i equal to 2 and the other equal to 1 provides two modes having the same frequency but different shapes.

Formulation of the analysis for other than simply supported plates, while not impossible, is rather cumbersome. Approximate methods are discussed in Chap. 5.

a. Example

It is desired to determine the maximum dynamic bending stress in a flat, simply supported, rectangular steel plate having the following dimensions and parameters:

$a = 60$ in.
$b = 40$ in.
$E = 30 \times 10^6$ psi
$\nu = 0.25$
$h = 1$ in. ($m = 0.00073$ lb-sec^2/in.^{3})

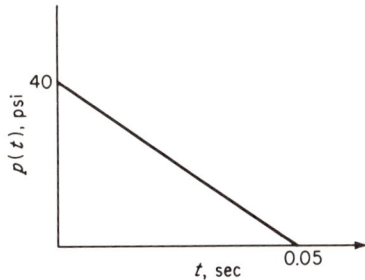

FIGURE 4.16 Flat-plate example—pressure-time function.

The plate is subjected to a uniformly distributed blast pressure, with the time variation shown in Fig. 4.16. Substitution of the numerical values into Eq. (4.65) provides

$$\ddot{A}_1 + 290{,}000 A_1 = 2{,}220 p(t) \tag{4.68}$$

Therefore
$$\omega_1 = \sqrt{290{,}000} = 539 \text{ rad/sec}$$
$$T_1 = \frac{2\pi}{\omega} = 0.0117 \text{ sec}$$

Proceeding as in Sec. 4.3 and noting that $p(t) = 40f(t)$, we recognize that the modal static deflection is the right side of Eq. (4.68), with the time function removed and divided by ω_1^2.

$$A_{1st} = \frac{2220 \times 40}{290{,}000} = 0.306 \text{ in.}$$

The maximum DLF for the load function specified is given by Fig. 2.7. Entering this figure with $t_d/T_1 = 0.05/0.0117 = 4.3$, we read

$$(\text{DLF})_{\max} = 1.88$$

The maximum dynamic deflection at the center of the plate is therefore

$$y_{\max} = (A_1)_{\max} = A_{1st}(\text{DLF})_{\max} = 0.306 \times 1.88 = 0.575 \text{ in.}$$

The bending stress in the z direction may be computed by

$$\sigma_z = -\frac{Eh}{2(1-\nu^2)}\left(\frac{\partial^2 y}{\partial z^2} + \nu \frac{\partial^2 y}{\partial x^2}\right)$$

From Eq. (4.61),

$$\frac{\partial^2 y}{\partial z^2} = -A_1 \frac{\pi^2}{b^2} \sin\frac{\pi x}{a} \sin\frac{\pi z}{b}$$
$$\frac{\partial^2 y}{\partial x^2} = -A_1 \frac{\pi^2}{a^2} \sin\frac{\pi x}{a} \sin\frac{\pi z}{b}$$

Therefore
$$\sigma_z = (A_1)\frac{Eh\pi^2}{2(1-\nu^2)}\left(\frac{1}{b^2} + \frac{\nu}{a^2}\right)\sin\frac{\pi x}{a} \sin\frac{\pi z}{b}$$

The maximum bending stress occurring at the center of the plate ($x = a/2$, $z = b/2$) is therefore

$$(\sigma_z)_{max} = (A_1)_{max} \frac{Eh\pi^2}{2(1 - \nu^2)} \left(\frac{1}{b^2} + \frac{\nu}{a^2}\right)$$

Substituting $(A_1)_{max} = 0.575$ in. and the other parameters as given, we find the maximum bending stress in the z direction to be 63,000 psi. The maximum dynamic stress in the x direction obtained by interchanging a and b is 39,500 psi.

An alternative method for computing the dynamic stresses would be that based on the classical solution for static loads. In the above example, the dynamic stress would equal the static stress due to a pressure of 40 psi multiplied by the DLF, or 1.88. A slight difference would be found between this result and that given above. Neither would be exact since, in the original analysis, we neglected the higher modes and, in the alternative approach, we assumed in effect that all modes have the same DLF as the first mode. In this example the first mode dominates both dynamic and static responses; therefore there is little error in either method.

4.8 Elasto-plastic Analysis of Beams

The determination of inelastic response for beams or other elements having distributed mass and load is extremely difficult. One possible approach is to conduct the usual elastic analysis up to the time when the ultimate bending capacity is attained at some point along the beam and then to assume that an idealized hinge has formed at this point, thus creating a new elastic system. The analysis would then be continued until a second hinge is formed or the rotation of the first hinge reverses in direction, thus indicating that elastic behavior should be restored at this point. This type of analysis is extremely cumbersome, because the system may change several times during the response and is still not exact since the plastic behavior is really not concentrated at a point as assumed.

An alternative approach is to replace the member by a lumped-parameter system as indicated in Fig. 4.17b. The mass and load are both considered to be concentrated at discrete points along the span, and it is assumed that plastic hinges may form only at these points. For the elastic range the stiffness coefficients would be determined in the usual manner by introducing unit deflections at each point. When a plastic hinge forms at a particular point, a new set of stiffness coefficients may be computed for the beam, with a pin inserted at that point. Although more practical than the first method mentioned, this approach is also

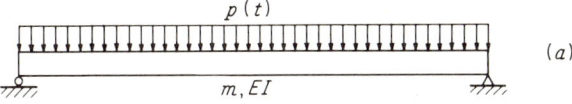
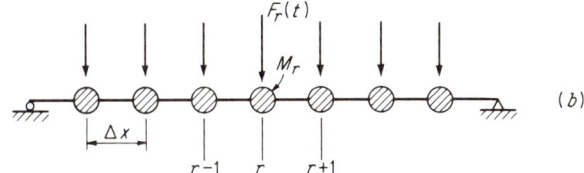
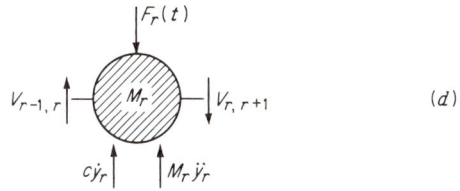
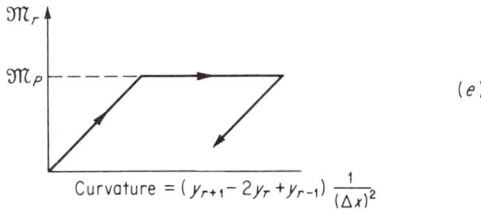

FIGURE 4.17 Finite-difference formulation of elasto-plastic beam analysis.

very cumbersome, because of the large number of potential hinge arrangements.

A simplification of the lumped-parameter approach is possible if use is made of finite-difference techniques. Consider a simple beam (Fig. 4.17) with uniform load $p(t)$, uniformly distributed mass m per unit length, and constant rigidity EI. The beam is divided into equal segments, so that the lumped parameters are

$$M = m\,(\Delta x) \qquad F(t) = (\Delta x)\,p(t)$$

The curvature at point r may be approximately expressed in terms of the *second central difference* as follows:

$$\left(\frac{d^2y}{dx^2}\right)_r \approx \frac{\delta^2 y_r}{(\Delta x)^2} = \frac{1}{(\Delta x)^2}(y_{r+1} - 2y_r + y_{r-1})$$

where y_r is the deflection at point r. The bending moment at this point is therefore

$$\mathfrak{M}_r = -EI\left(\frac{d^2y}{dx^2}\right)_r$$
$$\approx \frac{-EI}{(\Delta x)^2}(y_{r+1} - 2y_r + y_{r-1}) \qquad (4.69)$$

If we now consider the equilibrium of a segment of beam (assumed to be massless) as shown in Fig. 4.17c, the shear in that segment is obviously given by

$$V_{r-1,r} = \frac{\mathfrak{M}_r - \mathfrak{M}_{r-1}}{\Delta x}$$

Next consider the dynamic equilibrium of mass r as indicated in Fig. 4.17d. The equation of motion for this mass is

$$M_r \ddot{y}_r + V_{r-1,r} - V_{r,r+1} + c\dot{y}_r = F_r(t)$$

or

$$M_r \ddot{y}_r - \frac{\mathfrak{M}_{r-1} - 2\mathfrak{M}_r + \mathfrak{M}_{r+1}}{\Delta x} + c\dot{y}_r = F_r(t) \qquad (4.70)$$

where c is the damping coefficient and it is assumed that the damping force is proportional to the velocity at this point. At any time and for a given distortion of the system, Eq. (4.69) provides the bending moments at all points and Eq. (4.70) permits computation of the accelerations at these points. Thus numerical analysis may be executed in the usual manner for lumped-mass multidegree systems (Sec. 3.9).

When one of the moments reaches the ultimate bending capacity $\mathfrak{M}_P$, the moment is held constant at that value in subsequent computations. However, it is necessary to continue evaluation of Eq. (4.69) until a peak is reached and the hypothetical $\mathfrak{M}_r$ begins to decrease. This reversal indicates that the point has returned to elastic behavior, and subsequently the moment could be computed by

$$\mathfrak{M}_r = \mathfrak{M}_P - \frac{EI}{(\Delta x)^2}(y_{r+1} - 2y_r + y_{r-1}) - \mathfrak{M}_r^p \qquad (4.71)$$

where $\mathfrak{M}_r^p$ is the hypothetical peak previously computed. Equation (4.71) is based on the idealized moment-curvature relationship shown in Fig. 4.17e.

It should be apparent that the equations given above can easily be modified so as to be applicable to beams with nonuniform distributions of mass, rigidity, and loading.[25]

Structures with Distributed Mass and Load   195

In order to obtain acceptable accuracy by the method just outlined, a fairly large number of mass points must be used. The number depends upon the type of loading, but in a typical case it might be necessary to divide a simple beam span into 10 segments. Thus the lumped system would have nine degrees of freedom, and the time interval for the numerical analysis would have to be taken as a fraction of the smallest natural period of the elastic system. It is apparent that this type of analysis is not feasible for hand computation. However, it is ideally suited to electronic computation, and systems consisting of many lumped masses can be handled with relative ease.

If the loading on the beam is such that the hinge locations are known in advance and only one or two hinge arrangements occur, the procedure given above is unnecessarily complex. The much simpler procedures presented in Chap. 5 are quite adequate for such cases.

Problems

In Probs. 4.1 to 4.6 the beam is prismatic and has the following properties:
$m = 30$ lb-sec^2/ft per foot of span
$EI = 1.5 \times 10^8$ lb-ft^2
$l = 12$ ft

4.1 Determine the maximum dynamic values (in terms of p_1) of midspan deflection, midspan bending moment, and end shear for the beam of Fig. 4.18a due to a suddenly applied constant pressure of magnitude p_1 lb/ft.
Answer
$y = 2.82 \times 10^{-6} \, p_1$ ft
$\mathfrak{M} = 29.1 p_1$ lb-ft
$V = 7.63 p_1$ lb

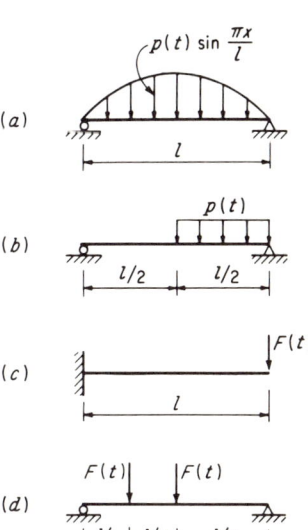

FIGURE 4.18 Problems 4.1 to 4.4.

4.2 *a.* Write the series (in terms of p_1 and DLF) for the dynamic midspan bending moment in the beam of Fig. 4.18b.

b. For the first two modes only, determine the maximum modal components of the midspan bending moment and end shear for the load-time function of Fig. 4.19. (*Note:* Figure 2.9 may be used.)

Answer

$\mathfrak{M}_1 = 12.3p_1$ lb-ft
$\mathfrak{M}_2 = 0$
$V_1 = 3.21p_1$ lb
$V_2 = 1.25p_1$ lb

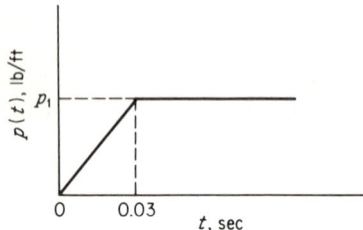

FIGURE 4.19 Problems 4.2 and 4.4. Load-time function.

4.3 Assuming that the static-deflection curve is the same as the first-mode shape and including only that mode, compute the maximum dynamic bending moment in the beam of Fig. 4.18c due to the load-time function of Fig. 4.20. (*Note:* Figure 2.7 may be used.)

Answer

$17.8F_1$ ft-lb

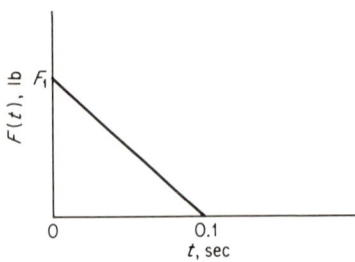

FIGURE 4.20 Problem 4.3. Load-time function.

4.4 *a.* Write the series expression (in terms of F_1 and DLF) for the deflection at each of the two load points on the beam of Fig. 4.18d.

b. For each of the first two modes, compute separately the maximum dynamic deflection at each load point due to the load-time function of Fig. 4.19, with p_1 replaced by F_1 lb.

Answer

Midspan: $y_1 = 5.6 \times 10^{-7} F_1$
$y_2 = 3.5 \times 10^{-9} F_1$ ft
Quarter point: $y_1 = 4.6 \times 10^{-7} F_1$
$y_2 = 6.8 \times 10^{-9} F_1$ ft

c. Repeat (*b*) for the left beam reaction.

Answer

$V_1 = 1.56F_1$
$V_2 = 0.15F_1$ lb

4.5 A simply supported beam is subjected to a uniformly distributed static load p_1 which is suddenly released. Write the series expression for the resulting free vibrations, and determine the amplitude of the first mode in terms of p_1.

4.6 A concentrated force given by $F_1 \sin 400t$ lb is applied to the quarter point of the simply supported span. If there is 10 percent of critical damping, what is the amplitude (in terms of F_1) of steady-state deflection at the loaded quarter-point in each of the first two modes?
Answer
$A_1 = 2.0 \times 10^{-8} F_1$ ft
$A_2 = 2.5 \times 10^{-8} F_1$ ft

4.7 Using the Rayleigh method, derive an approximate expression for the fundamental natural frequency in terms of EI, l, and m for the haunched beam in Fig. 4.21. Both EI and m vary linearly between midspan and support. Divide the beam into 10 equal segments.

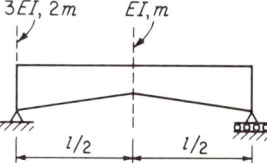

FIGURE 4.21 Problem 4.7. Haunched beam.

4.8 Determine the fundamental natural frequency of a two-span continuous beam in terms of EI, l, and m. The exterior supports are hinged, and the beam is prismatic over both spans, but one span is twice as long as the other.

4.9 Derive the frequency equation for a continuous beam of two identical spans with complete fixity at the exterior supports. By comparison with results given for single spans, write expressions for the natural frequencies of the first five modes.

4.10 Consider the two-span beam shown in Fig. 4.13. Suppose that, in addition to the concentrated force $F_1[f(t)]$, there is a uniformly distributed force of $p_1[f(t)]$ on the right span and that $f(t)$ is the same for both forces. Derive expressions similar to Eqs. (4.50) and (4.51) for the two midspan deflections.

4.11 *a.* Write the equations of motion for the three-beam system of Fig. 4.22, assuming the following shapes: $y_g = y_{go}[1 - \cos(\pi x_g/2l_g)]$ and $y_b = y_{go} + y_{bo} \sin(\pi x_b/l_b)$. y_{go} is the deflection at the end of the cantilever girder, and y_{bo} is the midspan beam deflection relative to the end of the girder.

b. Determine the natural frequencies of the two modes for the following parameters:
$m_b = 2m_g = 0.1$ lb-sec^2/in.^{2}
$EI_b = \tfrac{1}{2}EI_g = 10^{10}$ lb-in.^{2}
$l_b = 2l_g = 200$ in.

c. Write the modal equations of motion.

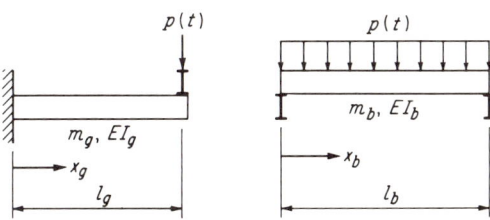

FIGURE 4.22 Problem 4.11. Beam-girder system.

4.12 A reinforced concrete slab simply supported on all edges is subjected to a pressure uniformly distributed over the central region indicated by the shaded area in Fig. 4.23. The pressure has the time function shown. Considering only the first mode, determine the maximum dynamic deflection at the middle of the slab. The thickness of the slab is 6 in., and the rigidity may be taken as

$$\frac{Eh^3}{12(1-\nu^2)} = 3 \times 10^7 \text{ lb-in.}^2 \text{ per inch of width}$$

Answer
$y_{max} = 0.476$ in.

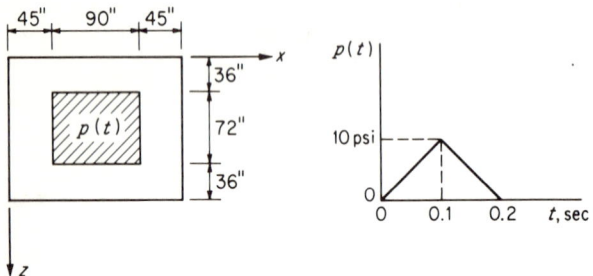

FIGURE 4.23 Problem 4.12. Two-way slab.

4.13 Consider the simply supported beam with uniformly distributed dynamic load as shown in Fig. 4.24. An elastic analysis is to be made, using the finite-difference technique, with three lumped masses as indicated.

  a. Write the two equations of motion (the third is eliminated by symmetry) in terms of y_1 and y_2.

  b. Using these equations, derive expressions for the natural frequencies and compare with the exact expressions for the first and third modes of the real beam.

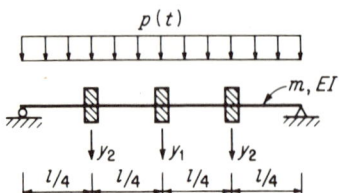

FIGURE 4.24 Problem 4.13. Finite-difference analysis.

5
Approximate Design Methods

5.1 Introduction

We must conclude from the developments in previous chapters that exact or rigorous dynamic analysis is possible only for relatively simple structures. This is particularly true of structures with continuous-mass distribution, the analysis of which is extremely complex for all but the most simple boundary conditions. We have also seen that rigorous solution is possible only when the load-time and the resistance-displacement variations are convenient mathematical functions. For these reasons it is often prudent, at least for practical design purposes, to adopt approximate methods which permit rapid analysis of even complex structures with reasonable accuracy. These methods usually require that both the structure and the loading be idealized in some degree. This chapter deals with methods by which the idealization may be accomplished.

It is frequently possible to reduce the system to one degree of freedom. For example, all three of the structural elements shown in Fig. 5.1 can, for practical purposes, be represented by an equivalent one-degree system having the parameters F_e, m_e, and k_e. Even though such elements are parts of a complete structure, it is often permissible to treat them independently.

One method of simplifying an analysis in the elastic range is to include only one or a few of the normal modes. For example, it was seen in the previous chapter that satisfactory solutions for simply supported, single-

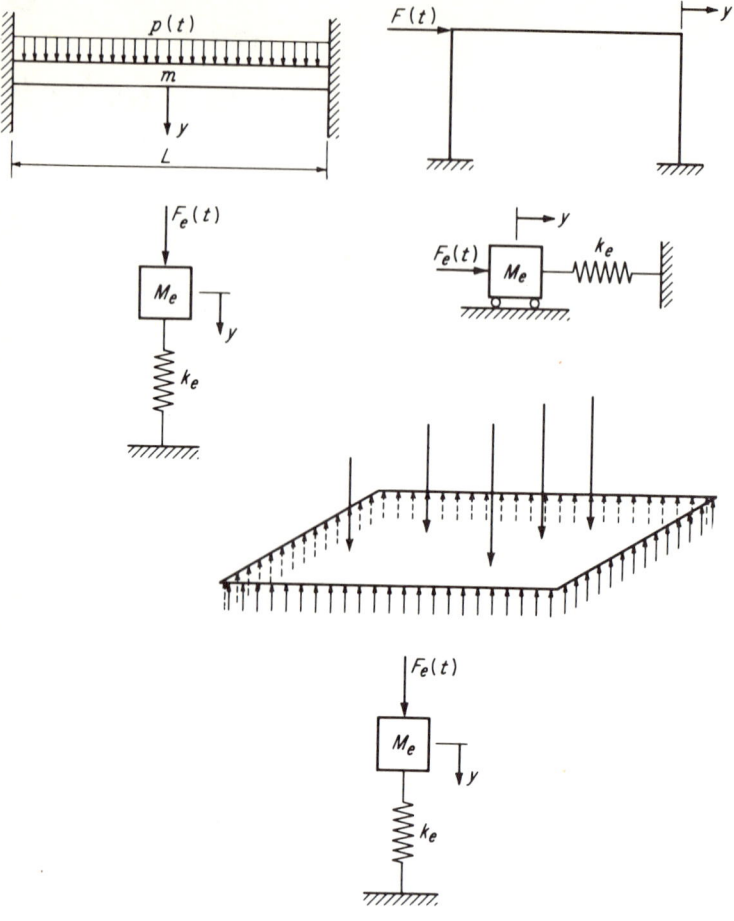

FIGURE 5.1 Equivalent one-degree systems.

span beams could be obtained by considering only the fundamental mode. This procedure is, in effect, an idealization of the actual dynamic system. However, for many structural elements this approach cannot be used, because it is too difficult to determine the modal shapes. Therefore even the fundamental mode must be approximated.

The approximate procedures developed herein are not based on the fundamental mode, even though the shape of that mode may be known. The methods presented are somewhat more accurate and are more general since the characteristic shape is not required.

Nonlinear resistance or nonmathematical load-time functions can be dealt with by numerical analysis. However, this is tedious and often

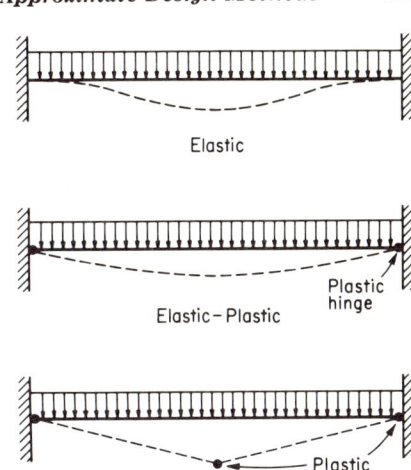

FIGURE 5.2 Stages of beam response.

not justified because of other uncertainties in the formulation of the problem which cannot be removed. This chapter deals primarily with direct, or closed, solutions based on approximate mathematical models.

Of particular importance is the difficulty in handling plastic behavior of structural elements. Rigorous solutions for such cases are not practical for design purposes. This difficulty is illustrated by the beam of Fig. 5.2, which goes through three stages of deformation. The first stage is purely elastic; the second, occurring after hinges have been formed at the supports, is a combination of elastic and plastic; and the third might be classified as purely plastic. Each stage has different characteristic shapes, and it is not possible to identify modal shapes which are meaningful throughout the response. The approximation used herein is to treat each stage as completely independent; e.g., the incremental distortions in the second stage of Fig. 5.2 are assumed to have the characteristic shapes of a simply supported beam.

From the viewpoint of practical design, the approximate methods presented here are extremely important. They should not be regarded as merely crude approximations, to be used for rough or preliminary analysis, nor should they be regarded as methods to be used only by engineers who lack the training or intellect to employ more sophisticated techniques. Problems in structural dynamics typically involve significant uncertainties, particularly with regard to loading characteristics. Such being the case, complex methods of analysis are often not justified. It is a waste of time to employ methods having precision much greater than that of the input of the analysis.

5.2 Idealized System

In order to define an equivalent one-degree system, it is necessary to evaluate the parameters of that system, namely, M_e, k_e, and F_e, as shown in Fig. 5.1. In addition, the load-time function $f(t)$ must be established in order to analyze the system. These parameters and loading are discussed below. With this representation of the actual structure and loading, the dynamic analysis becomes relatively simple by use of the methods for one-degree systems presented in Chaps. 1 and 2.

The equivalent system is usually selected so that the deflection of the concentrated mass is the same as that for some significant point on the structure, e.g., the midspan of a beam. It should be noted that stresses and forces in the idealized system are not directly equivalent to the same quantities in the structure. However, knowing the deflection, the stresses in the real structure may be readily computed. Since the time scale is not altered, the response of the equivalent system, defined in terms of displacement and time, is exactly the same as that of the significant point on the structure.

The constants of the equivalent system are evaluated on the basis of an assumed shape of the actual structure. This shape will be taken to be the same as that resulting from the static application of the dynamic loads. This concept is not quite the same as that using the first-mode shape, and for the types of elements considered here is somewhat more accurate, particularly with regard to stress computation. Furthermore, we are here dealing primarily with cases in which the modal shapes cannot be easily determined or, if determined, are expressible only by unmanageable mathematical functions.

It is convenient to introduce below certain *transformation factors*. These factors, denoted by K, are used to convert the real system into the equivalent system. When the total load, mass, resistance, and stiffness of the real structure are multiplied by the corresponding transformation factors, we obtain those parameters for the equivalent one-degree system.

a. Mass

As developed in Sec. 3.7a, the equivalent mass of a mode or, in the present case, of the equivalent one-degree system, is given by

$$M_e = \sum_{r=1}^{j} M_r \phi_r^2 \qquad (5.1a)$$

for a lumped-mass system or

$$M_e = \int m \phi^2(x)\, dx \qquad (5.1b)$$

for a structure with continuous mass. In Eqs. (5.1), ϕ_r or $\phi(x)$ is the assumed-shape function on which the equivalent system is based. For convenience, we now introduce the *mass factor* K_M, which is defined as the ratio of equivalent mass to the actual total mass of the structure.

$$K_M = \frac{M_e}{M_t} \qquad (5.2)$$

For example, in the case of the beam shown in Fig. 5.1, $M_t = mL$, where L is the span, and M_e is as given by Eq. (5.1b).

b. Load Distribution

As shown in Sec. 3.7a, the equivalent force on the idealized system is given by

$$F_e = \sum_{r=1}^{j} F_r \phi_r \qquad (5.3a)$$

for a structure with concentrated forces or

$$F_e = \int^L p(x)\phi(x)\,dx \qquad (5.3b)$$

for distributed loads. The *load factor* K_L is defined as the ratio of equivalent to actual total force.

$$K_L = \frac{F_e}{F_t} \qquad (5.4)$$

For the beam of Fig. 5.1, $F_t = pL$ and F_e is given by Eq. (5.3b). The above applies to magnitude of force, and both equivalent and real loads have the same time function.

c. Resistance Function

The resistance functions for actual structures may have a variety of forms. Three possible shapes are shown in Fig. 5.3a. Curve A corresponds to a structure of brittle material. Curve B would apply to a structure made of a ductile material with marked yielding such as steel or reinforced concrete. Curve C represents a situation in which resistance decreases above a certain deflection but before complete failure. The latter might occur in a structure of plain concrete or one in which failure results from instability. In order to simplify analysis, these resistance functions must be idealized. For most structures it is permissible to employ a bilinear function as indicated in Fig. 5.3b. The analyses discussed herein will be made on this basis.

The resistance of an element is the internal force tending to restore the element to its unloaded static position. For our present purpose we

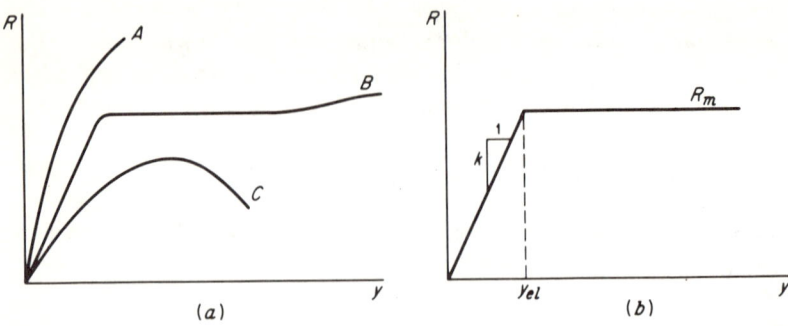

FIGURE 5.3 Resistance functions.

define resistance in terms of the load distribution for which the analysis is being made. Thus the maximum resistance is the total load having the given distribution which the element could support statically. The stiffness is numerically equal to the total load of the same distribution which would cause a unit deflection at the point where the deflection is equal to that of the equivalent system. By these definitions it is apparent that the *resistance factor* K_R must always equal the load factor K_L. Therefore

$$K_R = \frac{R_{me}}{R_m} = K_L \qquad (5.5a)$$

and

$$K_R = \frac{k_e}{k} = K_L \qquad (5.5b)$$

where K_L is given by Eq. (5.4), together with Eq. (5.3). Referring to the beam of Fig. 5.1 and to Fig. 5.3b, we see that R_m is the maximum value of pL, or the plastic-limit load which the beam could support statically, and k is the value of pL which would cause a unit elastic deflection at midspan. Resistance and deflection are related in the elastic range by $R = ky$ for the real structure and $R_e = k_e y$ for the equivalent system.

For the fixed-ended beam of Fig. 5.1, as well as in other cases, the resistance function is not bilinear even if ideal hinge formation is assumed. This is true because hinges first form at the supports and further deflection is required to form the midspan hinge and hence attain maximum resistance (Fig. 5.2). The result is the resistance function indicated by the solid line in Fig. 5.4. Since we require, for simplified analysis, a bilinear function, this solid line will be replaced by the dashed line shown in Fig. 5.4. The "effective" spring constant k_E will be selected so that the areas under the two curves are equal. Thus the energy absorbed will remain constant, and there will be little error in the maximum dynamic displacement computed. This concept is applied to an example in Sec. 5.6c.

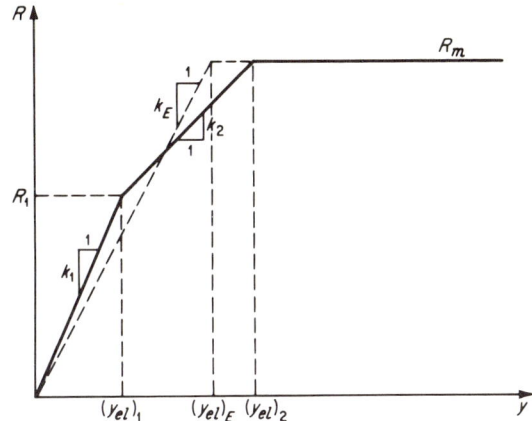

FIGURE 5.4 Effective bilinear resistance function.

d. Load-time Function

Any reasonable load-time function can be handled by numerical analysis. However, if a closed solution is desired, it is usually necessary to idealize the actual function by some simple mathematical shape. This can often be done without appreciable error in the final result. Furthermore, the actual function is usually not so well known that precision in its representation is justified.

Examples of this type of idealization are shown in Fig. 5.5. In each case the actual function (solid line) has been reduced to one of those for which solutions were given in Chap. 2 (Figs. 2.7 to 2.9 and 2.23 to 2.26). Thus the maximum response can be determined very easily by use of the charts provided.

The selection of the idealized load-time function requires judgment by the analyst. It is important to note that the actual and idealized functions need be similar only in the time range of interest. For example, in Fig. 5.5d, for the determination of maximum response, function A might be used if t_1 were the time of maximum response; but B might be used if that time were t_2. It is for this purpose that the times of maximum response given in Figs. 2.7 to 2.9 and 2.23 to 2.26 are particularly useful.

5.3 Transformation Factors

In this section transformation factors are tabulated for beams and slabs having various types of support conditions. These tables contain all information necessary to convert actual structural elements of these

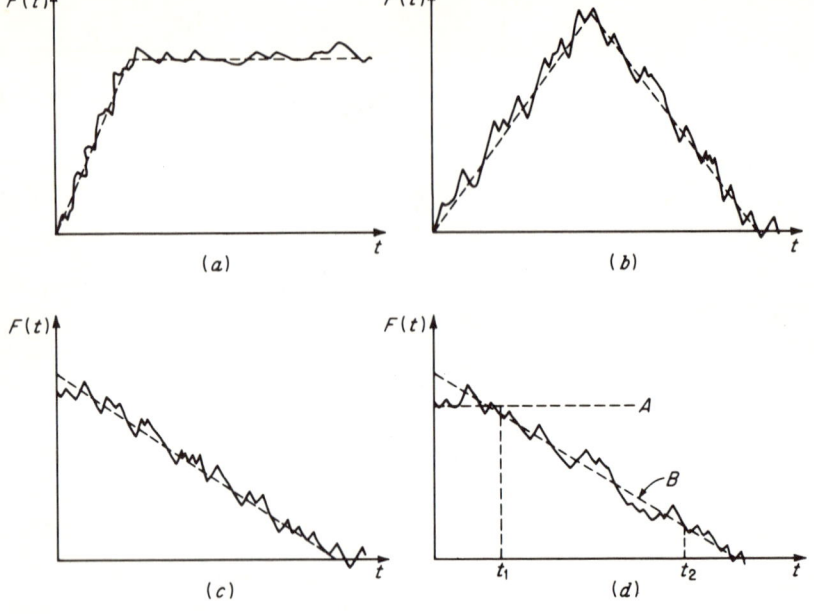

FIGURE 5.5 Idealized load-time functions.

types into equivalent one-degree systems for the purpose of approximate analysis.[9,10]

a. Beams and One-way Slabs

Table 5.1 is applicable to beams or one-way slabs having simple (hinged) supports. Factors are given for three different load distributions. As previously explained, these are based on the static deflected shape corresponding to the particular load distribution. These shapes are indicated in Fig. 5.6 and expressed by the following:

Uniform load:

$$\phi(x) = \frac{16}{5L^4}(L^3x - 2Lx^3 + x^4) \qquad \text{elastic}$$

$$\phi(x) = \frac{2x}{L} \qquad x < \frac{L}{2} \qquad \text{plastic}$$

Concentrated load at midspan:

$$\phi(x) = \frac{x}{L^3}(3L^2 - 4x^2) \qquad x < \frac{L}{2} \qquad \text{elastic}$$

$$\phi(x) = \frac{2x}{L} \qquad x < \frac{L}{2} \qquad \text{plastic}$$

Approximate Design Methods   207

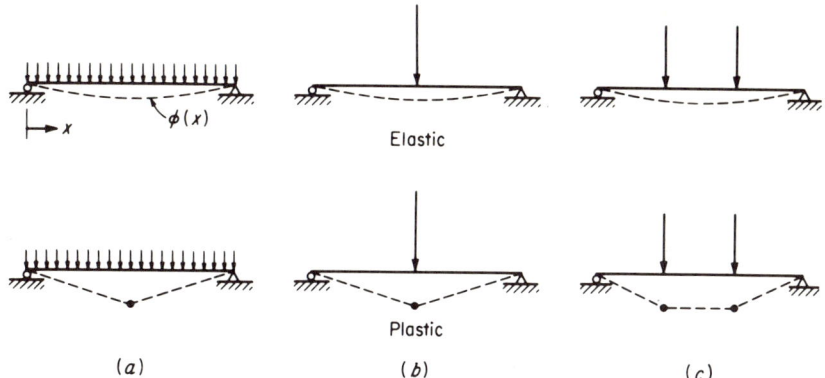

FIGURE 5.6 Assumed shapes.

Concentrated loads at third points:

$$\phi(x) = \frac{36}{23L^3}(2L^2 - x^2) \qquad x < \frac{L}{3} \qquad \text{elastic}$$

$$\phi(x) = \frac{36x}{23L^3}\left(3Lx - 3x^2 - \frac{L^2}{9}\right) \qquad \frac{L}{3} < x < \tfrac{2}{3}L \qquad \text{elastic}$$

$$\phi(x) = \frac{3x}{L} \qquad x < \frac{L}{3} \qquad \text{plastic}$$

$$\phi(x) = 1 \qquad \frac{L}{3} < x < \tfrac{2}{3}L \qquad \text{plastic}$$

The magnitude of these functions is set by the requirement that

$$\phi(L/2) = 1$$

Note that, in the plastic range, $\phi(x)$ is not the actual shape, but rather the shape of the incremental deflection after the fully plastic state has been attained. Use of this function implies a sudden change in the characteristics of the dynamic system. Although, clearly, this is not correct, the error resulting from this approximation is tolerable, and is in fact necessary if we are to obtain a practical solution.

According to Eqs. (5.1) and (5.2), the mass factors in Table 5.1 have been computed by evaluating

$$K_M = \frac{\int_0^L m\phi^2(x)\,dx}{mL}$$

where m is the mass per unit length, and the integration includes the entire span, using the appropriate expressions for $\phi(x)$ in each portion of the span. Also given in the table are factors for concentrated masses at the load points which frequently occur in practice. These are given by

$$K_M = \frac{\sum_r M_r \phi_r}{\sum_r M_r}$$

where ϕ_r is the ordinate of the assumed shape at mass r. To obtain the equivalent mass corresponding to these concentrations, K_M is multiplied by the sum of all concentrated masses. If both concentrated and distributed masses are being considered, the total equivalent value is simply the sum of those corresponding to the two types of masses.

The load factors in Table 5.1 have been computed by Eqs. (5.3) and (5.4). Thus

$$K_L = \frac{\int_0^L p\phi(x)\, dx}{pL}$$

for the distributed load and

$$K_L = \frac{\sum_r F_r \phi_r}{\sum_r F_r}$$

for concentrated loads.

Also given in Table 5.1 is the *load-mass factor*, which is merely the ratio of the mass and load factors. This is a convenience since the equation of motion may be written in terms of that factor alone. Furthermore, the natural frequency of both the real and idealized systems is given by

$$\omega = \sqrt{\frac{k_e}{M_e}} = \sqrt{\frac{K_L k}{K_M M_t}} = \sqrt{\frac{k}{K_{LM} M_t}}$$

The maximum resistances and spring constants tabulated are for the real system and are merely the conventional expressions for these quantities. When multiplied by the load factor, these become the corresponding quantities of the equivalent system. The term $\mathfrak{M}_P$ is the ultimate, or plastic, bending strength of the element. The resistances given are based upon flexure, and it is assumed herein that the element is designed so that shear failure is not critical.

The dynamic reactions given in Table 5.1 are discussed in Sec. 5.4.

Tables 5.2 and 5.3 contain transformation factors for fixed-ended beams and for beams with one end fixed and the other simply supported. These were derived in exactly the same manner as were those for simply supported beams, the only difference being in the assumed shapes. In these cases three ranges of behavior must be considered since the system changes but does not become fully plastic when the hinge is formed at the support. By the assumptions of the procedure being presented, the middle range, which we shall call elastic-plastic, is associated with the

Table 5.1 Transformation Factors for Beams and One-way Slabs

Simply-supported

Loading diagram	Strain range	Load factor K_L	Mass factor K_M		Load-mass factor K_{LM}		Maximum resistance R_m	Spring constant k	Dynamic reaction V
			Concentrated mass*	Uniform mass	Concentrated mass*	Uniform mass			
$F = pL$, L (uniform load)	Elastic	0.64	....	0.50	....	0.78	$\dfrac{8\mathfrak{M}_P}{L}$	$\dfrac{384EI}{5L^3}$	$0.39R + 0.11F$
	Plastic	0.50	....	0.33	....	0.66	$\dfrac{8\mathfrak{M}_P}{L}$	0	$0.38R_m + 0.12F$
F at midspan ($L/2$, $L/2$)	Elastic	1.0	1.0	0.49	1.0	0.49	$\dfrac{4\mathfrak{M}_P}{L}$	$\dfrac{48EI}{L^3}$	$0.78R - 0.28F$
	Plastic	1.0	1.0	0.33	1.0	0.33	$\dfrac{4\mathfrak{M}_P}{L}$	0	$0.75R_m - 0.25F$
$F/2$ at third points	Elastic	0.87	0.76	0.52	0.87	0.60	$\dfrac{6\mathfrak{M}_P}{L}$	$\dfrac{56.4EI}{L^3}$	$0.62R - 0.12F$
	Plastic	1.0	1.0	0.56	1.0	0.56	$\dfrac{6\mathfrak{M}_P}{L}$	0	$0.52R_m - 0.02F$

* Equal parts of the concentrated mass are lumped at each concentrated load.

Source: "Design of Structures to Resist the Effects of Atomic Weapons," U.S. Army Corps of Engineers Manual EM 1110-345-415, 1957.

Table 5.2 Transformation Factors for Beams and One-way Slabs

$\mathfrak{M}_{P_s}$ = ultimate moment capacity at support
$\mathfrak{M}_{P_m}$ = ultimate moment capacity at midspan

Fixed ends

Loading diagram	Strain range	Load factor K_L	Mass factor K_M		Load-mass factor K_{LM}		Maximum resistance R_m	Spring constant k	Effective spring constant k_E†	Dynamic reaction V
			Concentrated mass*	Uniform mass	Concentrated mass*	Uniform mass				
	Elastic	0.53	.....	0.41	.....	0.77	$\dfrac{12\mathfrak{M}_{P_s}}{L}$	$\dfrac{384EI}{L^3}$	.....	$0.36R + 0.14F$
	Elastic-plastic	0.64	.....	0.50	.....	0.78	$\dfrac{8}{L}(\mathfrak{M}_{P_s}+\mathfrak{M}_{P_m})$	$\dfrac{384EI}{5L^3}$	$\dfrac{307EI}{L^3}$	$0.39R + 0.11F$
	Plastic	0.50	.....	0.33	.....	0.66	$\dfrac{8}{L}(\mathfrak{M}_{P_s}+\mathfrak{M}_{P_m})$	0	.....	$0.38R_m + 0.12F$
	Elastic	1.0	1.0	0.37	1.0	0.37	$\dfrac{4}{L}(\mathfrak{M}_{P_s}+\mathfrak{M}_{P_m})$	$\dfrac{192EI}{L^3}$	.....	$0.71R - 0.21F$
	Plastic	1.0	1.0	0.33	1.0	0.33	$\dfrac{4}{L}(\mathfrak{M}_{P_s}+\mathfrak{M}_{P_m})$	0	.....	$0.75R_m - 0.25F$

* Concentrated mass is lumped at the concentrated load.
† See Fig. 5.4.
Source: "Design of Structures to Resist the Effects of Atomic Weapons," U.S. Army Corps of Engineers Manual EM 1110-345-415, 1957.

Table 5.3 Transformation Factors for Beams and One-way Slabs

$\mathfrak{M}_{Ps}$ = ultimate bending capacity at support
$\mathfrak{M}_{Pm}$ = ultimate positive bending capacity

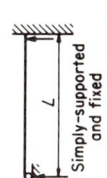
Simply-supported and fixed

Loading diagram	Strain range	Load factor K_L	Mass factor K_M		Load-mass factor K_{LM}		Maximum resistance R_m	Spring constant k	Effective spring constant k_E†	Dynamic reaction V
			Concentrated mass*	Uniform mass	Concentrated mass*	Uniform mass				
$F = pL$	Elastic	0.58	.....	0.45	.....	0.78	$\dfrac{8\mathfrak{M}_{Ps}}{L}$	$\dfrac{185EI}{L^3}$	$\dfrac{160EI}{L^3}$	$V_1 = 0.26R + 0.12F$ $V_2 = 0.43R + 0.19F$
	Elastic-plastic	0.64	.....	0.50	.....	0.78	$\dfrac{4}{L}(\mathfrak{M}_{Ps} + 2\mathfrak{M}_{Pm})$	$\dfrac{384EI}{5L^3}$		$V = 0.39R + 0.11F \pm \mathfrak{M}_{Ps}/L$
	Plastic	0.50	.....	0.33	.....	0.66	$\dfrac{4}{L}(\mathfrak{M}_{Ps} + 2\mathfrak{M}_{Pm})$	0		$V = 0.38R_m + 0.12F \pm \mathfrak{M}_{Ps}/L$
F	Elastic	1.0	1.0	0.43	1.0	0.43	$\dfrac{16MP_s}{3L}$	$\dfrac{107EI}{L^3}$	$\dfrac{106EI}{L^3}$	$V_1 = 0.54R + 0.14F$ $V_2 = 0.25R + 0.07F$
	Elastic-plastic	1.0	1.0	0.49	1.0	0.49	$\dfrac{2}{L}(\mathfrak{M}_{Ps} + 2\mathfrak{M}_{Pm})$	$\dfrac{48EI}{L^3}$		$V = 0.78R - 0.28F \pm \mathfrak{M}_{Ps}/L$
	Plastic	1.0	1.0	0.33	1.0	0.33	$\dfrac{2}{L}(\mathfrak{M}_{Ps} + 2\mathfrak{M}_{Pm})$	0		$V = 0.75R_m - 0.25F \pm \mathfrak{M}_{Ps}/L$
$F/2$, $F/2$	Elastic	0.81	0.67	0.45	0.83	0.55	$\dfrac{6MP_s}{L}$	$\dfrac{132EI}{L^3}$	$\dfrac{122EI}{L^3}$	$V_1 = 0.17R + 0.17F$ $V_2 = 0.33R + 0.33F$
	Elastic-plastic	0.87	0.76	0.52	0.87	0.60	$\dfrac{2}{L}(\mathfrak{M}_{Ps} + 3\mathfrak{M}_{Pm})$	$\dfrac{56EI}{L^3}$		$V = 0.62R - 0.12F \pm \mathfrak{M}_{Ps}/L$
	Plastic	1.0	1.0	0.56	1.0	0.56	$\dfrac{2}{L}(\mathfrak{M}_{Ps} + 3\mathfrak{M}_{Pm})$	......		$V = 0.52R_m - 0.02F \pm \mathfrak{M}_{Ps}/L$

* Equal parts of the concentrated mass are lumped at each concentrated load.
† See Fig. 5.4.
Source: "Design of Structures to Resist the Effects of Atomic Weapons," U.S. Army Corps of Engineers Manual EM 1110-345-415, 1957.

same shape as that of an elastic, simply supported beam, and therefore the factors are identical. Furthermore, in the plastic range, the shape is independent of the original support condition and therefore all beams have the same factors.

The maximum resistances given in Tables 5.2 and 5.3 are those which occur at the upper limit of each range. In addition to the spring constants associated with each range, the "effective" spring constant encompassing all ranges is also given. This concept was discussed in connection with Fig. 5.4.

b. Two-way and Flat Slabs

Transformation factors, as well as the constants defining the resistance function, are given in Tables 5.4 and 5.5, for two-way slabs with simple and fixed supports, respectively. These were obtained in the same manner as were those for beams. Values are given for several ratios of the lengths of the two sides.

After the deflected shapes for a slab have been assumed, the transformation factors are obtained by integration over the slab surface. The factors given were based upon approximations to the classical plate theory for deflection in the elastic range and yield-line theory in the plastic range. In the latter range the shape was assumed to be planar between yield lines. In the simply supported case there is obviously an elastic-plastic, or transition, range, but the behavior is exceedingly complex, and in order to simplify the procedure this range has been ignored. In the case of fixed supports, the elastic range has been terminated when the moment along most of the long edge has reached ultimate. This transition point cannot be determined precisely, and the limiting resistances given are estimates. For the elastic-plastic range, it has been assumed that the shape (but not the resistance), and hence the factors, are the same as those for a simply supported slab.

In Tables 5.4 and 5.5, the moment notation is as follows:

$\mathfrak{M}_{Pfa}$ = total positive ultimate moment capacity along midspan section parallel to short edge

$\mathfrak{M}_{Psa}$ = total negative ultimate moment capacity along short edge

$\mathfrak{M}_{Psb}^0$ = negative ultimate moment capacity per unit width at center of long edge

In the expressions for spring constant, I_a is the moment of inertia per unit width. Note that both maximum resistance and spring constant refer to the *total* load on the slab.

Transformation factors and resistance functions for square, interior flat slabs are given in Table 5.6. Various sizes of column capital are considered. These factors were based on approximate deflected shapes,

Table 5.4 Transformation Factors for Two-way Slabs: Simple Supports—Four Sides, Uniform Load

V_A = total dynamic reaction along short edge; V_B = total dynamic reaction along long edge.

Strain range	a/b	Load factor K_L	Mass factor K_M	Load-mass factor K_{LM}	Maximum resistance	Spring constant k	Dynamic reactions V_A	V_B
Elastic	1.0	0.45	0.31	0.68	$\dfrac{12}{a}(\mathfrak{M}_{Pfa} + \mathfrak{M}_{Pfb})$	$\dfrac{252 EI_a}{a^2}$	$0.07F + 0.18R$	$0.07F + 0.18R$
	0.9	0.47	0.33	0.70	$\dfrac{1}{a}(12\mathfrak{M}_{Pfa} + 11\mathfrak{M}_{Pfb})$	$\dfrac{230 EI_a}{a^2}$	$0.06F + 0.16R$	$0.08F + 0.20R$
	0.8	0.49	0.35	0.71	$\dfrac{1}{a}(12\mathfrak{M}_{Pfa} + 10.3\mathfrak{M}_{Pfb})$	$\dfrac{212 EI_a}{a^2}$	$0.06F + 0.14R$	$0.08F + 0.22R$
	0.7	0.51	0.37	0.73	$\dfrac{1}{a}(12\mathfrak{M}_{Pfa} + 9.8\mathfrak{M}_{Pfb})$	$\dfrac{201 EI_a}{a^2}$	$0.05F + 0.13R$	$0.08F + 0.24R$
	0.6	0.53	0.39	0.74	$\dfrac{1}{a}(12\mathfrak{M}_{Pfa} + 9.3\mathfrak{M}_{Pfb})$	$\dfrac{197 EI_a}{a^2}$	$0.04F + 0.11R$	$0.09F + 0.26R$
	0.5	0.55	0.41	0.75	$\dfrac{1}{a}(12\mathfrak{M}_{Pfa} + 9.0\mathfrak{M}_{Pfb})$	$\dfrac{201 EI_a}{a^2}$	$0.04F + 0.09R$	$0.09F + 0.28R$
Plastic	1.0	0.33	0.17	0.51	$\dfrac{12}{a}(\mathfrak{M}_{Pfa} + \mathfrak{M}_{Pfb})$	0	$0.09F + 0.16R_m$	$0.09F + 0.16R_m$
	0.9	0.35	0.18	0.51	$\dfrac{1}{a}(12\mathfrak{M}_{Pfa} + 11\mathfrak{M}_{Pfb})$	0	$0.08F + 0.15R_m$	$0.09F + 0.18R_m$
	0.8	0.37	0.20	0.54	$\dfrac{1}{a}(12\mathfrak{M}_{Pfa} + 10.3\mathfrak{M}_{Pfb})$	0	$0.07F + 0.13R_m$	$0.10F + 0.20R_m$
	0.7	0.38	0.22	0.58	$\dfrac{1}{a}(12\mathfrak{M}_{Pfa} + 9.8\mathfrak{M}_{Pfb})$	0	$0.06F + 0.12R_m$	$0.10F + 0.22R_m$
	0.6	0.40	0.23	0.58	$\dfrac{1}{a}(12\mathfrak{M}_{Pfa} + 9.3\mathfrak{M}_{Pfb})$	0	$0.05F + 0.10R_m$	$0.10F + 0.25R_m$
	0.5	0.42	0.25	0.59	$\dfrac{1}{a}(12\mathfrak{M}_{Pfa} + 9.0\mathfrak{M}_{Pfb})$	0	$0.04F + 0.08R_m$	$0.11F + 0.27R_m$

Source: "Design of Structures to Resist the Effects of Atomic Weapons," U.S. Army Corps of Engineers Manual EM 1110-345-415, 1957.

Table 5.5 Transformation Factors for Two-way Slabs: Fixed Four Sides, Uniform Load

V_A = total dynamic reaction along short edge; V_B = total dynamic reaction along long edge.

Strain range	a/b	Load factor K_L	Mass factor K_M	Load-mass factor K_{LM}	Maximum resistance	Spring constant k	Dynamic reactions V_A	V_B
Elastic	1.0	0.33	0.21	0.63	$29.2 \mathfrak{M}^0 P_{ab}$	$810 EI_a/a^2$	$0.10F + 0.15R$	$0.10F + 0.15R$
	0.9	0.34	0.23	0.68	$27.4 \mathfrak{M}^0 P_{ab}$	$742 EI_a/a^2$	$0.09F + 0.14R$	$0.10F + 0.17R$
	0.8	0.36	0.25	0.69	$26.4 \mathfrak{M}^0 P_{ab}$	$705 EI_a/a^2$	$0.08F + 0.12R$	$0.11F + 0.19R$
	0.7	0.38	0.27	0.71	$26.2 \mathfrak{M}^0 P_{ab}$	$692 EI_a/a^2$	$0.07F + 0.11R$	$0.11F + 0.21R$
	0.6	0.41	0.29	0.71	$27.3 \mathfrak{M}^0 P_{ab}$	$724 EI_a/a^2$	$0.06F + 0.09R$	$0.12F + 0.23R$
	0.5	0.43	0.31	0.72	$30.2 M^0 P_{ab}$	$806 EI_a/a^2$	$0.05F + 0.08R$	$0.12F + 0.25R$
Elastic-plastic	1.0	0.46	0.31	0.67	$(1/a)[12(\mathfrak{M}_{Pfa} + \mathfrak{M}_{Psa}) + 12(\mathfrak{M}_{Pfb} + \mathfrak{M}_{Psb})]$	$252 EI_a/a^2$	$0.07F + 0.18R$	$0.07F + 0.18R$
	0.9	0.47	0.33	0.70	$(1/a)[12(\mathfrak{M}_{Pfa} + \mathfrak{M}_{Psa}) + 11(\mathfrak{M}_{Pfb} + \mathfrak{M}_{Psb})]$	$230 EI_a/a^2$	$0.06F + 0.16R$	$0.08F + 0.20R$
	0.8	0.49	0.35	0.71	$(1/a)[12(\mathfrak{M}_{Pfa} + \mathfrak{M}_{Psa}) + 10.3(\mathfrak{M}_{Pfb} + \mathfrak{M}_{Psb})]$	$212 EI_a/a^2$	$0.06F + 0.14R$	$0.08F + 0.22R$
	0.7	0.51	0.37	0.73	$(1/a)[12(\mathfrak{M}_{Pfa} + \mathfrak{M}_{Psa}) + 9.8(\mathfrak{M}_{Pfb} + \mathfrak{M}_{Psb})]$	$201 EI_a/a^2$	$0.05F + 0.13R$	$0.08F + 0.24R$
	0.6	0.53	0.39	0.74	$(1/a)[12(\mathfrak{M}_{Pfa} + \mathfrak{M}_{Psa}) + 9.3(\mathfrak{M}_{Pfb} + \mathfrak{M}_{Psb})]$	$197 EI_a/a^2$	$0.04F + 0.11R$	$0.09F + 0.26R$
	0.5	0.55	0.41	0.75	$(1/a)[12(\mathfrak{M}_{Pfa} + \mathfrak{M}_{Psa}) + 9.0(\mathfrak{M}_{Pfb} + M_{Psb})]$	$201 EI_a/a^2$	$0.04F + 0.09R$	$0.09F + 0.28R$
Plastic	1.0	0.33	0.17	0.51	$(1/a)[12(\mathfrak{M}_{Pfa} + \mathfrak{M}_{Psa}) + 12(\mathfrak{M}_{Pfb} + \mathfrak{M}_{Psb})]$	0	$0.09F + 0.16R_m$	$0.09F + 0.16R_m$
	0.9	0.35	0.18	0.51	$(1/a)[12(\mathfrak{M}_{Pfa} + \mathfrak{M}_{Psa}) + 11(\mathfrak{M}_{Pfb} + \mathfrak{M}_{Psb})]$	0	$0.08F + 0.15R_m$	$0.09F + 0.18R_m$
	0.8	0.37	0.20	0.54	$(1/a)[12(\mathfrak{M}_{Pfa} + \mathfrak{M}_{Psa}) + 10.3(M_{Pfb} + \mathfrak{M}_{Psb})]$	0	$0.07F + 0.13R_m$	$0.10F + 0.20R_m$
	0.7	0.38	0.22	0.58	$(1/a)[12(\mathfrak{M}_{Pfa} + \mathfrak{M}_{Psa}) + 9.8(\mathfrak{M}_{Pfb} + \mathfrak{M}_{Psb})]$	0	$0.06F + 0.12R_m$	$0.10F + 0.22R_m$
	0.6	0.40	0.23	0.58	$(1/a)[12(\mathfrak{M}_{Pfa} + \mathfrak{M}_{Psa}) + 9.3(M_{Pfb} + \mathfrak{M}_{Psb})]$	0	$0.05F + 0.10R_m$	$0.10F + 0.25R_m$
	0.5	0.42	0.25	0.59	$(1/a)[12(\mathfrak{M}_{Pfa} + \mathfrak{M}_{Psa}) + 9.0(\mathfrak{M}_{Pfb} + \mathfrak{M}_{Psb})]$	0	$0.04F + 0.08R_m$	$0.11F + 0.27R_m$

Source: "Design of Structures to Resist the Effects of Atomic Weapons," U.S. Army Corps of Engineers Manual EM 1110-345-415, 1957.

Approximate Design Methods

Table 5.6 *Transformation Factors for Flat Slabs: Square Interior, Uniform Load*

d = width of column capital
a = column spacing, in.
E = compressive modulus of elasticity of concrete, ksi
I_a = average of gross and transformed moments of inertia per unit width, equal in both directions, in.^{4}/in. (Sec. 5.6a)
F = total load on one slab panel, excluding capitals
R = total resistance of one slab panel, excluding capitals
$\Sigma \mathfrak{M}_P = \mathfrak{M}_{Pmp} + \mathfrak{M}_{Pmn} + \mathfrak{M}_{Pcp} + \mathfrak{M}_{Pcn}$ in.-kips/in.

Strain range	d/a	Load factor K_L	Mass factor K_M	Load-mass factor K_{LM}	Spring constant k kips/in.	Maximum resistance R_m kips	Dynamic column load V_c kips
Elastic	0.05	8/15	0.34	0.64	$208 E I_a / a^2$	$4.2 \Sigma \mathfrak{M}_P$	
	0.10	8/15	0.34	0.64	$230 E I_a / a^2$	$4.4 \Sigma \mathfrak{M}_P$	$0.16F + 0.84R$
	0.15	8/15	0.34	0.64	$252 E I_a / a^2$	$4.6 \Sigma \mathfrak{M}_P$	+ load on capital
	0.20	8/15	0.34	0.64	$276 E I_a / a^2$	$4.8 \Sigma \mathfrak{M}_P$	
	0.25	8/15	0.34	0.64	$302 E I_a / a^2$	$5.0 \Sigma \mathfrak{M}_P$	
Plastic	0.05	1/2	7/24	7/12	0	$4.2 \Sigma \mathfrak{M}_P$	
	0.10	1/2	7/24	7/12	0	$4.4 \Sigma \mathfrak{M}_P$	$0.14F + 0.86R_m$
	0.15	1/2	7/24	7/12	0	$4.6 \Sigma \mathfrak{M}_P$	+ load on capital
	0.20	1/2	7/24	7/12	0	$4.8 \Sigma \mathfrak{M}_P$	
	0.25	1/2	7/24	7/12	0	$5.0 \Sigma \mathfrak{M}_P$	

Source: "Design of Structures to Resist the Effects of Atomic Weapons," U.S. Army Corps of Engineers Manual EM 1110-345-415, 1957.

as in the case of two-way slabs. Because of the complexity of the behavior, the elastic-plastic transition range has been ignored. Note that both load and resistance are for the surface area *outside* the column capital. Load applied directly to the column capital is assumed to be carried by the column and not to affect the slab response. In Table 5.6 the following notation is used:

$\mathfrak{M}_{Pmp}, \mathfrak{M}_{Pmn}$ = positive and negative ultimate moment capacity per unit width in middle strip of width $a/2$

$\mathfrak{M}_{Pcp}, \mathfrak{M}_{Pcn}$ = positive and negative ultimate moment capacity per unit width in column strip of width $a/2$

The data given for slabs in Tables 5.4 to 5.6 are obviously approximate. The transformation factors are based on assumed shapes, and the ultimate resistance on yield-line theory. More refined data could probably be developed, but in view of the many uncertainties in the dynamic analysis of slabs, it is doubtful that this would be warranted for design purposes.

c. Frames

Consider now the simple rigid frame shown in Fig. 5.7. The mass is distributed along both the roof and the sides, and the dynamic load

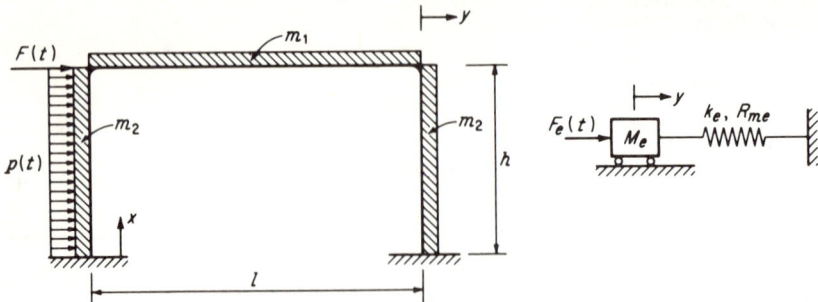

FIGURE 5.7 Equivalent one-degree system for rigid frame.

consists of a concentrated force at roof level plus a distributed load on one wall surface. Only horizontal motion is considered. We wish to define the equivalent one-degree system. The displacement of this system y is taken to be that at the top of the frame.

It will be assumed that the walls remain straight; that is, $y_x = xy/h$, where y_x is the displacement of a point at a distance x from the base, and y is that at the top of the frame. Although the columns of the frame and probably the walls of the building do not in fact remain straight, this is an acceptable approximation.

For the assumption stated above, the equivalent mass is given by [Eq. (5.1)]

$$M_e = m_1 l + 2 \int_0^h m_2 \left(\frac{x}{h}\right)^2 dx$$
$$= m_1 l + \tfrac{2}{3} m_2 h \tag{5.6}$$

where m_1 and m_2 are the masses per unit length (along the frame) of the roof and walls, respectively.

The equivalent load is [Eq. (5.3)]

$$F_e(t) = F(t) + \int_0^h p(t) \left(\frac{x}{h}\right) dx$$
$$= F(t) + p(t) \frac{h}{2} \tag{5.7}$$

Since we have taken the deflection of the equivalent system to be the same as that at roof level, the equivalent stiffness k_e is equal to the actual stiffness of the frame referred to a horizontal load at the top. This stiffness may be computed by conventional frame analysis and is merely the concentrated force required to produce a unit horizontal deflection of the roof.

The maximum resistance of both the real and idealized systems is simply the maximum horizontal force which can be carried by the frame.

If the columns have less bending strength than the girder (the usual case),

$$R_m = R_{me} = \frac{4\mathfrak{M}_{Pc}}{h}$$

where $\mathfrak{M}_{Pc}$ is the ultimate column bending strength.

On the basis of the foregoing, the equivalent system is established and then analyzed in the usual manner. The analysis yields the frame deflection, and stresses within the frame are directly related to this quantity. If a closed solution is to be made, both $F(t)$ and $p(t)$ must have the same time variation. Otherwise numerical analysis is necessary.

For multistory frames with mass and load distributed over the walls, it is in most cases sufficiently accurate to lump the mass and load at floor levels on the basis of tributary wall area.

d. Other Structural Elements

Transformation and resistance factors such as those discussed above could be derived for a wide assortment of structural elements. The only requirements are assumed shapes in the possible strain ranges and the maximum resistance for each range. Following the principles developed above, the general procedure should be apparent.

In some cases shear distortion may be important, and should be considered when determining the deflected shape, spring constant, and maximum resistance. For example, the shear effect may be significant in deep beams of reinforced concrete.

Trusses may be handled in the same manner as beams either by expressing the properties in terms of equivalent beam parameters or by computing the deflected shape on the basis of truss analysis. Because of instability of individual compression members, it may not be possible to design for plastic behavior of trusses.

Analysis of continuous, reinforced concrete T beams or continuous, composite steel beams involves special problems because the effective rigidity EI varies along the span and depends upon the magnitude and sign of the bending moment. However, approximations can obviously be developed. Analysis of beams supporting two-way slabs presents similar problems, since both the load intensity and the mass vary along the span because of the nature of the slab behavior.[10]

5.4 Dynamic Reactions

It is important to recognize that the dynamic reactions of the real structural element have no direct counterpart in the equivalent one-degree system. In other words, the reaction of the equivalent system, i.e., the spring force, is not the same as the real reaction. This is true because the

218   *Introduction to Structural Dynamics*

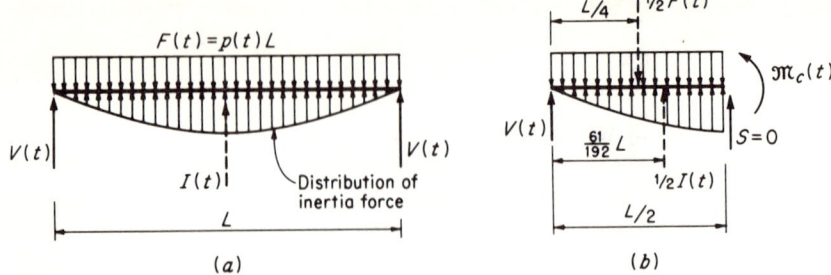

FIGURE 5.8 Determination of dynamic reactions.

simplified system was deliberately selected so as to have the same dynamic deflection as the real element, rather than the same force or stress characteristics. We are of course very much interested in the reactions, since these are usually related to the maximum shear in the element and also because they are necessary for the design of the supporting structure.

Expressions for the real reactions may be obtained by considering the dynamic equilibrium of the complete element. For example, consider the elastic, simply supported beam with uniformly distributed load and mass as shown in Fig. 5.8. For dynamic equilibrium we must include the inertia force, which at all times has a distribution identical with the assumed deflected shape of the beam. This distribution exists because the motion is harmonic and at any point along the span

$$\ddot{y}(x) = -\omega^2 y(x) = A\phi(x)$$

where A is a constant. Thus, at any point, the intensity of the inertia force is proportional to the ordinate of the deflected shape. It is apparent (Fig. 5.8) that the dynamic reaction $V(t)$ depends upon both the load $F(t)$ and the total inertia force $I(t)$.

Consider now the dynamic equilibrium of the left half of the beam as indicated in Fig. 5.8b. By symmetry, it is known that the shear S at midspan must be zero. Furthermore, the location of the resultant inertia force (for the half-span) may be computed from the assumed shape, which for this case is

$$\phi(x) = \frac{16}{5L^4}(L^3x - 2Lx^3 + x^4)$$

Thus Fig. 5.8b represents a set of forces in equilibrium. It is convenient to take moments about the resultant inertia force, thus obtaining the following equation:

$$V(^{61}\!/_{192})L - \mathfrak{M}_c - \tfrac{1}{2}F(^{61}\!/_{192}L - \tfrac{1}{4}L) = 0$$

where $\mathfrak{M}_c$ is the dynamic bending moment at midspan. Consistent with our definition of resistance, we have

$$\mathfrak{M}_c = \frac{RL^2}{8}$$

where R is the resistance which varies with time. Substituting for $\mathfrak{M}_c$ in the equilibrium equation and solving for V, we obtain

$$V = 0.39R + 0.11F \qquad (5.8)$$

Thus the dynamic reaction is a function of load and resistance, both of which are functions of time. Since this equation must also hold for static loading, and since in that case $R = F$, the sum of the two coefficients must equal ½. Note that all terms in Eq. (5.8) are for the real, rather than the equivalent, system.

For the plastic range or for beams with other loading or support conditions, the same procedure as above is followed, the only difference being in the assumed deflected shape. On this basis the expressions given for dynamic reaction in Tables 5.1 to 5.3 were obtained. Edge reactions for two-way slabs and dynamic column loads for flat slabs are given in Tables 5.4 to 5.6.

5.5 Response Calculations

Having established the equivalent one-degree system by the procedures outlined above and with the aid of the factors given in the tables presented, we may write the equation of motion

$$M_e \ddot{y} + k_e y = F_e(t) \qquad (5.9)$$

or in terms of the real system,

$$K_M M_t \ddot{y} + K_L k y = K_L F(t) \qquad (5.10)$$

where M_t is the total mass of the beam, slab, or other element. Equation (5.10) may also be written as

$$K_{LM} M_t \ddot{y} + ky = F(t) \qquad (5.11)$$

indicating that it is possible to use only one factor, K_{LM}, modifying one parameter of the real system, rather than three. The natural period of the system is given by

$$T = 2\pi \sqrt{\frac{K_{LM} M_t}{k}} \qquad (5.12)$$

Since the transformation factors change as the element progresses through the different stress ranges, i.e., elastic, elastic-plastic, and

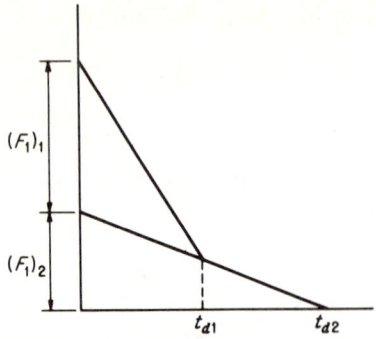

FIGURE 5.9 Load function idealized by two triangles.

plastic, a complete solution requires that each range be treated separately. For example, if a closed solution were being obtained, the response in each range would be computed, using the factors for that range and beginning with initial conditions equal to the final conditions in the preceding range. However, such refinements are probably not justified in the majority of practical cases. It is therefore suggested that, when the response is expected to extend beyond the elastic range, an approximate weighted average of the various factors be employed. This approximation would permit the use of charts for maximum response such as those given in Chap. 2. If a numerical analysis is being made, no difficulty exists since the factors may simply be changed at the appropriate time.

a. Combined Load-time Functions

As discussed in Sec. 5.2d, it is often possible to idealize the real load-time function by a simple relationship, thus making possible the use of available charts for maximum response. For example, in Fig. 5.5c, the real function is replaced by a triangular load pulse. If this is deemed to be excessively crude, it may be desirable to replace the load function by the superposition of two (or more) triangles as shown in Fig. 5.9. An approximate solution for this case, based upon the response due to each triangle separately, has been developed by Newmark.[26] This method is expressed by the empirical relationship

$$\frac{(F_1)_1}{R_m} C_1(\mu) + \frac{(F_1)_2}{R_m} C_2(\mu) = 1 \qquad (5.13)$$

where $C_1(\mu)$ and $C_2(\mu)$ are values of the ratio R_m/F_1 corresponding to a certain value of the ductility ratio μ and the ratios of duration to period t_{d1}/T and t_{d2}/T, respectively. $C_1(\mu)$ and $C_2(\mu)$ may be taken directly from Fig. 2.24, the inelastic response curve for triangular pulses of this type.

This procedure is restricted to systems undergoing plastic response and cannot be extended to other types of load-time functions.

Equation (5.13) provides an upper bound of response; i.e., it overestimates the maximum deflection or the required strength for a desired deflection. The maximum error in required strength which occurs when one load duration is very short and the other very long is of the order of 20 percent. Any number of triangles may be combined in this way by adding terms to the left side of the equation.

Equation (5.13) must be solved by trial and error. For analysis, i.e., when the natural period and maximum resistance are given, values of μ must be assumed, $C_1(\mu)$ and $C_2(\mu)$ read from Fig. 2.24, and the process repeated until a μ value satisfying Eq. (5.13) is found. For design, in which case the required μ is given, a natural period is assumed, so that $C_1(\mu)$ and $C_2(\mu)$ may be obtained and the required R_m is then computed. However, selection of the structural properties for this R_m will result in a new natural period, and hence the process must be repeated until R_m and T are consistent.

To illustrate application of the above to analysis, consider a given system and loading for which

$$\frac{(F_1)_1}{R_m} = 1.2 \qquad \frac{t_{d1}}{T} = 0.5$$

$$\frac{(F_1)_2}{R_m} = 0.8 \qquad \frac{t_{d2}}{T} = 1.0$$

We desire the maximum deflection in terms of the ductility ratio μ. The trial-and-error procedure is demonstrated in the following table, where $C_1(\mu)$ and $C_2(\mu)$ are read from Fig. 2.24 for each assumed μ.

Assumed μ	$C_1(\mu)$	$C_2(\mu)$	$\dfrac{(F_1)_1}{R_m} C_1(\mu)$	$\dfrac{(F_1)_2}{R_m} C_2(\mu)$	Σ
7	0.37	0.59	0.44	0.47	0.91
6	0.39	0.63	0.47	0.50	0.97
5	0.43	0.67	0.52	0.54	1.06

Since, for the correct solution, the summation would be unity, μ is between 5 and 6; i.e., the maximum deflection is between five and six times the elastic-limit deflection. This should not be regarded as an accurate solution, since deflections in the plastic range are always difficult to compute and the approximation in the method may lead to sizable error. From the more important design viewpoint, however, the corresponding error in required strength is not great.

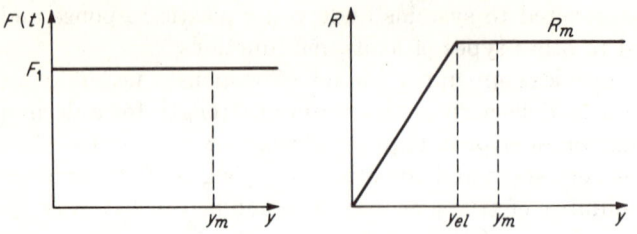

FIGURE 5.10 Long-duration loading and bilinear resistance.

b. Limiting Cases of Inelastic Response

For competence in design, one must have an intuitive feeling for the dynamic response of structural elements. To aid in the development of this intuition, it may be useful to consider two limiting cases of inelastic response: (1) the maximum deflection of a system subjected to a long-duration loading relative to the natural period, and (2) the deflection due to a load of very short duration.

The first case is indicated in Fig. 5.10, where the force F_1 changes so slowly that the variation up to the time of maximum response, y_m, is negligible. Such being the case, the work done by the force, and hence the total energy imparted to the system, is

$$W_e = F_1 y_m$$

If the resistance function is bilinear as indicated in Fig. 5.10, the strain energy at maximum deflection, or the total energy absorbed by the system, is

$$\mathcal{U} = R_m(y_m - \tfrac{1}{2} y_{el})$$

The latter expression is merely the area under the resistance curve. Since the work done by the force must equal the energy absorbed at maximum deflection ($\dot{y} = 0$), we may write

$$F_1 y_m = R_m(y_m - \tfrac{1}{2} y_{el})$$

Introducing the ductility ratio, $\mu = y_m/y_{el}$, and rearranging,

$$\text{Required } R_m = F_1 \left(\frac{1}{1 - 1/2\mu} \right) \tag{5.14}$$

or

$$\mu = \frac{1}{2(1 - F_1/R_m)} \tag{5.15}$$

Equation (5.15) may be used for the analysis of a given system, and (5.14) for the design of an element given the required ductility ratio. The equations are of course valid only if response extends beyond the elastic limit.

Approximate Design Methods 223

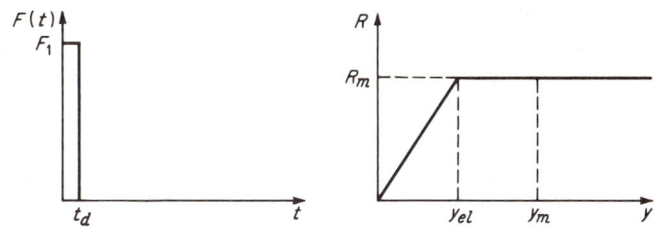

FIGURE 5.11 Short-duration loading and bilinear resistance.

Equation (5.14) indicates that, if μ is permitted to be very large, R_m may be made equal to F_1. However, for purely elastic response ($\mu = 1$), R_m must be twice F_1. Thus, by allowing large plastic deformation, the designer may reduce the strength of the structure by almost one-half. It should be noted that, in this limiting case of long load duration, the natural period is not a factor in the design.

The other extreme case, that involving a short-duration load, is shown in Fig. 5.11. In this situation it is convenient to consider the loading to be a pure impulse (i.e., the load duration is small compared with the time of maximum response; see Sec. 2.3a), denoted by

$$i = \text{impulse} = F_1 t_d$$

This impulse gives the mass of the system an initial velocity $\dot{y} = i/M$, and the resulting kinetic energy is

$$\mathcal{K} = \frac{i^2}{2M}$$

At maximum deflection this energy is completely absorbed by the spring (since the kinetic energy is zero), and hence we may write

$$\frac{i^2}{2M} = R_m(y_m - \tfrac{1}{2}y_{el}) = \frac{R_m^2}{k}(\mu - \tfrac{1}{2})$$

Noting that $\omega^2 = k/M$, we rearrange to obtain

$$\text{Required } R_m = \frac{i\omega}{\sqrt{2\mu - 1}} \qquad (5.16)$$

or

$$\mu = \tfrac{1}{2}\left(\frac{i^2\omega^2}{R_m^2} + 1\right) \qquad (5.17)$$

In this case we may observe that, if $\mu = 1$ (elastic response), R_m must equal $i\omega$, but as μ becomes very large, the required strength approaches zero. The potential design economy associated with large permitted deflection is apparent. Equations (5.16) and (5.17) may be used with-

224 *Introduction to Structural Dynamics*

out appreciable error if the natural period is five or more times the load duration. In such cases, the actual shape of the load pulse is of no importance, because the load effect is expressed entirely by impulse, or the area under the load-time curve.

Equations (5.14) and (5.16) are useful for design purposes because, even for load durations which are not extremely long or short, one or the other of these equations provides a good basis for preliminary design. This statement does not apply if the load begins near zero and increases slowly to its peak value.

5.6 Design Examples

This section contains several very simple examples of the dynamic design of beams and slabs, based on the approximate methods described above. In each case the dynamic load is given as a simple linear time function, with the implication that these may be idealized forms of the actual time variation.

It must be recognized that dynamic design can seldom be direct since the response depends upon mass and stiffness, which are both related to strength. Therefore we must apply the technique of trial and error.

In most of these examples the design criterion is given in terms of the ductility ratio μ. This is often done in practice because in some respects μ is a better indication of structural damage than is the actual deflection. A proper μ value for design depends upon the function of the element and the amount of damage that can be tolerated. It also depends upon how many times the design load is expected to occur. If the purpose of design is merely to prevent collapse under one load application, μ values as high as 20 may be permissible. On the other hand, if there are to be many load applications or if no damage can be tolerated, the element must remain elastic; that is, $\mu = 1$. A commonly used criterion for moderate damage is $\mu = 3$, which implies considerable yielding of steel or cracking of concrete but no significant impairment of the resistance to future loading.

a. Reinforced Concrete Beam

Consider the problem of designing a simply supported, reinforced concrete beam as shown in Fig. 5.12. It is subjected to a uniformly distributed load, having the magnitude and time function shown. In addition to the weight of the beam itself, there is an attached dead weight (perhaps a slab) of 1200 lb/ft. We are required to design a beam which will remain elastic; i.e., the response will attain, but not exceed, the elastic limit.

Approximate Design Methods

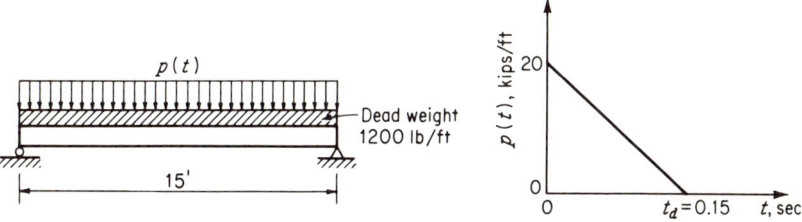

FIGURE 5.12 Design example. Reinforced concrete beam.

The material properties are specified as follows:

Dynamic yield strength of steel = 50,000 psi = σ_{dy}
Dynamic concrete compressive strength = 4000 psi = σ'_{dc}
Steel ratio = ρ_s = 0.015

With regard to the material strengths, it should be noted that these are time-sensitive; i.e., the strength increases with rate of strain.[9] Although the amount of increase varies, for the type of elements and loading considered herein, the dynamic strengths are approximately 25 per cent greater than static values. The steel ratio is specified for this example to simplify the problem by reducing the number of variables.

The ultimate bending strength of the beam will be taken as that given by the well-known formula

$$\mathfrak{M}_P = \rho_s bd^2 \sigma_{dy} \left(1 - \frac{\rho_s \sigma_{dy}}{1.7 \sigma'_{dc}}\right)$$

where b and d are the width and effective depth, respectively. For the material properties given, this may be written

$$\mathfrak{M}_P = 666bd^2 \text{ in.-lb}$$

The maximum resistance of the beam (Table 5.1) is

$$R_m = \frac{8\mathfrak{M}_P}{L} = \frac{8 \times 666bd^2}{15 \times 12} = 29.6bd^2 \quad \text{lb}$$

By previous experience it is known that the natural period of such elements is usually of the order of 0.05 sec. Therefore it may be expected that t_d/T will be about 3, since t_d = 0.15 sec. Reference to Fig. 2.7 leads us to estimate that the maximum DLF will be approximately 1.8. The estimated required strength of the beam is therefore

$$\text{Required } R_m = 1.8 \times 20,000 \times 15 + 2,000(\pm) \times 15$$
$$= 570,000 \text{ lb}$$

where 2000 lb/ft is the dead weight, including an estimate for the beam itself. The dead load must be added, because this part of the resistance is not available to oppose the dynamic load. Since

$$\text{Required } bd^2 = \frac{570{,}000}{29.6} = 19{,}300 \text{ in.}^3$$

the trial beam size will be taken as $b = 18$ in. and $d = 32.8$ in. We now proceed to analyze this particular beam.

In order to determine the stiffness of a reinforced concrete beam, an estimate must be made of the effective moment of inertia. This is difficult to accomplish because the amount of cracking, which has a very appreciable effect, varies along the span and depends upon the amount of deflection due to loading, as well as other phenomena such as shrinkage, creep, etc. Some designers have used the fully cracked transformed section for computing moment of inertia. It has also been suggested that, as an approximation, the average of the uncracked and cracked transformed sections be used.[9] The latter appears reasonable, and the result can be closely approximated by the expression

$$I_a = \frac{bd^3}{2}(5.5\rho_s + 0.083)$$

which will be used below for the sake of simplicity. Therefore

$$I_a = \frac{18(32.8)^3}{2}(5.5 \times 0.015 + 0.083) = 52{,}500 \text{ in.}^4$$

and, from Table 5.1,

$$k = \frac{384EI}{5L^3} = \frac{384(3 \times 10^6)52{,}500}{5(15 \times 12)^3} = 2.08 \times 10^6 \text{ lb/in.}$$

With an allowance for steel cover, the total weight of the beam selected is 660 lb/ft and the total weight and mass of the system are

$$\text{Weight} = (660 + 1200)15 = 28{,}000 \text{ lb}$$
$$M_t = 28{,}000/386 = 72.3 \text{ lb-sec}^2/\text{in.}$$

From Table 5.1, $K_{LM} = 0.78$, and hence the natural period is

$$T = 2\pi\sqrt{\frac{K_{LM}M_t}{k}} = 2\pi\sqrt{\frac{0.78 \times 72.3}{2.08 \times 10^6}} = 0.033 \text{ sec}$$

Approximate Design Methods

The required resistance is obtained with the aid of Fig. 2.7.

$$\frac{t_d}{T} = \frac{0.15}{0.033} = 4.5$$

Therefore $\quad (DLF)_{max} = 1.89$

and

Required $R_m = 1.89 \times 20{,}000 \times 15 + 28{,}000$
$= 567{,}000 + 28{,}000 = 595{,}000$ lb

The actual strength is 570,000 lb, and therefore the beam is slightly underdesigned. If desired, a second cycle of design would be executed in exactly the same manner as demonstrated above.

In order to determine the maximum shear, we compute the dynamic reaction, using the expression given in Table 5.1. However, the load at the time of maximum response must first be determined. From Fig. 2.7b we obtain $t_m/T = 0.487$, or $t_m = 0.016$ sec. At this time the dynamic load is 17.9 kips/ft (Fig. 5.12) and the dynamic reaction, or maximum beam shear, is

$V = 0.39R_m \text{ (live)} + 0.11F + \text{dead}$
$= 0.39 \times 567{,}000 + 0.11 \times 17{,}900 \times 15 + \frac{1}{2} \times 28{,}000$
$= 264{,}000$ lb

The web reinforcement of the beam (stirrups) would be designed for this value.

As an example of elasto-plastic design, consider the same problem as above, except that now the design criterion is to be $\mu = 3$; that is, the desired maximum deflection is three times the elastic limit.

Since the load duration in this example is relatively long, we use Eq. (5.14) to obtain a first trial value of required resistance. Thus

Required $R_m = F_1 \left(\dfrac{1}{1 - 1/2\mu} \right) = 20{,}000 \times 15 \times 1.2$
$= 360{,}000$ lb

This is, of course, an upper bound [since Eq. (5.14) is based on an infinite duration], so we might arbitrarily reduce this value, say, to 300,000 lb. If the beam weight is estimated to be 300 lb/ft,

Required $R_m = 300{,}000 + 1500 \times 15 = 322{,}000$ lb

and

Required $bd^2 = 322{,}000/29.6 = 10{,}900$ in.^{3}

Let us try

$b = 15$ in. $\quad$ and $\quad d = 27.0$ in.

Therefore

$$\text{Beam weight} = 450 \text{ lb/ft}$$
$$\text{Total weight} = (450 + 1200)15 = 25{,}000 \text{ lb}$$
$$M_t = 25{,}000/386 = 64.0 \text{ lb-sec}^2/\text{in.}$$
$$I_a = \frac{bd^3}{2}(5.5\rho_s + 0.083) = \frac{15(27)^3}{2}(5.5 \times 0.015 + 0.083)$$
$$= 24{,}400 \text{ in.}^4$$
$$k = \frac{384EI_a}{5L^3} = \frac{384(3 \times 10^6)24{,}400}{5(15 \times 12)^3} = 0.96 \times 10^6 \text{ lb/in.}$$

By Table 5.1, the load-mass factors are 0.78 and 0.66 for the elastic and plastic ranges, respectively. Since the behavior is to be more plastic than elastic ($\mu = 3$), we estimate that the proper K_{LM} is 0.70. Thus

$$T = 2\pi\sqrt{\frac{K_{LM}M_t}{k}} = 2\pi\sqrt{\frac{0.70 \times 64.0}{0.96 \times 10^6}} = 0.043 \text{ sec}$$
$$\frac{t_d}{T} = \frac{0.15}{0.043} = 3.5$$
$$\frac{R_m \text{ (net)}}{F_1} = \frac{300{,}000}{20{,}000 \times 15} = 1.0$$

Using the last two values computed, we obtain from Fig. 2.24 $\mu = 3.7$. This is somewhat more than desired, and the beam should be slightly strengthened to obtain $\mu = 3$. However, inspection of Fig. 2.24 indicates that an increase of less than 10 percent would be required, and this step will not be included here.

In order to determine accurately the maximum beam shear, or end reaction, a more complete analysis would be required. The maximum shear occurs at the end of the elastic range (since at later times R is constant and F decreases), but there is no convenient way to obtain the corresponding time without making a rigorous analysis of the elastic range. As a slightly conservative estimate, we may combine the maximum load with the maximum resistance, using the expression for dynamic reaction given in Table 5.1. Thus

$$V_{\max} = 0.39R_m \text{ (net)} + 0.11F + \text{dead}$$
$$= 0.39 \times 300{,}000 + 0.11 \times 20{,}000 \times 15 + \tfrac{1}{2} \times 25{,}000$$
$$= 162{,}000 \text{ lb}$$

The error in this value is small, since the elastic limit is attained in something less than half the natural period, at which time there has been little decrease in the applied load.

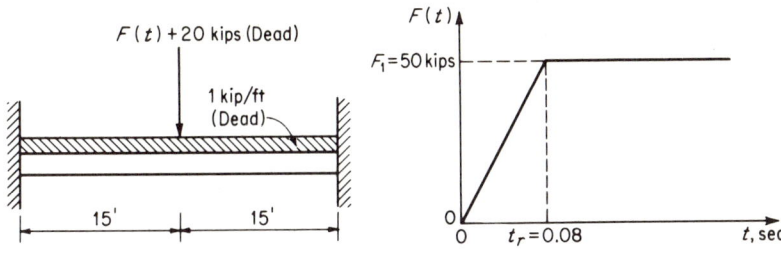

FIGURE 5.13 Design example. Steel beam.

b. Steel Beam

The fixed-ended beam shown in Fig. 5.13 is to be designed using a standard steel section. It is required that the bending stress in the beam not exceed 30,000 psi. For simplicity, it will be assumed that the dead load of 1 kip/ft includes an allowance for beam weight. It will also be assumed that the beam is so braced that the possibility of buckling is eliminated.

Inspection of the appropriate response chart, Fig. 2.9, indicates that the DLF will probably not exceed 1.4, since t_r/T will probably not be less than about $\frac{2}{3}$. Adopting this value of $(\text{DLF})_{\max}$ and combining dead and dynamic effects, we estimate the maximum moment at the support to be

$$\mathfrak{M}_{\max} = \frac{w_D L^2}{12} + \frac{F_D L}{8} + \frac{F_1 L}{8} (\text{DLF})_{\max}$$

$$= 1 \times \frac{30^2}{12} + 20 \times {}^{30}\!/_{8} + 50 \times {}^{30}\!/_{8} \times 1.4$$

$$= 75 + 75 + 262 = 412 \text{ kip-ft}$$

$$\text{Required section modulus} = \frac{\mathfrak{M}}{\sigma} = \frac{412 \times 12}{30} = 165 \text{ in.}^3$$

Most economical section, 24WF76
$I = 2096$ in.^{4}
$S = 175.4$ in.^{3}

Table 5.2:
$$k = \frac{192EI}{L^3} = \frac{192(30 \times 10^3)2096}{(30 \times 12)^3}$$
$$= 258 \text{ kips/in.}$$

In order to establish the equivalent system, the concentrated and distributed masses must be dealt with separately, using the factors given in Table 5.2. A dead load must of course represent a mass which is part of the system. The computations leading to the actual stress intensity are as follows:

$$K_L = 1.0 \qquad K_M = 1.0 \qquad \text{concentrated mass}$$
$$K_M = 0.37 \qquad \text{distributed mass}$$

$$M_e = \sum K_M M = \frac{20 \times 1.0 + 1 \times 30 \times 0.37}{386} = 0.081 \text{ kip-sec}^2/\text{in.}$$

$$k_e = K_L k = 258 \times 1.0 = 258 \text{ kips/in.}$$

$$T = 2\pi \sqrt{\frac{M_e}{k_e}} = 2\pi \sqrt{\frac{0.081}{258}} = 0.111 \text{ sec}$$

$$\frac{t_r}{T} = \frac{0.08}{0.111} = 0.72$$

Fig. 2.9:
$$(\text{DLF})_{max} = 1.35$$

$$\mathfrak{M}_{max} = 1 \times \frac{30^2}{12} + 20 \times {}^{30}\!/_{8} + 50 \times {}^{30}\!/_{8} \times 1.35$$

$$= 403 \text{ kip-ft}$$

$$\max \sigma = \frac{\mathfrak{M}}{S} = \frac{403 \times 12}{175.4} = 27.6 \text{ ksi}$$

Thus a slightly smaller beam size could be used, but we shall not consider further trials here.

To determine the maximum beam shear we first observe, in Fig. 2.9b, that t_m/t_r is always greater than unity, which indicates that, under any circumstances, the maximum load, 50 kips, would be acting at the time of maximum response. At this same time, that part of the beam resistance associated with the dynamic load is $1.35 \times 50 = 67.5$ kips. Therefore, using the expression for dynamic reaction in Table 5.2, the maximum shear, or reaction, is found to be

$$V_{max} = 0.71 \times 67.5 - 0.21 \times 50 + \text{dead}$$
$$= 48.0 - 10.5 + \tfrac{1}{2} \times 20 + 1 \times 30 \times \tfrac{1}{2} = 62.5 \text{ kips}$$

c. Reinforced Concrete Slab

The two-way slab shown in Fig. 5.14 is fixed on all four edges and subjected to a uniformly distributed pressure, having the triangular time function indicated. The design is to be based on an allowable deflection of 6 in. at the center of the slab. Although this is a very large deflection, it would be an appropriate criterion if the aim of the design were merely to prevent collapse under a rare loading condition. Deflection, rather than ductility ratio, is specified, presumably because clearance or some other functional consideration, rather than structural damage, controls the design.

The design factors for this case are given in Table 5.5. In order to simplify the procedure, we shall make use of the "effective" resistance function as indicated in Fig. 5.4. To simplify further, let us assume

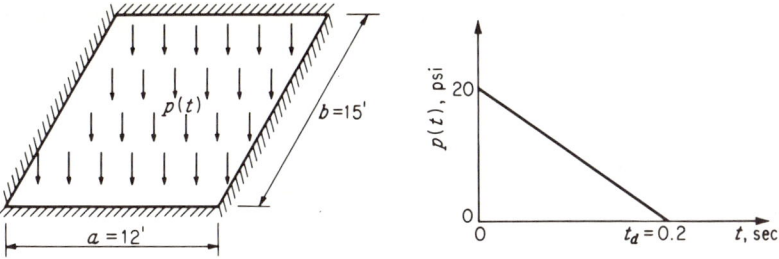

FIGURE 5.14 Design example. Two-way reinforced concrete slab.

that the bending strength of the slab is uniform around the complete perimeter and along the two midspan axes. Thus, by the notation of Sec. 5.3b and Table 5.5,

$$\mathfrak{M}_{Pfa} = \mathfrak{M}_{Psa} = \mathfrak{M}_{Pfb} = \mathfrak{M}_{Psb} = b\mathfrak{M}_{Psb}^0$$

According to Table 5.5 for $a/b = 12/15 = 0.8$ and with the notation of Fig. 5.4,

$$R_1 = 26.4\mathfrak{M}_{Psb}^0$$

$$R_m = \frac{1}{a}[12(M_{Pfa} + M_{Psa}) + 10.3(M_{Pfb} + M_{Psb})]$$

$$= \frac{1}{12}[12(2) + 10.3(2)]\,15M_{Psb}^0 = 55.8M_{Psb}^0$$

$$k_1 = \frac{705EI_a}{a^2}$$

$$k_2 = \frac{212EI_a}{a^2}$$

$$(y_{el})_1 = \frac{R_1}{k_1} = 0.0375\,\frac{\mathfrak{M}_{Psb}^0 a^2}{EI_a}$$

$$(y_{el})_2 = (y_{el})_1 + \frac{R_m - R_1}{k_2} = 0.176\,\frac{M_{Psb}^0 a^2}{EI_a}$$

Referring to Fig. 5.4, if we equate the area under the trilinear resistance curve up to $(y_{el})_2$ to the area under the bilinear (dashed) curve up to the same point, we obtain

$$(y_{el})_E = 0.130\,\frac{\mathfrak{M}_{Psb}^0 a^2}{EI_a}$$

and

$$k_E = \frac{R_m}{(y_{el})_E} = \frac{430EI_a}{a^2}$$

which define the "effective" bilinear resistance function.

Referring to Table 5.5, the load-mass factors for the three ranges of behavior are found to be 0.69, 0.71, and 0.54. Since, in this case, the behavior is primarily plastic, we select a value near that for the plastic range, say, $K_{LM} = 0.57$.

As a first trial, let us investigate a slab with a resistance just equal to the peak value of the applied load; i.e.,

$$R_m = 20 \text{ psi} \times 144 \times 12 \times 15 = 5.2 \times 10^5 \text{ lb}$$

Therefore

$$\text{Required } \mathfrak{M}_{Psb}^0 = \frac{R_m}{55.8} = \frac{5.2 \times 10^5}{55.8} = 9300 \text{ lb-in./in.}$$

where the dead weight has been neglected.

If the dynamic material strengths are $\sigma_{dy} = 50{,}000$ psi and $\sigma'_{dc} = 4000$ psi and if we arbitrarily set the steel ratio as $\rho_s = 0.01$, the bending strength per inch of slab width is

$$\mathfrak{M}_{Psb}^0 = \rho_s b d^2 \sigma_{dy}\left(1 - \frac{\rho_s \sigma_{dy}}{1.7 \sigma'_{dc}}\right) = 464 d^2 \text{ lb-in./in.}$$

The required effective depth of slab is therefore

$$\text{Required } d = \sqrt{\frac{\text{Required } \mathfrak{M}_{Psb}^0}{464}} = \sqrt{\frac{9300}{464}} = 4.5 \text{ in.}$$

and the total depth might be 6 in. Using the empirical expression for moment of inertia,

$$I_a = \frac{bd^3}{2}(5.5\rho_s + 0.083) = \frac{(1)(4.5)^3}{2}(5.5 \times 0.01 + 0.083)$$
$$= 6.30 \text{ in.}^4/\text{in.}$$

we compute the parameters of the equivalent system.

$$k_E = \frac{430 E I_a}{a^2} = \frac{430(3 \times 10^6)6.30}{(12 \times 12)^2} = 392{,}000 \text{ lb/in.}$$

$$M_t = \frac{150 \times \frac{6}{12} \times 12 \times 15}{386} = 35.0 \text{ lb-sec}^2/\text{in.}$$

$$T = 2\pi\sqrt{\frac{K_{LM} M_t}{k_E}} = 2\pi\sqrt{\frac{0.57 \times 35.0}{392{,}000}} = 0.045 \text{ sec}$$

$$\frac{t_d}{T} = \frac{0.20}{0.045} = 4.4$$

Entering Fig. 2.24 with this value of t_d/T and with $R_m/F_1 = 1$, we read $\mu = 4.3$. Since

$$(y_{el})_E = \frac{R_m}{k_E} = \frac{5.2 \times 10^5}{392{,}000} = 1.33 \text{ in.}$$

the maximum deflection at midslab is

$$y_{max} = \mu(y_{el})_E = 4.3 \times 1.33 = 5.7 \text{ in.}$$

This deflection is close to the value desired (6 in.), and the design is therefore deemed satisfactory.

5.7 Approximate Design of Multidegree Systems

Thus far in this chapter, we have dealt with the analysis and design of single elements with distributed mass which could be analyzed on the basis of an equivalent one-degree system. Of course, most structures consist of combinations of such elements. However, in many cases it is permissible to assume that the individual elements act independently of the others. For example, if we consider a beam supported by girders, it may be possible to analyze the former as though it were on rigid supports and to design the girders for the dynamic reactions of the beam so determined. This is permissible provided that the natural periods of the elements are sufficiently different. If the girders have a much longer period, they will respond slowly and the inertia forces along the beam due to the girder motion will be small compared with those due to the vibration of the beam itself, and hence will have little effect on beam response. On the other hand, if the girder period is short compared with that of the beam, the girder deflection will be small relative to beam deflection, and again there will be little effect on beam response. As an approximate rule of thumb, it may be said that two such elements may be treated separately if the periods differ by a factor of 2 or more. An example of this type of uncoupled system is given in Sec. 5.7a.

If the periods of the elements are not sufficiently different, it is necessary to analyze the multidegree system as such. Generally, this can be done by assuming a single shape for each element. Thus the number of degrees of freedom equals the number of elements. This procedure was demonstrated for a beam-girder system in Sec. 4.6 and is further demonstrated in Sec. 5.7(b), using the methods of this chapter.

a. Rigid Frame with Vertical Load

Suppose it is required to determine the maximum dynamic stresses in the frame of Fig. 5.15a. The analysis of such a rigid frame involves consideration of two separable effects of the applied load: (1) the flexural stresses in the girder and columns, and (2) the axial stresses in the columns. The periods of the natural modes associated with the first effect are much larger than those associated with the second. Stated differently, the vertical displacements due to change in column length

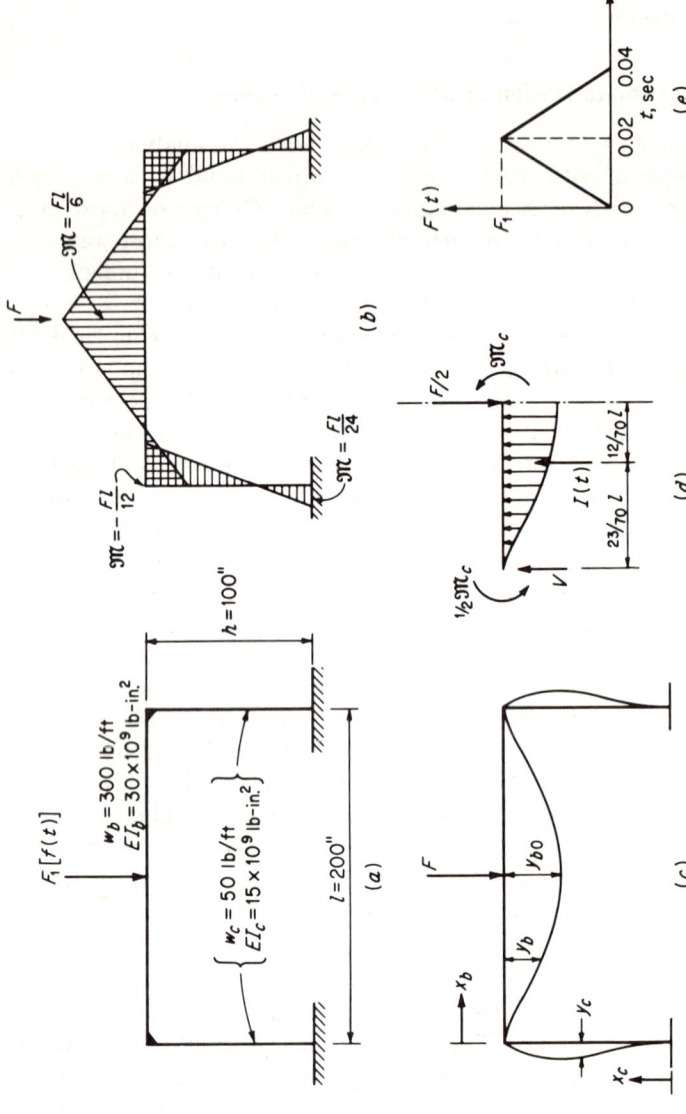

FIGURE 5.15 Example. Rigid-frame analysis.

are too small to affect the flexural response. Furthermore, the girder response develops so slowly that the load on the column may be considered static. Therefore our procedure will be to analyze the flexural response, ignoring the change in column length, and then to design the column for the maximum girder reaction applied statically.

It will be sufficiently accurate to base the flexural analysis on an equivalent one-degree system. As in previous cases, the deflected shape associated with the predominant mode will be taken as that due to static application of the load. In Fig. 5.15b are shown the moments due to this load obtained by conventional frame analysis. The deflections corresponding to these moments as obtained by the moment-area method or some other standard procedure are given by

$$y_b = \frac{F x_b}{48 E I_b} (l^2 + 2l x_b - 4 x_b^2)$$

$$y_c = \frac{F l}{48 E I_c} \left(x_c^2 - \frac{x_c^3}{h} \right)$$

for which the coordinates are defined in Fig. 5.15c. The midspan girder deflection is given by

$$y_{bo} = \frac{F l^3}{96 E I_b}$$

If we select this deflection as the modal coordinate, the characteristic shape is defined as follows:

(a) $$\phi_b(x_b) = \frac{2 x_b}{l^2} \left(l + 2 x_b - 4 \frac{x_b^2}{l} \right) \quad x < \frac{l}{2}$$

(b) $$\phi_c(x_c) = \frac{2 I_b}{l^2 I_c} \left(x_c^2 - \frac{x_c^3}{h} \right)$$

(5.18)

since, by definition,

$$y_b = y_{bo} \phi_b(x_b) \quad \text{and} \quad y_c = y_{bo} \phi_c(x_c)$$

Referring to Eq. (5.1), we see that the equivalent mass of the system is

$$M_e = 2 \int_0^{l/2} m_b [\phi_b(x_b)]^2 \, dx_b + 2 \int_0^{h} m_c [\phi_c(x_c)]^2 \, dx_c$$

where m_b and m_c are the mass per unit length of the girder and column, respectively. Executing the integrations, we obtain

$$M_e = 0.443 m_b l + 0.076 m_c h \left(\frac{h^4 I_b^2}{l^4 I_c^2} \right) \quad (5.19)$$

and substituting the numerical values given in Fig. 5.15a,

$$M_e = 5.74 + 0.02 = 5.76 \text{ lb-sec}^2/\text{in.}$$

It may be observed that the contribution of the column mass (0.02) is very small and might well have been ignored. Furthermore, the mass factor for the girder (0.443) lies between that for a simply supported beam (0.49) and that for a fixed beam (0.37) as given by Tables 5.1 and 5.2. The latter observation suggests that we might have estimated the mass factor by approximate interpolation between these two extremes without introducing appreciable error.

Since we have selected the deflection of the load point y_{bo} to be the deflection of the equivalent system, the load factor K_L is unity. Thus

$$k = k_e = \frac{F}{y_{bo}} = \frac{96EI_b}{l^3} = 360{,}000 \text{ lb/in.}$$

and
$$F_e(t) = F(t) = F_1[f(t)]$$

The natural period of the system is

$$T = 2\pi\sqrt{\frac{M_e}{k_e}} = 2\pi\sqrt{\frac{5.76}{360{,}000}} = 0.0251 \text{ sec}$$

For the given load-time function shown in Fig. 5.15e, the maximum DLF is given by Fig. 2.8a.

$$\frac{t_d}{T} = \frac{0.04}{0.0251} = 1.59 \qquad (\text{DLF})_{max} = 1.23$$

The time of maximum response is given by Fig. 2.8b.

$$\frac{t_m}{t_d} = 0.55 \qquad t_m = 0.55 \times 0.04 = 0.022 \text{ sec}$$

The maximum dynamic girder deflection is therefore

$$(y_{bo})_{max} = (y_{bo})_{st}(\text{DLF})_{max} = \frac{F_1}{k}(\text{DLF})_{max}$$
$$= \frac{F_1}{360{,}000} \times 1.23 = (3.42 \times 10^{-6})F_1 \quad \text{in.}$$

and the maximum bending moment is

$$\mathfrak{M}_{max} = \frac{F_1 l}{6}(\text{DLF})_{max} = \frac{F_1 \times 200}{6} \times 1.23 = 41F_1 \quad \text{lb-in.}$$

In order to obtain the maximum column load, we must first derive the expression for the dynamic girder reaction. Following the procedure of Sec. 5.4, we consider the dynamic equilibrium of the half girder span as indicated in Fig. 5.15d. The location of the resultant inertia force is the

Approximate Design Methods

centroid of the area under the characteristic shape defined by Eq. (5.18a). Taking moments about this point and noting that the girder end moment is always one-half the midspan moment, we obtain the equation

$$V \times {}^{23}\!\!/_{70}l + \frac{F}{2} \times {}^{12}\!\!/_{70}l - {}^{3}\!\!/_{2}\mathfrak{M}_c = 0$$

R is by definition the static load of the type applied corresponding to a given deformation or stress condition of the frame, and it follows that $\mathfrak{M}_c = Rl/6$. Using this expression to eliminate $\mathfrak{M}_c$ from the last equation and solving, we obtain

$$V = 0.76R - 0.26F \qquad (5.20)$$

which is also the dynamic column load. Comparison with the corresponding expressions in Tables 5.1 and 5.2 indicates that the coefficients in Eq. (5.20) lie between those for pinned and fixed-ended beams, as would be expected.

In the present case, the maximum resistance occurs at $t = 0.022$ sec at which time $F = 0.9F_1$ and $R = 1.23F_1$. Thus the maximum column load is

$$V = 0.76(1.23F_1) - 0.26(0.9F_1) = 0.70F_1$$

As far as column axial stress is concerned, the load is applied very slowly relative to the period of the mode corresponding to this stress. Therefore the column may be designed as though the load just computed were applied statically. The maximum column bending moment which occurs simultaneously is equal to $(F_1 l/12)1.23$.

b. Elasto-plastic Beam-girder System

In this section the analysis of the beam-girder system and loading shown in Fig. 5.16 is considered. It is known from preliminary considerations that the girders and/or beam will deflect into the plastic range. The maximum deflections of both beam and girders are desired.

In order to determine whether or not the beam and girders may be treated as uncoupled, the natural periods of the separate elements should be compared. Using the elastic factors in Table 5.1, we get for the beam

$$M_{be} = K_{Mb}M_b = 0.50(200 \times 0.1) = 10 \text{ lb-sec}^2/\text{in.}$$

$$k_{be} = K_{Lb}k_b = K_{Lb}\frac{384EI_b}{5l_b^3} = 0.64\frac{384 \times 2 \times 10^{10}}{5(200)^3} = 123{,}000 \text{ lb/in.}$$

$$T_b = 2\pi\sqrt{\frac{M_{be}}{k_{be}}} = 2\pi\sqrt{\frac{10}{123{,}000}} = 0.057 \text{ sec}$$

For one of the girders, the effective mass is that of the girder itself plus

238 Introduction to Structural Dynamics

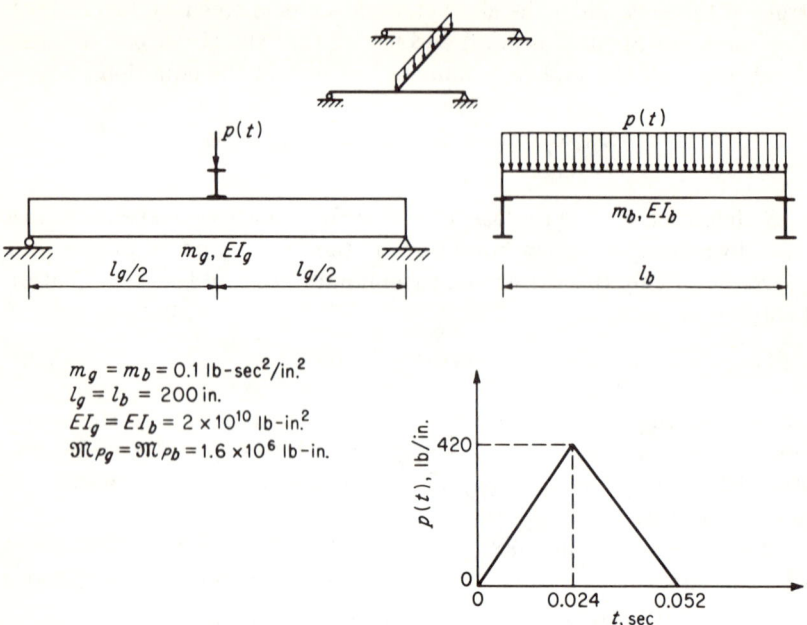

FIGURE 5.16 Example. Elasto-plastic analysis of beam-girder system.

one-half the beam mass (since this "rides" along with the girder) with a mass factor of unity.

$$M_{ge} = K_{M_g}M_g + \tfrac{1}{2}M_b = 0.49(200 \times 0.1) + \tfrac{1}{2}(200 \times 0.1)$$
$$= 19.8 \text{ lb-sec}^2/\text{in.}$$

$$k_{ge} = K_{L_g}k_g = K_{L_g}\frac{48EI_g}{l_g{}^3} = 1.0\frac{48(2 \times 10^{10})}{(200)^3}$$
$$= 120{,}000 \text{ lb/in.}$$

$$T_g = 2\pi\sqrt{\frac{M_{ge}}{k_{ge}}} = 2\pi\sqrt{\frac{19.8}{120{,}000}} = 0.081 \text{ sec}$$

The two periods differ by less than a factor of 2, and therefore we must consider the dynamic interaction; i.e., we must analyze as a two-degree system. The symmetry of structure and loading eliminates participation of the third mode.

In writing the equation of motion for the beam, we must include as additional load the inertia forces due to the rigid-body acceleration associated with the support motion as indicated in Fig. 5.17a. Thus

$$K_{Mb}M_b\ddot{y}_b + K_{Lb}k_b y_b = K_{Lb}[p(t)l_b - m_b l_b \ddot{y}_g]$$

where y_b and y_g are the relative beam midspan deflection and the midspan girder deflection, respectively. In this equation $k_b y_b$ is the resistance R_b,

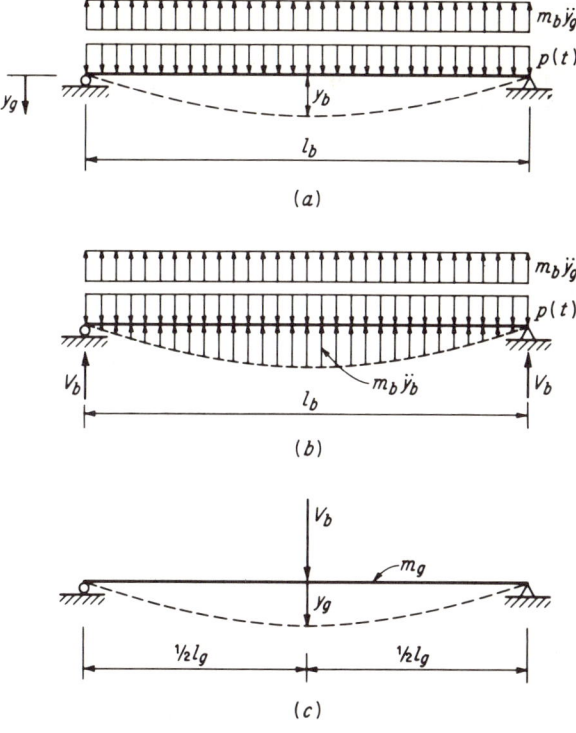

FIGURE 5.17 Example. Force distributions.

which is limited by $R_{bm} = 8\mathfrak{M}_{Pb}/l_b = 64{,}000$ lb. With the load and mass factors given for the elastic range in Table 5.1 ($K_{Lb} = 0.64$, $K_{Mb} = 0.50$) and the numerical properties given in Fig. 5.16, the equation of motion becomes

$$10\ddot{y}_b + (123{,}000 y_b \text{ or } 41{,}000) = 128 p(t) - 12.8 \ddot{y}_g \quad (5.21)$$

The beam reaction, which is the dynamic load on the girder, is simply one-half the total net force (including inertia effects) on the beam, as indicated in Fig. 5.17b. This is given by

$$\begin{aligned} V_b &= \tfrac{1}{2}[p(t) l_b - m_b l_b \ddot{y}_g - 0.64 m_b l_b \ddot{y}_b] \\ &= 100 p(t) - 10 \ddot{y}_g - 6.4 \ddot{y}_b \end{aligned} \quad (5.22)$$

The factor 0.64 is derived from the condition that the distribution curve of relative inertia force is the deflected shape. The equation of motion for the girder is therefore (Fig. 5.17c)

$$K_{Mg} M_g \ddot{y}_g + K_{Lg} k_g y_g = K_{Lg} V_b$$

The girder resistance is $k_g y_g$, and when the girder is in the plastic range, this term is replaced by the constant $R_{gm} = 4\mathfrak{M}_{Pg}/l_g = 32{,}000$ lb. With the insertion of the elastic load and mass factors for a beam with concentrated load at midspan (Table 5.1: $K_{Lg} = 1.0$, $K_{Mg} = 0.49$) and the numerical values of the girder properties (Fig. 5.16), the girder equation of motion becomes

$$9.8\ddot{y}_g + (120{,}000 y_g \text{ or } 32{,}000) = 100 p(t) - 10\ddot{y}_g - 6.4 \ddot{y}_b \quad (5.23)$$

Equations (5.21) and (5.23) will form the basis for analysis, which must be done by numerical integration, since we anticipate plastic behavior. The student may wish to compare the elastic equations given above with those derived for the same beam-girder system in Sec. 4.6. It will be found that they are practically identical, there being only slight differences in the numerical coefficients, which result from differences in the assumed shapes. In Sec. 4.6 the assumed shapes were sine waves (i.e., the fundamental-mode shapes), while here they are the static-load-deflection curves upon which Table 5.1 is based.

Strictly speaking, Eqs. (5.21) and (5.23) should be altered in the plastic range by changing the load and mass factors to reflect the change in deflected shape. However, if this were done, four conditions would have to be considered: both elements elastic, both plastic, girder elastic/beam plastic, and girder plastic/beam elastic. Thus four sets of equations would have to be used. This greatly complicates the analysis, and the increased accuracy is probably not enough to warrant the effort.

As the equations of motion now stand, each step in the analysis would have to be done by trial, since both accelerations appear in each equation. To avoid this, we may solve Eqs. (5.21) and (5.23) simultaneously to obtain two new equations, each involving only one acceleration. The result is

(a) $\quad \ddot{y}_b = 10.8 p(t) + (13{,}200 y_g \text{ or } 3{,}530) - (21{,}000 y_b \text{ or } 6{,}990)$
(b) $\quad \ddot{y}_g = 1.56 p(t) + (6{,}790 y_b \text{ or } 2{,}260) - (10{,}300 y_g \text{ or } 2{,}760)$ $\quad (5.24)$

The alternative constant terms within the parentheses are merely upper limits of the resistance. For example, in Eq. (5.24a), 3,530 should be used in place of $13{,}200 y_g$ if the latter exceeds the former. Rebound into the elastic range is handled in the usual way (Sec. 1.5).

The natural periods of the two-degree system may be derived from the elastic equations of motion. This was done in Sec. 4.6 with the result $T_1 = 0.090$ and $T_2 = 0.039$ sec. The time interval for the numerical analysis is taken as approximately one-tenth of the shorter period, or 0.004 sec. The computations leading to maximum displacements are shown in Table 5.7. Reference is made to Chap. 1 for the details of the numerical procedure.

Table 5.7 Numerical Integration; Beam-girder System of Fig. 5.16

$y^{(s+1)} = 2y^{(s)} - y^{(s-1)} + \ddot{y}^{(s)} (\Delta t)^2$

t	$p(t)$	$10.8p(t)$	$13{,}200y_a$ or 3530	$21{,}000y_b$ or 6990	$\ddot{y}_b$ Eq. (5.24a)	$\ddot{y}_b (\Delta t)^2$	y_b	$1.56p(t)$	$6790y_b$ or 2260	$10{,}300y_a$ or 2760	$\ddot{y}_a$ Eq. (5.24b)	$\ddot{y}_a (\Delta t)^2$	y_a
0	0	0	0	0	0	0.0019*	0	0	0	0	0	0.0003*	0
0.004	70	760	0	40	720	0.0115	0.0019	110	10	0	120	0.0019	0.0003
0.008	140	1510	30	320	1220	0.0195	0.0153	220	100	30	290	0.0046	0.0025
0.012	210	2270	120	1010	1380	0.0220	0.0482	330	330	100	560	0.0090	0.0093
0.016	280	3020	330	2160	1190	0.0190	0.1031	440	700	260	880	0.0141	0.0251
0.020	350	3780	730	3720	790	0.0127	0.1770	550	1200	570	1180	0.0189	0.0550
0.024	420	4530	1370	5530	370	0.0059	0.2636	660	1790	1070	1380	0.0221	0.1038
0.028	360	3890	2300	3990	-800	-0.0128	0.3561	560	2260	1800	1020	0.0163	0.1747
0.032	300	3240	3460	6990	-290	-0.0046	0.4358	470	2260	2700	30	0.0005	0.2619
0.036	240	2590	3530	6990	-870	-0.0139	0.5109	370	2260	2760	-130	-0.0021	0.3496
0.040	180	1940	3530	6990	-1520	-0.0243	0.5721	280	2260	2760	-220	-0.0035	0.4352
0.044	120	1300	3530	6990	-2160	-0.0346	0.6090	190	2260	2760	-310	-0.0050	0.5173
0.048	60	650	3530	6990	-2810	-0.0450	**0.6113**	90	2260	2760	-410	-0.0066	0.5944
0.052	0	0	3530	6090	-2560	-0.0410	0.5686	0	1970	2760	-790	-0.0127	0.6649
0.056	0	0	3530	4330	-800	-0.0128	0.4849	0	1400	2760	-1360	-0.0172	0.7227
0.060	0	0	3530	2300	+1230	+0.0197	0.3884	0	750	2760	-2010	-0.0322	0.7633
0.064	0	0	3530	700	2830	0.0453	0.3116	0	230	2760	-2530	-0.0405	**0.7717**
0.068	0						0.2801						0.7396

* Determined by trial as $\frac{1}{6}\ddot{y}(\Delta t)^2$ at $t = 0.004$ (Sec. 1.2b).

With reference to the computations, the beam deflection exceeds the elastic limit at 0.028 sec and the girder goes plastic at 0.036 sec. The beam rebounds elastically after 0.048 sec. The maximum beam deflection (relative to the girder) is 0.61 in., or 1.8 times the elastic limit, and the maximum girder deflection is 0.77 in., or 2.9 times its elastic limit.

Problems

5.1 *a.* Determine the transformation factors K_L and K_M for a cantilever beam under uniformly distributed load. Use the elastic-static-deflection curve as the assumed shape.
   b. Compare the natural frequency based on these factors with the exact expression for the first normal mode.
   c. Derive the expressions for dynamic reactions (shear and moment at the fixed end).
Answer
   a. $K_L = 0.40$; $K_M = 0.257$
   c. $V = 0.3F + 0.7R$; $\mathfrak{M} = Rl/2$

5.2 A simply supported beam has a uniformly distributed mass m and, in addition, lumped masses equal to ml at the third points of the span. It is subjected to a uniformly distributed dynamic load $p(t)$. Compute the mass, stiffness, and load of the equivalent one-degree system.

5.3 Determine the transformation factors for a square, uniformly loaded plate fixed on two adjacent edges and free at the other edges. The shape may be assumed to be $y = A[1 - \cos(\pi x/2l)][1 - \cos(\pi z/2l)]$, where l is the length of the sides and x and z are perpendicular coordinates with origin at the intersection of the fixed edges.
Answer
   $K_M = 0.0515$
   $K_L = 0.1325$

5.4 A fixed-ended beam with concentrated load at midspan has a negative moment capacity equal to one-half the positive moment capacity. Plot the trilinear resistance function, and determine the effective stiffness of the equivalent bilinear function.

5.5 A one-degree system defined by the parameters given in Fig. 5.18 is subjected to the load-time function shown. Using the approximate method of analysis for multiple triangles (Sec. 5.5a):
   a. Compute the maximum deflection.
   b. Compute the value of R_m which would result in $\mu = 2$.

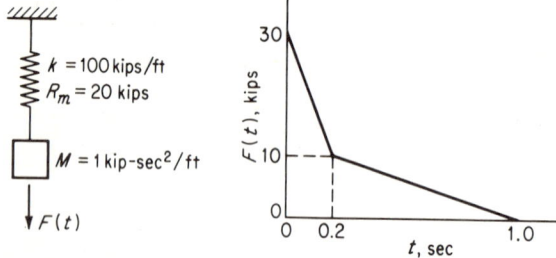

FIGURE 5.18 Problem 5.5.

Answer
  a. $\mu = 2.5$
  b. $R_m = 22.2$ kips

5.6 A triangular load pulse with zero rise time is applied to a one-degree system. If the load duration is sufficiently long, Eq. (5.14) gives the required resistance with only slight error. To investigate the range in which Eq. (5.14) might be used, plot the ratio of R_m given thereby to the correct value versus t_d/T for $\mu = 3$. How would the value of μ affect your conclusion?

5.7 Referring to Prob. 5.6, Eq. (5.16) gives the required resistance with only slight error if the load duration is sufficiently short. To investigate the range in which Eq. (5.16) might be used, plot the ratio of R_m given thereby to the correct value versus t_d/T for $\mu = 3$. How would the value of μ affect your conclusion?

5.8 A square two-way slab is fixed on all edges and loaded by a uniformly distributed pressure which is applied suddenly with an intensity of 1 kip/ft² and which then decays linearly to zero at 0.2 sec. The slab properties are:
  Span = 20 ft
  Weight = 100 lb/ft²
  $I_a = 40$ in.⁴/ft
  $E_c = 3 \times 10^6$ lb/in.²
  $\mathfrak{M}_{Pfa} = \mathfrak{M}_{Pfb} = \mathfrak{M}_{Psa} = \mathfrak{M}_{Psb} = 2000$ kip-in.

Using the appropriate transformation factors, compute the maximum central deflection:
  a. By the response chart (Fig. 2.24), based on an average load-mass factor and an effective stiffness.
  b. By numerical integration, taking into account all three stress ranges.

5.9 A reinforced concrete beam is fixed at one end, simply supported at the other, and subjected to a uniformly distributed load, with the time function shown in Fig. 5.19. The beam properties are:
  Span = 20 ft
  $b = 12$ in.
  $d = 24$ in. (total depth = 26 in.)
  $\rho_s = 0.015$
  $\sigma_{dy} = 60,000$ psi, $\sigma'_{dc} = 5000$ psi
  $E_c = 4 \times 10^6$ psi

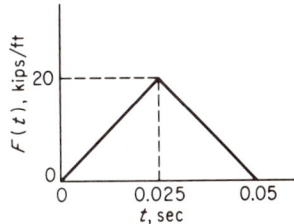

FIGURE 5.19 Problem 5.9.

The steel ratio given applies to both positive and negative moment capacities. Compute by approximate methods the maximum dynamic deflection. Assume no appreciable shear distortion.

Answer
  $y_{\max} = 2.9$ in.

5.10 Design a fixed-ended one-way slab to resist a suddenly applied constant pressure of 5 psi with duration of 0.3 sec. The following properties are given:
Span = 15 ft
$p_s = 0.01$
$\sigma_{dy} = 50{,}000$ psi, $\sigma'_{dc} = 5000$ psi
Design for $\mu = 10$.

5.11 A simply supported 24WF94 steel beam has a span of 20 ft, a total dead weight of 2000 lb/ft, and a dynamic yield strength of 45,000 psi. Determine the maximum dynamic deflection due to loads at the third points of the span, each having a rise time of 0.04 sec and a constant value thereafter of 125 kips. What is the maximum dynamic beam reaction?

5.12 Select the most economical steel WF beam which will resist elastically a rectangular-pulse load of magnitude 100 kips and duration 0.1 sec. The load is concentrated at midspan, and there is a uniformly distributed weight (in addition to the beam itself) of 3000 lb/ft. The span is 18 ft. It may be assumed that buckling is prevented, but the shear capacity of the beam selected should be checked. $\sigma_{dy} = 60{,}000$ psi.

5.13 Referring to Prob. 4.11, make an elasto-plastic numerical analysis (two degrees of freedom) to determine maximum beam and girder deflections for the following additional data: $p(t)$ is a triangular pulse with zero rise time, an initial value of 400 lb/in., and a duration of 0.25 sec, and the ultimate bending capacities are 200 kip-ft for the beam and 350 kip-ft for the cantilever girder.

6

Earthquake Analysis and Design

6.1 Introduction

The problems associated with the design of structures to withstand earthquakes have long been of great interest to engineers. This is due not only to the catastrophic nature of the possible failure but also to the fact that the difficulties encountered are technically intriguing. The literature on the subject is voluminous and is increasing rapidly. Considerable progress is being made, but in spite of this fact, a really satisfactory method of design still does not exist.[27–29]

The major difficulty lies in the prediction of the character and intensity of the earthquakes to which a structure might be subjected during its life. Strong earthquakes are rare events, and the number of actual measurements which have been made is in a statistical sense very small. Design must therefore be based on rather crude estimates of the expected ground motion.

Another major difficulty lies in the fact that a realistic analysis for earthquake should account for inelastic behavior of the structure. Very few structures could withstand a strong earthquake without some plastic deformation. In fact, it would be uneconomical to design a structure so that it would remain completely elastic. The inherent difficulty of inelastic analysis of multidegree systems, coupled with the irregular and uncertain nature of the ground motion, makes a rigorous solution to the problem impractical.[31]

246   Introduction to Structural Dynamics

It is obvious that in this brief chapter only an introduction to the subject can be offered. Attention will be restricted to a few basic concepts which will provide a foundation for further study.

6.2 Response of Multidegree Systems to Support Motion

In Sec. 2.6 the response to support motion of one-degree systems such as that shown in Fig. 6.1 was investigated. It was found there that the equation of motion could be written as

$$M\ddot{u} + ku + c\dot{u} = -M\ddot{y}_{so}f_a(t)$$

or
$$\ddot{u} + \omega^2 u + 2\beta\dot{u} = -\ddot{y}_{so}f_a(t) \qquad (6.1)$$

and the response could be expressed by

$$u(t) = -\frac{\ddot{y}_{so}}{\omega^2}(\text{DLF})_a \qquad (6.2)$$

where $u(t)$ = relative displacement of the mass with respect to the support, or

$$u(t) = y - y_s$$

and $(\text{DLF})_a = \omega \int_0^t f_a(\tau) e^{-\beta(t-\tau)} \sin \omega(t-\tau)\, d\tau$

$\ddot{y}_{so}$ = maximum support acceleration
$f_a(t)$ = time function for support acceleration
$\ddot{y}_s = \ddot{y}_{so}f_a(t)$
β = damping coefficient = $c/2M$

$(\text{DLF})_a$ is the dynamic load factor as previously defined, except that, in this case, it is based on the time function for the support acceleration. Any of the DLF expressions developed in Chap. 2 for a specific time function could therefore be used if this function represented the variation in ground acceleration. It should be noted that the expression given above for $(\text{DLF})_a$ is valid only if the initial ground velocity is zero.

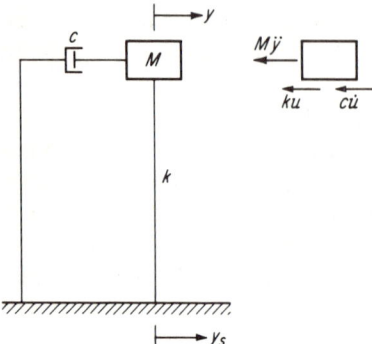

FIGURE 6.1 Damped one-degree system with support motion.

Earthquake Analysis and Design

It is a significant fact that the maximum relative displacement u_m and the maximum absolute acceleration $\ddot{y}_m$ are directly related. To demonstrate, we consider the equation of motion (Fig. 6.1)

$$M\ddot{y} + c\dot{u} + ku = 0$$

which, at maximum relative displacement ($\dot{u} = 0$), becomes

$$M\ddot{y}_m + ku_m = 0$$

Observe that $\ddot{y}_m$ and u_m must occur simultaneously. Therefore

$$\ddot{y}_m = -\frac{k}{M} u_m = -\omega^2 u_m = \ddot{y}_{so}(\text{DLF})_{a,\max} \qquad (6.3)$$

which indicates that the maximum values of *absolute* acceleration and *relative* displacement are related in the same way as the corresponding terms for a pure harmonic motion. This is in the nature of a coincidence, and it should not be concluded that the response due to support motion is harmonic. The fictitious velocity associated with this apparent harmonic motion is sometimes called the *spectral velocity*, and its maximum value is given by

$$v_{sm} = \omega u_m = -\frac{\ddot{y}_{so}}{\omega}(\text{DLF})_{a,\max} \qquad (6.4)$$

In the analysis of multidegree systems with support motion, we may apply the modal method as developed in Sec. 3.7.* The modal equation of motion (Eq. 3.46) for a lumped-mass system with external forces is

$$\ddot{A}_n + \omega_n^2 A_n + 2\beta_n \dot{A}_n = \frac{f(t) \sum_{r=1}^{j} F_{r1} \phi_{rn}}{\sum_{r=1}^{j} M_r \phi_{rn}^2}$$

where the applied force at mass r is $F_{r1}f(t)$. It was shown in Sec. 2.6 that the response due to support motion is equivalent to the response due to applied forces equal to $-M\ddot{y}_s$. For the multidegree case, the equivalent force at mass r would therefore be $-M_r\ddot{y}_s = -M_r\ddot{y}_{so}f_a(t)$, and if this is used to replace $F_{r1}f(t)$ in the above, the resulting modal equation for support motion is

$$\ddot{A}_n + \omega_n^2 A_n + 2\beta_n \dot{A}_n = -\frac{f_a(t)\ddot{y}_{so} \sum_{r=1}^{j} M_r \phi_{rn}}{\sum_{r=1}^{j} M_r \phi_{rn}^2} \qquad (6.5)$$

---
* The equations which follow are also presented in matrix form in the Appendix.

It must be remembered that A_n is the *relative* modal displacement with respect to the support. Comparison of this equation with Eq. (6.1) for a one-degree system reveals that the only difference is the presence of the two summations in the right side of Eq. (6.5). Since these are constants for a mode and merely modify the input acceleration, the same modification may be applied to the response of the one-degree system to obtain the modal response. If we define the *modal participation factor* by

$$\Gamma_n = \frac{\sum_{r=1}^{j} M_r \phi_{rn}}{\sum_{r=1}^{j} M_r \phi_{rn}^2} \qquad (6.6)$$

then the modal displacement is given by

$$A_n(t) = \Gamma_n u_n{}^0(t) \qquad (6.7)$$

where $u_n{}^0(t)$ is the response of the one-degree system having circular frequency ω_n and is given by Eq. (6.2), or

$$u_n{}^0(t) = -\frac{\ddot{y}_{so}}{\omega_n^2} (\text{DLF})_{na} \qquad (6.8)$$

where $(\text{DLF})_{na}$ is the dynamic load factor in the usual sense, associated with the frequency ω_n and the time function $f_a(t)$. The modal displacement at mass r is

$$u_{rn}(t) = \Gamma_n u_n{}^0(t) \phi_{rn} \qquad (6.9)$$

and the total relative displacement is

$$u_r(t) = \sum^{n} \Gamma_n u_n{}^0(t) \phi_{rn} \qquad (6.10)$$

If the ground motion had been given in terms of displacement rather than acceleration, the absolute response of the one-degree system could be obtained by (Sec. 2.6a)

$$y_n{}^0(t) = y_{so}(\text{DLF})_n$$

and the relative response by

$$u_n{}^0(t) = y_{so}(\text{DLF})_n - y_s(t) \qquad (6.11)$$

where $y_s(t) = y_{so} f(t)$, and $(\text{DLF})_n$ is based on the time function $f(t)$ rather than $f_a(t)$. In the analysis of a multidegree system, Eq. (6.11) would be used in place of Eq. (6.8) if y_s rather than $\ddot{y}_s$ were specified as input.

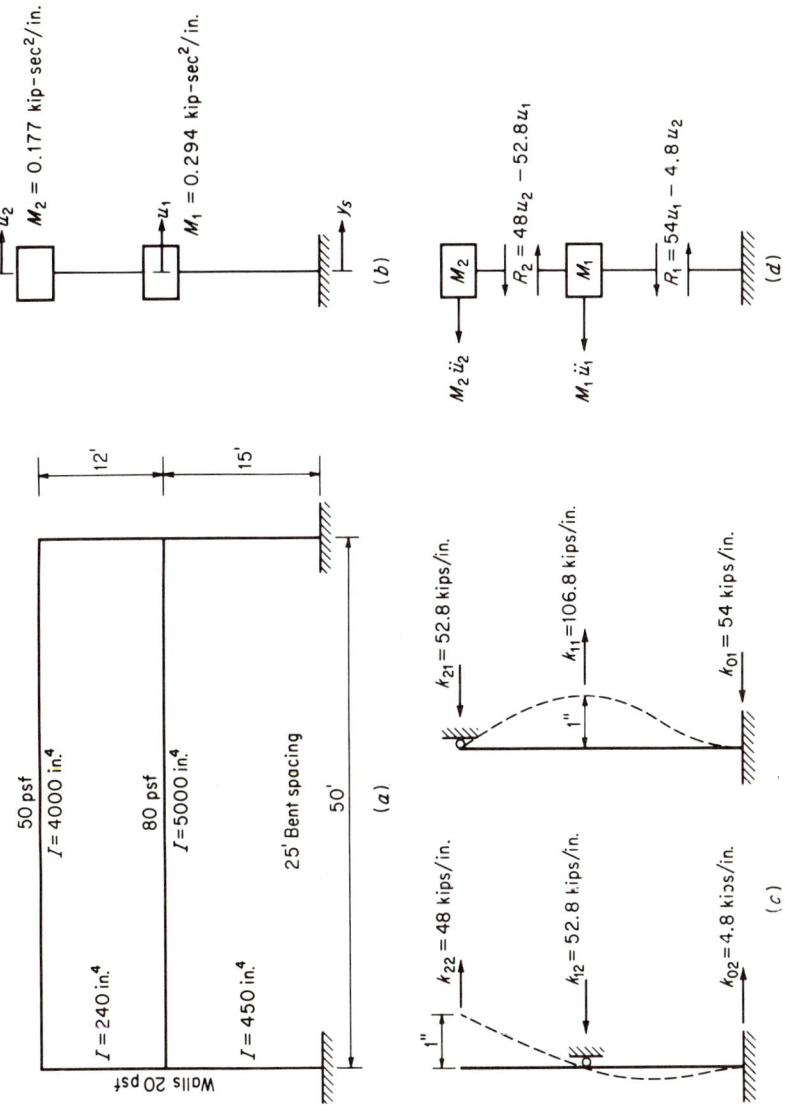

FIGURE 6.2 Example. Two-story steel frame.

6.3 Multistory-building Analysis

To illustrate application of the modal analysis developed in the preceding section, we consider the two-story steel frame shown in Fig. 6.2. The weights of the roof, first floor, and walls are indicated in Fig. 6.2a, as are the moments of inertia of the structural elements. The lumped masses at the floor levels which are given in Fig. 6.2b have been computed on the basis of the floor weights plus that of the tributary wall areas. The elastic-stiffness coefficients shown in Fig. 6.2c have been computed by conventional frame analysis. If there were no support motion, the condition of dynamic equilibrium would be as indicated in Fig. 6.2d.

Before considering the actual input, we shall establish the normal modes of the system. Based on Fig. 6.2d, the equations of motion are

$$M_2 \ddot{y}_2 + 48 y_2 - 52.8 y_1 = 0$$
$$M_1 \ddot{y}_1 - 52.8 y_2 + 106.8 y_1 = 0$$

When computing normal modes no ground motion is considered, and therefore the u's in Fig. 6.2d are both absolute displacements and those relative to the ground. Operating on these equations in the usual way, we find that the natural frequencies are

$$\omega_1 = 9.0 \text{ rad/sec} \qquad \omega_2 = 23.5 \text{ rad/sec}$$

and that the characteristic shapes may be defined by

First mode:
$$\phi_{11} = +1.00$$
$$\phi_{21} = +1.57$$

Second mode:
$$\phi_{12} = +1.00$$
$$\phi_{22} = -1.06$$

We now wish to make an analysis of the building frame for the ground motion given in Fig. 6.3. This sinusoidal motion is not intended to be

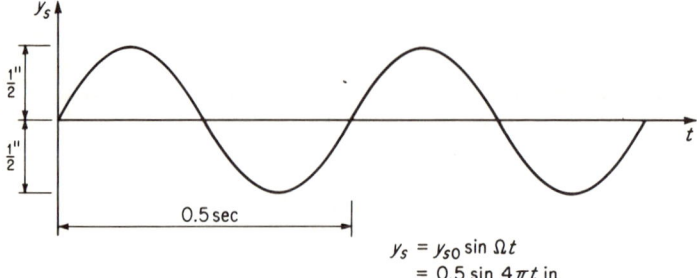

FIGURE 6.3 Example. Assumed support motion.

typical of earthquakes, but will serve to illustrate analysis for ground motion. We shall first make a completely elastic analysis and then obtain an inelastic solution for the case in which the column strengths are limited so as to cause such behavior. A comparison of these two solutions will serve to demonstrate the significance of plastic behavior in earthquake design.

a. Elastic Solution

Making use of Eq. (6.6), the modal participation factors are computed as follows:

$$\Gamma_n = \frac{\sum_{r} M_r \phi_{rn}}{\sum_{r} M_r \phi_{rn}^2}$$

$$\Gamma_1 = \frac{0.294(1.00) + 0.177(1.57)}{0.294(1.00)^2 + 0.177(1.57)^2} = 0.784$$

$$\Gamma_2 = \frac{0.294(1.00) + 0.177(-1.06)}{0.294(1.00)^2 + 0.177(-1.06)^2} = +0.215$$

The dynamic load factor (ignoring damping) is given by

$$(\text{DLF})_n = \omega_n \int_0^t f(\tau) \sin \omega_n(t - \tau) \, d\tau$$

where $f(\tau) = \sin \Omega \tau$. Integration yields

$$(\text{DLF})_n = \frac{1}{1 - \Omega^2/\omega_n^2} \left(\sin \Omega t - \frac{\Omega}{\omega_n} \sin \omega_n t \right)$$

which is the same as Eq. (2.34b). The modal values of this quantity are

$$(\text{DLF})_1 = -1.05(\sin 4\pi t - 1.40 \sin 9.0t)$$
$$(\text{DLF})_2 = +1.40(\sin 4\pi t - 0.535 \sin 23.5t)$$

By Eq. (6.11), we have for the two modes

$$u_1^0(t) = 0.5(\text{DLF})_1 - 0.5 \sin 4\pi t$$
$$= -1.025 \sin 4\pi t + 0.735 \sin 9.0t$$
$$u_2^0(t) = 0.5(\text{DLF})_2 - 0.5 \sin 4\pi t$$
$$= 0.2 \sin 4\pi t - 0.375 \sin 23.5t$$

Finally, the relative displacements of the two masses as given by Eq. (6.10) are

$$u_1(t) = (0.784)[u_1^0(t)](1.00) + (0.215)[u_2^0(t)](1.00)$$
$$= 0.784[u_1^0(t)] + 0.215[u_2^0(t)]$$
$$u_2(t) = 0.784[u_1^0(t)](1.57) + (0.215)[u_2^0(t)](-1.06)$$
$$= 1.230[u_1^0(t)] - 0.228[u_2^0(t)]$$

252   Introduction to Structural Dynamics

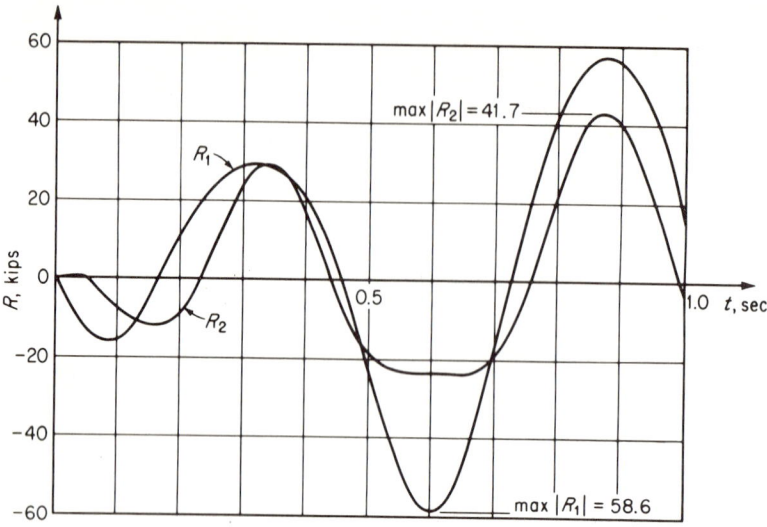

FIGURE 6.4 Example. Response of two-story frame.

The first of these is also the first-story distortion, i.e., the motion of the floor above relative to the floor below. The second-story distortion would be given by

$$u_2(t) - u_1(t) = 0.446[u_1{}^0(t)] - 0.443[u_2{}^0(t)]$$

The story shears which determine the magnitude of column bending would be computed by (Fig. 6.2d)

$$R_2 = 48u_2(t) - 52.8u_1(t)$$
$$R_1 = 54u_1(t) - 4.8u_2(t)$$

The magnitude of these shears, computed after evaluating the above expressions for $u_n{}^0(t)$ and $u_n(t)$ at various times, are plotted in Fig. 6.4 through two cycles of ground motion.

The story shears computed above are very large; e.g., the maximum in the top story is about 0.6 of the roof weight and that in the bottom story is about one-third of the total building weight. These are much larger than would normally be used for earthquake design, even though the assumed ground motion is not particularly severe as compared with actual earthquakes. There are several reasons for this difference: (1) earthquake motions are not perfectly harmonic; (2) damping was ignored in the analysis; and (3) some plastic deformation of the columns is permissible, and the column strengths need not be as great as the shears

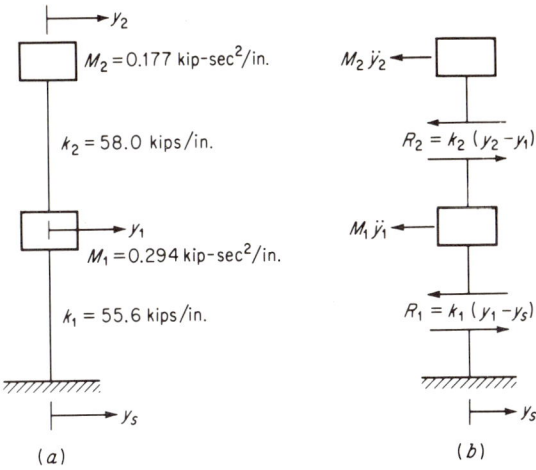

FIGURE 6.5 Idealized two-story frame.

computed above indicate. The validity of the last point will now be demonstrated.

b. Elasto-plastic Solution

Suppose that the definition of the problem is now modified by setting elastic limits for the story shears. It will arbitrarily be assumed that these values are approximately one-half of the maximum shears computed in the elastic analysis above. Thus the maximum plastic story resistances are taken to be

$$R_{m1} = 29.5 \text{ kips} \qquad R_{m2} = 21 \text{ kips}$$

In all other respects, including column stiffness, the structure remains as shown in Fig. 6.2a and the ground motion remains as shown in Fig. 6.3.

As discussed in Sec. 3.9, it is difficult, and probably not worthwhile, to include the effect of girder flexibility in an elasto-plastic analysis of a building frame. We shall therefore assume the girders to be rigid, with the result that there are only two springs in the idealized system. The stiffness of each spring corresponding to a story is

$$k = \frac{12E(2I)}{h^3}$$

where h is the story height, and I is the moment of inertia of one column in the story. Thus the idealized dynamic system is as shown in Fig. 6.5. The resistance function for each story is assumed to be bilinear.

Table 6.1 Inelastic Analysis of Two-story Building Frame (Fig. 6.5)

$y^{(s+1)} = 2y^{(s)} - y^{(s-1)} + \ddot{y}^{(s)} (\Delta t)^2$

t	y_s	$y_1 - y_s$	R_1 (Eq. 6.12d)	$y_2 - y_1$	R_2 (Eq. 6.12c)	$\ddot{y}_1$ (Eq. 6.12b)	y_1	$\ddot{y}_2$ (Eq. 6.12a)	y_2
0	0	0	0	0	0	0	0	0	0
0.025	+0.154	−0.151	−8.4	−0.003	−0.2	+27.9	+0.003*	+1.1	0*
0.050	+0.294	−0.270	−15.0	−0.023	−1.3	+46.6	+0.024	+7.3	+0.001
0.075	+0.405	−0.331	−18.4	−0.067	−3.9	+49.3	+0.074	+22.0	+0.007
0.100	+0.475	−0.320	−17.8	−0.128	−7.4	+35.4	+0.155	+41.8	+0.027
0.125	+0.500	−0.242	−13.5	−0.185	−10.7	+9.5	+0.258	+60.4	+0.073
0.150	+0.475	−0.108	−6.0	−0.210	−12.2	−21.1	+0.367	+69.0	+0.157
0.175	+0.405	+0.058	+3.2	−0.179	−10.4	−46.2	+0.463	+58.7	+0.284
0.200	+0.294	+0.236	+13.1	−0.082	−4.8	−60.9	+0.530	+27.2	+0.448
0.225	+0.154	+0.405	+22.5	+0.070	+4.1	−62.5	+0.559	−23.2	+0.629
0.250	0	+0.549	+29.5	+0.247	+14.3	−51.6	+0.549	−80.6	+0.796
0.275	−0.154	+0.661	+29.5	+0.406	+21.0	−28.9	+0.507	−118.7	+0.913
0.300	−0.294	+0.741	+29.5	+0.509	+21.0	−28.9	+0.447	−118.7	+0.956

*At $t = 0.025$, $y = \frac{1}{6}(\ddot{y}$ at $t = 0.025)(\Delta t)^2$ (established by trial).

The equations of motion in terms of absolute displacements, as derived from the condition of dynamic equilibrium shown in Fig. 6.5b, are

(a) $$M_2 \ddot{y}_2 + R_2 = 0$$
(b) $$M_1 \ddot{y}_1 + R_1 - R_2 = 0$$

where (6.12)

(c) $$R_2 = k_2(y_2 - y_1), \text{ or 21 max}$$
(d) $$R_1 = k_1(y_1 - y_s), \text{ or 29.5 max}$$

Since the system is nonlinear, numerical analysis provides the easiest method of solution, and the first few steps are shown in Table 6.1. Since the smaller natural period is 0.27 sec $(2\pi/\omega_2)$, the time interval is taken as 0.025 sec. Analyses such as this can be executed with the aid of a computer for buildings of many stories. For very tall buildings, it may be desirable to lump the masses at every second or third floor rather than at each floor in order to reduce machine time.

The result of the inelastic analysis is plotted in Fig. 6.6 in the form of story distortions. Also shown are the elastic responses computed in Sec. 6.3a. The important point to be noted is that the distortions of the inelastic structure are actually *smaller*, even though the maximum story resistances are only one-half of the resistances developed in the elastic case. This is true primarily because the plastic deformation tends to eliminate the resonant effect of the sinusoidal input. Even though earthquake motions are not harmonic, such resonant effects do occur, and the

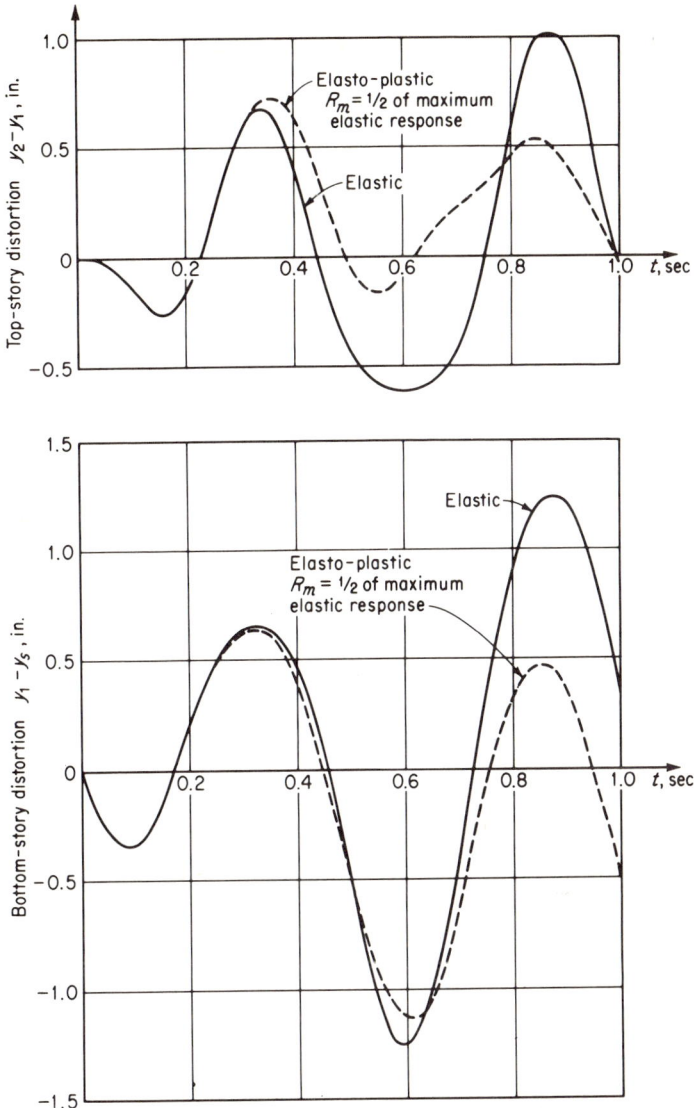

FIGURE 6.6 Example. Elastic and inelastic responses.

general conclusion reached here is still valid. Since the cost of the structural frame is related to the resistances provided, the economic advantage of permitting plastic deformation is apparent. This does, of course, result in some permanent distortion, but if kept within reasonable limits, does not imply serious damage.

256   Introduction to Structural Dynamics

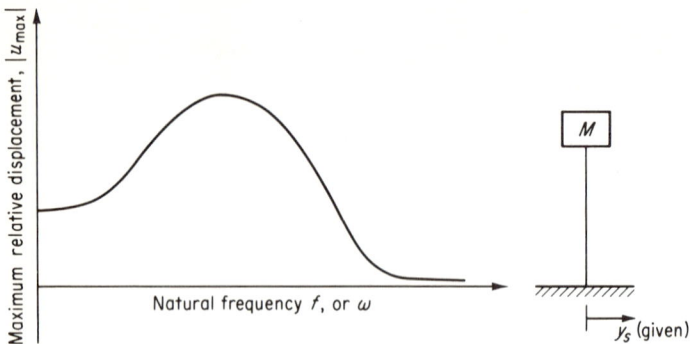

FIGURE 6.7 Typical response spectrum.

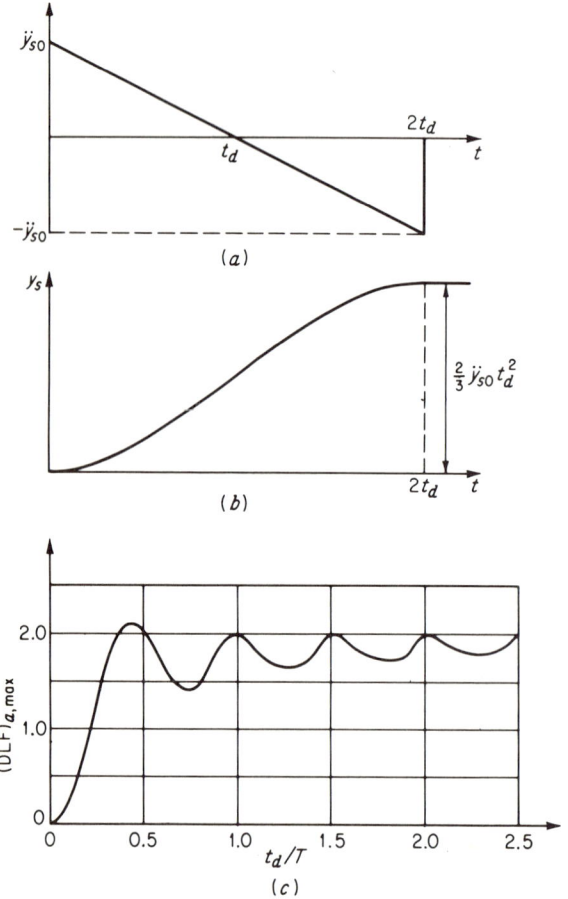

FIGURE 6.8 Response spectra for given support motion.

6.4 Response Spectra

Before proceeding to the actual earthquake problem, it will be useful to introduce the concept of a *response spectrum*. This is a plot giving the maximum responses (in terms of displacement or stress or acceleration, etc.) of all possible linear one-degree systems due to a given input, which in the present case is a ground motion. The abscissa of the spectrum is the natural frequency (or period) of the system, and the ordinate is the maximum response. Such a plot is shown in Fig. 6.7. Thus, in order to determine response for the particular input, we need know only the natural frequency of the responding system.[30,32]

To illustrate the construction of a response spectrum, consider the ground acceleration shown in Fig. 6.8a, which corresponds to the ground displacement shown in Fig. 6.8b. We wish to plot spectra for maximum relative displacement and maximum absolute acceleration of undamped

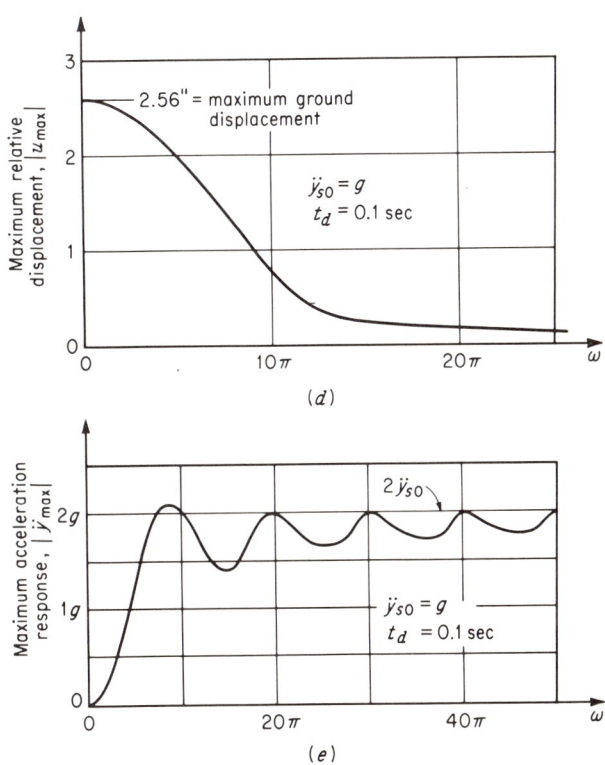

FIGURE 6.8 (*Continued*)

linear systems. According to Eqs. (6.2) and (6.3), the desired values are given by

(a) $$|u_{max}| = \frac{\ddot{y}_{so}}{\omega^2}(\text{DLF})_{a,max}$$
(b) $$|\ddot{y}_{max}| = \ddot{y}_{so}(\text{DLF})_{a,max}$$
(6.13)

The time function for the acceleration is

$$f_a(t) = 1 - \frac{t}{t_d} \qquad t \leqslant 2t_d$$
$$f_a(t) = 0 \qquad t > 2t_d$$

and the dynamic load factor is

$$(\text{DLF})_a = \omega \int_o^t \left(1 - \frac{\tau}{t_d}\right) \sin \omega(t - \tau)\, d\tau \qquad t \leqslant 2t_d$$

which, when integrated, gives

$$(\text{DLF})_a = 1 - \cos \omega t + \frac{\sin \omega t}{\omega t_d} - \frac{t}{t_d} \qquad t \leqslant 2t_d \qquad (6.14)$$

After time $2t_d$, we may express the DLF as

$$(\text{DLF})_a = (\text{DLF})_{t=2t_d} \cos \omega(t - 2t_d) + (\text{DLF})_{t=2t_d} \frac{\sin \omega(t - 2t_d)}{\omega} \qquad (6.15)$$

which is based upon consideration of the DLF and its derivative at $t = 2t_d$ as an initial displacement and velocity for the following motion. The maximum value of $(\text{DLF})_a$ which may be obtained from Eqs. (6.14) and (6.15) depends only on the ratio t_d/T. This is plotted in Fig. 6.8c.

We may now determine the maximum responses by Eqs. (6.13). To take a numerical example, let $\ddot{y}_{so} = g$ (the acceleration of gravity) and $t_d = 0.1$ sec. For these input parameters the spectrum for the absolute value of maximum relative displacement is as shown in Fig. 6.8d and that for maximum acceleration, in Fig. 6.8e. In connection with the use of Eq. (6.13a) for relative displacement, it should be noted that, as $\omega \to 0$, $(\text{DLF})_{a,max}$ also approaches zero and the value of u_{max} is indeterminate. However, it may be shown that, as $\omega \to 0$, u_{max} approaches the maximum ground motion. In fact, the ordinates to the spectra at the extremes of natural frequency are intuitively obvious. For example, if ω is very large, i.e., the structure is very stiff, the acceleration is applied suddenly with a corresponding DLF of 2, and hence the maximum acceleration of the mass is twice the initial ground acceleration. At the same time the spring distortion is negligible (since the stiffness is great), and hence the

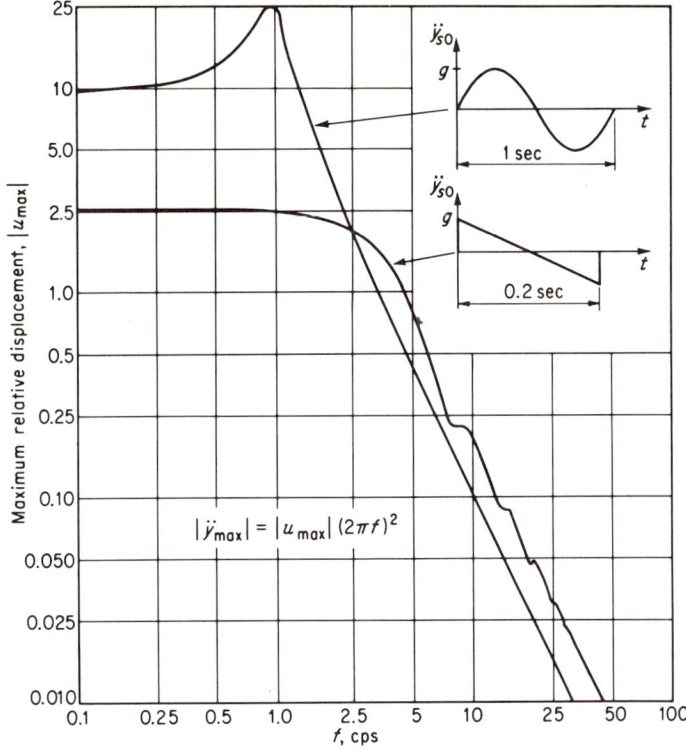

FIGURE 6.9 Log-log response spectra.

motion of the mass is the same as that of the ground; i.e., the relative motion is very small. On the other hand, if ω is very small, the spring is flexible, and hence the mass remains stationary while the ground moves beneath it. Thus the relative displacement equals the ground displacement, and the acceleration of the mass is zero.

An alternative representation of the response spectrum is the log-log plot in Fig. 6.9, where the result of the preceding example is repeated. This type of plot has the advantage that all spectra for the kinds of input considered herein consist essentially of two straight lines, except that there is some distortion near the intersection. This is true because, regardless of the details of the input, the displacement response in the small-frequency range is a constant equal to the maximum ground displacement, and, at large frequencies, the response acceleration is a constant equal to a multiple of the maximum ground acceleration. In the latter frequency range the maximum relative displacement is given by

$$|u_{\max}| = \alpha \frac{\ddot{y}_{so}}{\omega^2} = \alpha \frac{\ddot{y}_{so}}{(2\pi f)^2}$$

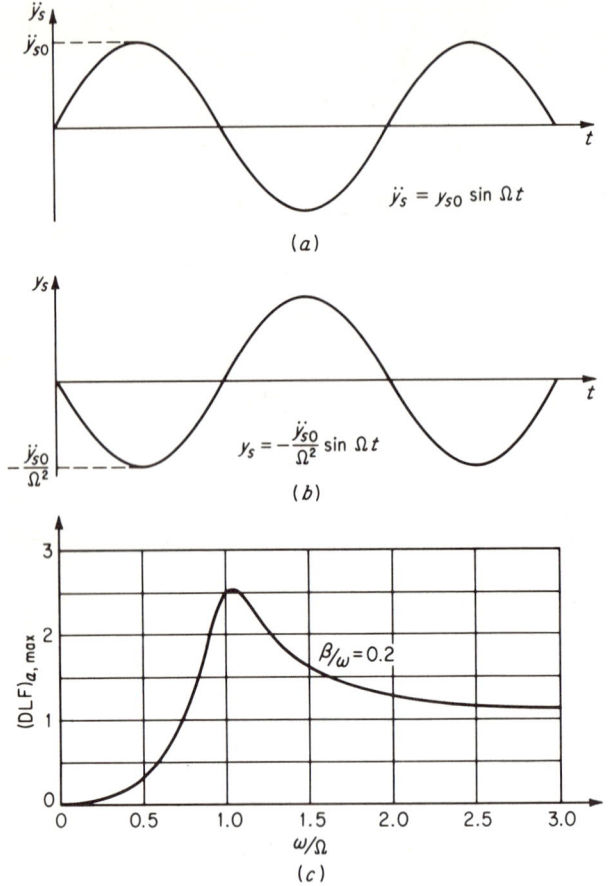

FIGURE 6.10 Response spectra for sinusoidal input.

where α is a constant which depends on the type of input ($\alpha = 2$ in the example of Fig. 6.8). The last equation is, of course, a straight line on the log-log plot. The fact that all spectra have the same general form on a log-log plot makes the estimation of a spectrum for a poorly defined input somewhat easier. If the spectrum gives relative displacement as in Fig. 6.8, the absolute acceleration of the response can be computed directly therefrom and a second plot is not really necessary.

As a second example we consider the spectrum for a sinusoidal variation of ground motion. The acceleration is shown in Fig. 6.10a, and the corresponding ground displacement is shown in Fig. 6.10b, with the assumption that there is an initial velocity of $-\ddot{y}_{so}/\Omega$. The maximum DLF for the damped steady-state response (which does not depend on the initial

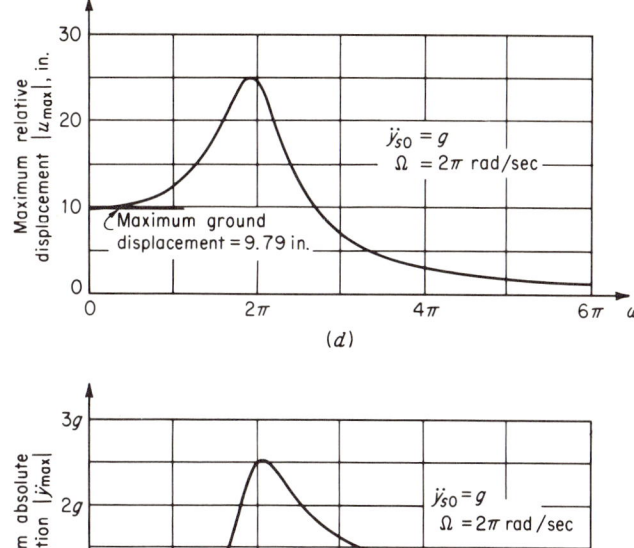

FIGURE 6.10 (*Continued*)

conditions) was developed for this input in Sec. 2.5b and is given by Eq. (2.41):

$$(\text{DLF})_{a,\max} = \frac{1}{\sqrt{(1 - \Omega^2/\omega^2)^2 + 4(\beta\Omega/\omega^2)^2}}$$

This is plotted in Fig. 6.10c for 20 percent of critical damping; that is, $\beta/\omega = 0.2$. The separate response spectra for relative displacement and absolute acceleration, computed by Eqs. (6.13), are plotted in Figs. 6.10d and e for the specific parameters $\ddot{y}_{so} = g$ and $\Omega = 2\pi$ rad/sec. Both curves display the expected peak near the point of resonance. At small frequency the maximum relative displacement equals the maximum ground displacement, and at large frequency u approaches zero as the maximum acceleration response approaches the maximum ground acceleration. In the latter case, the mass of the system simply "rides" along with the ground and the two motions are identical. The log-log plot of the spectrum for the sinusoidal input with 20 per cent damping is also shown in Fig. 6.9, where the two straight lines mentioned previously

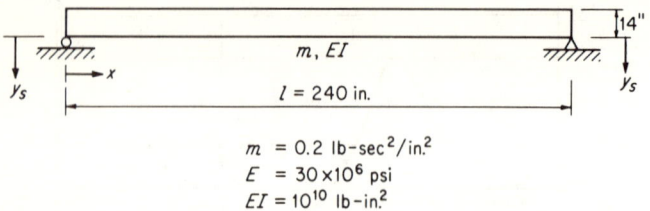

$m = 0.2 \text{ lb-sec}^2/\text{in}^2$
$E = 30 \times 10^6 \text{ psi}$
$EI = 10^{10} \text{ lb-in}^2$

FIGURE 6.11 Example. Beam with support motion.

correspond exactly to the maximum ground displacement and the maximum ground acceleration; that is, $\alpha = 1$.

To illustrate the application of the response-spectrum technique to an actual structural element, consider the simple beam shown in Fig. 6.11. Suppose that both supports move vertically in the manner indicated in Fig. 6.8a. We wish to determine the maximum bending stress in the beam resulting from this support motion. Only the fundamental mode will be considered since higher modes are of little importance.

In order to use the response spectrum, we first compute the natural frequency by Eq. (4.7).

$$f = \frac{\pi}{2l^2}\sqrt{\frac{EI}{m}} = \frac{\pi}{2(240)^2}\sqrt{\frac{10^{10}}{0.2}} = 6.1 \text{ cps}$$

Reading either Fig. 6.9 or 6.8d, we find the maximum relative displacement of the equivalent one-degree system to be

$$u^0_{\max} = 0.44 \text{ in.}$$

This is not the beam deflection we seek, since the motion of the beam support is not directly equivalent to the motion of the support of the equivalent system. For the present example involving distributed mass, it is apparent that Eq. (6.6), which gives the modal participation factor, should be written

$$\Gamma = \frac{\int_0^l m\phi(x)\,dx}{\int_0^l m[\phi(x)]^2\,dx}$$

Substituting $\phi(x) = \sin(\pi x/l)$ for the fundamental mode and evaluating the integrals,* we find

$$\Gamma = \frac{4}{\pi}$$

* Values of these integrals are given in Table 4.1 for beams with various support conditions and uniform mass.

Therefore, by Eq. (6.7), we obtain

$$A_{max} = \Gamma u^0_{max} = \frac{4}{\pi} \times 0.44 = 0.56 \text{ in.}$$

where A_{max} is the maximum displacement of the beam at midspan relative to the supports.

Since we are considering only the first mode, the bending moment is computed as follows:

$$\mathfrak{M} = -EI \frac{\partial^2 u}{\partial x^2}$$

where

$$u = A \sin \frac{\pi x}{l}$$

$$\frac{\partial^2 u}{\partial x^2} = -A \frac{\pi^2}{l^2} \sin \frac{\pi x}{l}$$

Therefore

$$\mathfrak{M}_{max} = \frac{EI\pi^2}{l^2} A_{max} \quad \text{at } x = \frac{l}{2}$$

The maximum bending stress is

$$\sigma_{max} = \mathfrak{M}_{max} \frac{c}{I} = \frac{Ec\pi^2}{l^2} A_{max}$$

where c is half the beam depth. Therefore

$$\sigma_{max} = \frac{(30 \times 10^6)(7)\pi^2}{(240)^2} \times 0.56 = 20{,}100 \text{ psi}$$

which is the desired result.

6.5 Earthquake Ground Motions

The motions of the ground during an earthquake are essentially random; i.e., the peaks of acceleration, both positive and negative, have various amplitudes and occur at various time intervals without any regular pattern. These occur in all directions, but our attention will be restricted to horizontal motions, which are the most damaging. Up to the present time (1964) very few records of actual strong-motion earthquakes have been obtained, and there is therefore little statistical basis for the prediction of future earthquakes. In any event, we shall never be able to predict the exact nature of the earthquake for which a given structure should be designed.

There are two possible solutions to the problem of defining the input. First, we could adopt a "standard" earthquake, with a certain amplitude of acceleration and time variation. This would be comparable with the approach commonly used with other types of loads (e.g., standard design truck loads), but would make even less sense in the case of earthquake, because not only the magnitude of the input, but also the time variation, has an important effect on structural response. A second and more promising approach is to treat the ground motion as a random variable. By this method the motion could be envisaged as the superposition of many sine waves of various frequencies, the amplitudes of which have certain probability distributions. Alternatively, the motion could be represented by a series of random impulses. If the intensity, or "power," of the random input is selected so as to be equivalent to that of actual measured earthquakes, the structural response should be similar. This type of analysis produces a probabilistic result; e.g., it would give the probabilities that the relative displacement of the structure would not exceed certain values. Such an approach has not been developed sufficiently for direct application, and further discussion is beyond the scope of this text.[33,34]

The most practical approach to the problem involves the use of response spectra such as discussed in Sec. 6.4. Although this method is perhaps an oversimplification and certainly approximate, it appears justified, in view of the limited data available on earthquake motions. Many investigators have computed spectra from actual earthquake records. This is accomplished by the direct application of Eq. (6.2). The DLF may be computed by numerical integration of the measured acceleration time function $f_a(t)$. The input is normally in the form of acceleration rather than displacement, because field instruments measure the former. Response spectra derived in this way are of course limited in usefulness since they apply only to the particular earthquake which happens to have been recorded. However, they are very useful in establishing the general nature of response to earthquake.

Based upon spectra computed for actual earthquakes, it is possible to estimate a proper spectrum for general purposes. It is here that the log-log plot for spectra (Fig. 6.9) is particularly useful. It was noted in Sec. 6.4 that the general forms of all spectra are similar, and this fact is helpful in estimating the earthquake response spectrum. Blume, Newmark, and Corning[29] have suggested that the spectrum can be approximated by three straight lines as follows: (1) a line of constant acceleration equal to twice the maximum ground acceleration, (2) a line of constant spectral velocity equal to 1.5 times the maximum ground velocity, and (3) a line of constant displacement equal to the maximum ground displacement. A spectrum constructed in this way is shown in Fig. 6.12 as curve a. It is based on the May 18, 1940, El Centro, Calif., earthquake (N-S component), for which the recorded maximum quantities were $0.33g$, 13.7 in./sec, and 8.3 in. It is intended to apply to elastic systems having between 5 and 10 percent of critical damping. For the El Centro input given above, the

Earthquake Analysis and Design

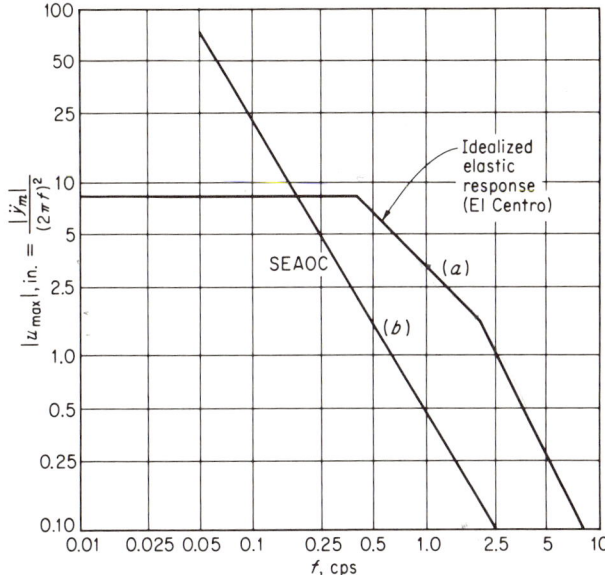

FIGURE 6.12 Idealized response spectrum for El Centro earthquake, May 18, 1940, N-S component and SEAOC recommendation.

three straight lines are defined by

(1) $\quad u_{\max} = (y_{so})_{\max} = 8.3 \text{ in.} \quad$ small f

(2) $\quad u_{\max} = \dfrac{1.5(\dot{y}_{so})_{\max}}{2\pi f} = \dfrac{3.3}{f} \text{ in.} \quad$ intermediate f

(3) $\quad u_{\max} = \dfrac{2(\ddot{y}_{so})_{\max}}{(2\pi f)^2} = \dfrac{6.6}{f^2} \text{ in.} \quad$ large f

The simplified spectrum thus defined and shown in Fig. 6.12 is based on a particular earthquake and may not be appropriate for general design purposes. Additional data which will become available in the future may indicate that the amplitudes of ground motion and the multiplying coefficients should be altered.

6.6 Earthquake Spectrum Analysis of Multidegree Systems

In Sec. 6.2 we developed a procedure of modal analysis for structures subjected to ground motion, and in Sec. 6.5 we discussed response spectra for earthquake. We now combine these two concepts in order to make an approximate elastic multidegree analysis for earthquake.

Since the response spectra give only maximum response, we shall obtain the maximum values for each mode, which must then be superimposed to give total response. The actual time variation of the design earthquake motion is unknown, and therefore it is impossible to compute the time variation of response either for a modal component or for the total. A conservative upper bound for the total response may be obtained by adding numerically the maximum modal components. However, this is excessively conservative, and it has been suggested that the "probable" value of the maximum response is approximately the square root of the sum of the squares (root mean square) of the modal maxima.[36] This is based on the assumption that the modal components are random variables, which is consistent with the random nature of the input. The accuracy of this approach increases with the number of degrees of freedom.

The elastic analysis demonstrated below is not intended to represent a proper method for earthquake design. A design on this basis would be too conservative and inconsistent with the observed behavior of structures during earthquake. The primary reason for this discrepancy is that most structures can undergo some plastic deformation without excessive damage. As demonstrated in Sec. 6.3, this capability results in a considerable decrease in the required strength and stiffness of the structure. Unfortunately, inelastic analysis of multidegree systems subjected to random earthquake motion cannot easily be accomplished. Considerable research is being conducted on the problem (1964), and in time satisfactory methods will probably be developed. In the meantime, the empirical approach as discussed in Sec. 6.7 is the most practical design procedure. The example of elastic analysis given below is presented to give the student a better insight into the general problem and to demonstrate a method which would be applicable to those special cases in which a truly elastic response is desired.

a. Building-frame Example

To illustrate modal spectrum analysis, we consider the three-story building frame shown in Fig. 6.13a. If this is considered to be a "shear building" (Sec. 3.8), it may be represented by the close-coupled system shown in Fig. 6.13b. Assuming the masses and stiffnesses shown, the natural frequencies and characteristic shapes have been computed, with the results tabulated in Fig. 6.13c (Sec. 3.4).

The next step is to compute the modal participation factors as given by Eq. (6.6). These computations are shown in Table 6.2.

It will be assumed that the input can be represented by the response spectrum shown in Fig. 6.12 (curve a). It will be recalled that this is an empirical representation of a particular earthquake and assumes a moderate amount of damping in the system. Its validity for general design purposes has not been proved. The responses of the equivalent

Earthquake Analysis and Design 267

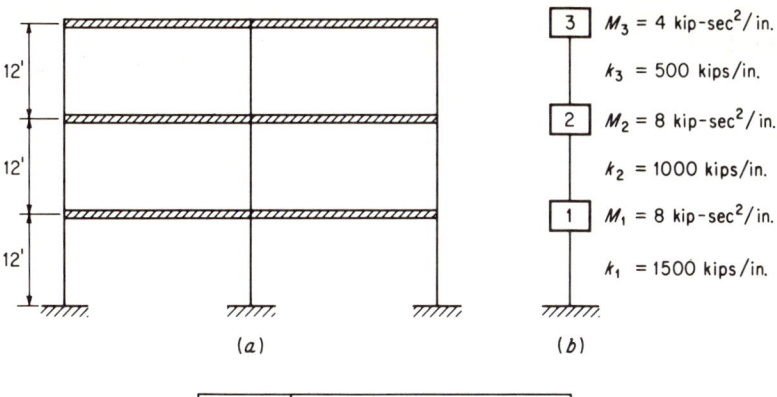

	Mode		
	1	2	3
f_n, cps	1.00	2.18	3.18
ϕ_{1n}	+0.314	−0.511	3.18
ϕ_{2n}	+0.686	−0.489	−2.18
ϕ_{3n}	1.000	1.000	1.00

(c)

FIGURE 6.13 Example. Three-story shear building.

Table 6.2 Modal Participation Factors (Structure Shown in Fig. 6.13)

Mass	M_r	First mode		Second mode		Third mode	
		$M_r\phi_{r1}$	$M_r\phi_{r1}^2$	$M_r\phi_{r2}$	$M_r\phi_{r2}^2$	$M_r\phi_{r3}$	$M_r\phi_{r3}^2$
1	8	2.51	0.79	−4.09	2.09	25.4	80.9
2	8	5.49	3.77	−3.91	1.91	−17.4	38.0
3	4	4.00	4.00	4.00	4.00	4.0	4.0
	Σ	+12.00	8.56	−4.00	8.00	12.0	122.9

$$\Gamma_n = \frac{\Sigma M_r \phi_{rn}}{\Sigma M_r \phi_{rn}^2} \quad \Gamma_1 = +1.40 \quad \Gamma_2 = -0.50 \quad \Gamma_3 = +0.098$$

one-degree systems in terms of relative displacement $u^0_{n,\max}$ are read directly from Fig. 6.12 and depend only on natural frequency. The modal responses $A_{n,\max}$ are simply those values multiplied by the corresponding participation factor [Eq. (6.7)].

Mode	f	$u^0_{n,\max}$	$A_{n,\max} = \Gamma_n u^0_{n,\max}$
1	1.00 cps	3.3 in.	4.6 in.
2	2.18	1.4	0.70
3	3.18	0.66	0.065

↳ Figure 6.12(a)

268　Introduction to Structural Dynamics

Having the maximum modal amplitudes, any other function such as displacement or acceleration at a point or a certain force or stress may be computed by Eq. (6.9). For example, the maximum modal components of first-floor deflection (relative to the ground) are obtained by multiplying the modal amplitude by the characteristic-shape factor for that floor:

Mode	ϕ_{1n}	$(u_{1n})_{max}$
1	0.314	1.44 in.
2	−0.511	0.36
3	3.18	0.21

$\Sigma = 2.01$

Of course, signs should not be attached to these modal deflections, since each could be in either direction. The upper bound for the maximum deflection of this floor is the numerical sum of the modal components, or 2.01 in. The "probable" maximum, or root mean square, is 1.50 in. The latter may not be appropriate in this case since there are only three degrees of freedom. In fact, the maximum may be expected to lie somewhere between the two values computed.

Of primary interest in earthquake design are the maximum values of the story shears. To obtain these, we first multiply the modal amplitudes $A_{n,max}$ by the relative story displacements corresponding to the modal shapes, ϕ_Δ. This provides the maximum story displacement Δ_n of the response, which, when multiplied by the spring constant, gives the story shear. The computations leading to these values are shown in Table 6.3.

Table 6.3　Maximum Story Shears by Spectrum Analysis; System in Fig. 6.13

Mode	First story		Second story		Third story	
	$\phi_{\Delta 1} = \phi_{1n}$	$\phi_{1n} = A_n \phi_{\Delta 1}$	$\phi_{\Delta 2} = \phi_{2n} - \phi_{1n}$	$\Delta_{2n} = A_n \phi_{\Delta 2}$	$\phi_{\Delta 3} = \phi_{3n} - \phi_{2n}$	$\Delta_{3n} = A_n \phi_{\Delta 3}$
1	0.314	1.44	0.372	1.71	0.314	1.44
2	0.511	0.36	0.022	0.02	1.489	1.04
3	3.18	0.21	5.36	0.35	3.18	0.21
Absolute max story displacement		2.01 in.		2.08 in.		2.69 in.
"Probable" max story displacement		1.50 in.		1.74 in.		1.79 in.
Absolute max story shear		3020 kips		2080 kips		1345 kips
"Probable" max story shear		2250 kips		1740 kips		895 kips

It may be observed that the first mode makes the major contribution in all cases, although in the top story the second-mode effect is also significant.

It may also be of interest to compute the maximum expected horizontal acceleration of a floor. In any mode the maximum absolute acceleration of a mass is simply the maximum displacement times the square of the natural frequency; that is, $(\ddot{y}_{rn})_{\max} = A_{n,\max} \phi_{rn} \omega_n^2$. For example, the acceleration of the top floor is computed as follows:

Mode	ω_n^2	$A_{n,\max}$	ϕ_{3n}	$(\ddot{y}_{3n})_{\max}$
1	39.2	4.6	+1.00	180 in./sec²
2	188	0.70	+1.00	132
3	398	0.065	+1.00	26

Third floor absolute maximum acceleration, 338 in./sec²
Third floor "probable" maximum acceleration, 225 in./sec²

As would be expected, the higher modes are relatively more important in the case of acceleration. The "probable" value computed above is about $0.6g$, which is very severe with regard to nonstructural damage to the building and its contents.

6.7 Practical Design for Earthquake

As discussed in the foregoing sections, we are not presently capable of applying rigorous methods of analysis to the design of actual structures to withstand earthquake. This is not meant to imply that we are unable to execute satisfactory designs. Although the detailed behavior of a given structure cannot be accurately predicted, we can ensure with reasonable confidence that it will survive. Survival requires that the structure be able to withstand a moderate earthquake, such as might occur several times during its life, with only slight damage. It should also be able to withstand the most severe earthquake without collapse. The latter requirement can be met if advantage is taken of the ability of most structures to absorb energy by inelastic response. It is apparent that, in earthquake-resistant design, the structure should be proportioned and detailed so as to ensure the ductility necessary for inelastic behavior.

Current practice in earthquake design is embodied in design codes, one of which is discussed below. Ideally, these are based on experience gained by the observation of structures which have undergone earthquake conditions, coupled with an understanding of the nature of dynamic response to support motion. Although special structures may justify more

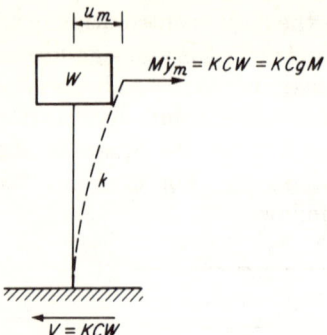

FIGURE 6.14 Seismic coefficients.

thorough investigation, the better codes provide a simple yet adequate method of design.

a. Design Codes

As an example of a commonly used code, we shall consider the Recommended Lateral Force Requirements (1959) of the Structural Engineers Association of California (SEAOC). Essentially, the procedure specified therein is based on only the first mode of the structure, which is justified by the belief that the higher modes are of secondary importance. By assuming a characteristic shape for the first mode, it is possible to convert the maximum condition of response into a set of equivalent static forces. The actual design may then be executed on the basis of static analysis.

The basic concept of the SEAOC recommendation is contained in the two formulas

$$V = KCW \tag{6.16}$$

$$C = \frac{0.05}{T^{1/3}} \tag{6.17}$$

where V = total dynamic base shear
W = total weight of building
T = natural period of first mode
K = coefficient, varying between 0.67 and 1.50

In the above, C, the seismic coefficient, is equivalent to the maximum acceleration expressed as a fraction of g, since, when multiplied by the weight, it gives the maximum horizontal inertia force (Fig. 6.14). The coefficient K is intended to reflect the ability of the structure to deform into the plastic range. For example, the smallest value (0.67) applies to moment-resisting frames which are relatively ductile, while higher values apply to less ductile arrangements such as those making use of concrete shear walls ($K = 1.33$).

The expression for C may be interpreted in the light of the discussion of Sec. 6.5. By reference to Fig. 6.14 and with $K = 1$, it is apparent that

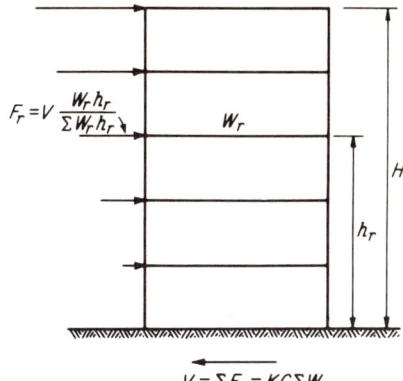

FIGURE 6.15 SEAOC recommendation for earthquake forces on buildings.

the spring distortion and the seismic coefficient are related by

$$ku_{max} = V = CgM$$

Therefore
$$u_{max} = Cg\frac{M}{k} = \frac{Cg}{\omega^2}$$

Substituting Eq. (6.17) for C, $1/f$ for T, and $2\pi f$ for ω, we obtain

$$u_{max} = \frac{0.49}{f^{5/3}} \text{ in.} \tag{6.18}$$

where f is in cycles per second. This equation represents a response spectrum as plotted in Fig. 6.12 (curve b), where it appears as a straight line. When compared with the idealized elastic spectrum previously used, Eq. (6.18) indicates a less severe response over most of the frequency range. This is to be expected, since, as discussed previously, the elastic spectrum is unduly conservative. The important point being demonstrated is that Eq. (6.17), and hence (6.18), takes into account the effect of the natural period in a rational manner.

The response spectrum represented by Eq. (6.18) could of course not be used for elastic modal analysis as in Sec. 6.6 because it implies inelastic response. However, the question should not arise, since the code considers only one mode of response. In the mode considered, the maximum inertia force on mass r of a lumped-mass system is given by

$$F_r = M_r \ddot{A}_m \phi_r$$

where $\ddot{A}_m$ is the maximum modal acceleration, and ϕ_r is the coordinate of the characteristic shape at mass r. Furthermore, for dynamic equilibrium, the sum of all inertia forces on a building must equal the base shear. Thus

$$\ddot{A}_m \sum^r M_r \phi_r = V$$

Eliminating $\ddot{A}_m$ from the last two equations, we obtain, for the inertia force on mass r,

$$F_r = \frac{M_r \phi_r}{\sum^r M_r \phi_r} V$$

The SEAOC recommendation implies that $\phi_r = h_r/H$, where h_r is the height aboveground of the rth mass and H is the total height of the structure. By this assumption

$$F_r = \frac{M_r(h_r/H)}{\sum^r M_r(h_r/H)} V$$

or
$$F_r = \frac{W_r h_r}{\sum^r W_r h_r} V \qquad (6.19)$$

which is the SEAOC recommendation for the distribution of lateral force. The resulting set of forces in dynamic equilibrium is shown in Fig. 6.15. For design purposes the building may be analyzed as though these forces were applied statically.

As indicated above, Eq. (6.19) is based on the assumption that the characteristic shape of the fundamental mode is a straight line from the foundation to the top of the building. This is of course an approximation, but is reasonable for typical buildings. The justification for the assumption may be understood if it is recognized that the total distortion of a typical building is the sum of two effects: (1) the shear distortion in the stories of the frame, and (2) the change in length of the columns due to overall bending of the building. The former tends to produce a deflected shape which is concave to the left, and the latter a shape concave to the right. The combined effect results in a shape which approaches a straight line.

To illustrate application of the SEAOC code, we consider again the three-story frame shown in Fig. 6.13, which was analyzed in Sec. 6.6, using the elastic response spectrum. The weights of the three floors are $W_1 = W_2 = 3090$ kips and $W_3 = 1545$ kips, and the total weight is 7725 kips. The fundamental mode has a natural frequency of 1.00 cps, or a period of 1.00 sec. Using Eq. (6.17), we find $C = 0.05$, and if K is taken to be 0.67 (moment-resisting frame), the total base shear is

$$V = KCW = 0.67 \times 0.05 \times 7725 = 259 \text{ kips}$$

According to Eq. (6.19), the floor forces are

$$F_1 = \frac{3090 \times 12}{3090 \times 12 + 3090 \times 24 + 1545 \times 36} \times 259 = \frac{9.6 \times 10^6}{166{,}700} = 58 \text{ kips}$$

$$F_2 = \frac{3090 \times 24}{166{,}700} \times 259 = 115 \text{ kips}$$

$$F_3 = \frac{1545 \times 36}{166{,}700} \times 259 = 86 \text{ kips}$$

Thus the design values of the story shears are:

First story, 259 kips
Second story, 201 kips
Third story, 86 kips

Comparison of these values with those tabulated in Table 6.3 reveals that they are approximately in the same proportion but that the code values are roughly one-tenth of those given by the elastic analysis. The fact that the relation between the three story shears is similar in the two cases indicates that the SEAOC recommendations are a reasonable representation of the dynamic response. The fact that the code values are very much smaller should not be alarming because the elastic analysis is known to be excessively conservative. The response spectrum on which the latter analysis was based (Fig. 6.12, curve a) is an approximation for a rather strong earthquake, and a structure should be expected to undergo considerable plastic distortion under that condition. Furthermore, the two sets of shears are not exactly comparable because the code values would be used with allowable design stresses rather than yield values.

Problems

6.1 The following data are given for an undamped one-degree system with sinusoidal support motion (Fig. 6.16): $k = 100$ kips/in., $M = 0.50$ kip-sec^2/in., and $\ddot{y}_{so} = 125$ in/sec^2. Determine the maximum relative motion of the mass with respect to the

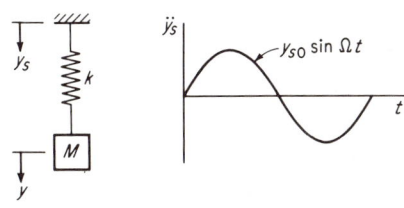

FIGURE 6.16 Problem 6.1.

support and the maximum absolute acceleration of the mass for the following two cases: a. $\Omega = 20$ rad/sec; b. $\Omega = \omega$, the natural circular frequency. Note that the support acceleration continues for only one cycle (Sec. 2.5).

Answer
a. $u_{max} = 0.88$ in.
   $\ddot{y}_{max} = 177$ in./sec^2
b. $u_{max} = 1.96$ in.
   $\ddot{y}_{max} = 392$ in./sec^2

6.2 A cantilever beam is described by $EI = 5 \times 10^{10}$ lb-in.^{2}, $m = 0.1$ lb-sec^2/in.^{2}, and $l = 200$ in. If the support motion (transverse to the beam) is the same as in Prob. 6.1, what is the maximum relative motion of the end of the beam when $\Omega = 50$ rad/sec? Consider only the first beam mode and assume the shape of that mode to be the same as the static dead-weight deflected shape.

Answer
$u_{max} = 0.17$ in.

6.3 The support of the one-degree system in Prob. 6.1 moves as indicated in Fig. 6.17. What are the maximum relative displacement and the maximum absolute acceleration?

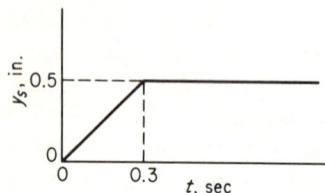

FIGURE 6.17 Problem 6.3.

6.4 The support of the two-degree system shown in Fig. 6.18 moves as indicated. Plot the relative deflection of M_2 up to $t = 0.2$ sec. The natural frequencies of this system were determined in Prob. 3.1.

6.5 Referring to Prob. 6.4, compute the maximum values of the modal components of the absolute deflection of M_2.

Answer
$A_{1,max} = 0.196$ ft
$A_{2,max} = 0.044$ ft

6.6 If the strengths of the springs in Prob. 6.4 are $R_{m1} = 80$ lb and $R_{m2} = 40$ lb, the response will reach the plastic range. Using a numerical analysis, compute the permanent distortion caused in each spring.

6.7 Plot an elastic response spectrum as in Fig. 6.9 for the ground motion shown in Fig. 6.18.

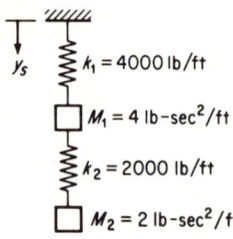 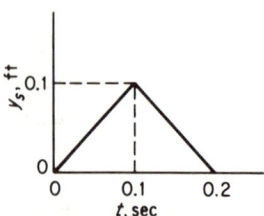

FIGURE 6.18 Problem 6.4.

6.8 Using the idealized earthquake response spectrum of Fig. 6.12, determine the maximum modal amplitudes for the two-degree system of Prob. 6.4. Assuming that the modes may be added numerically, what would be the maximum relative displacement of M_2? The maximum force in spring 1?

Answer

$A_{1,\text{max}} = 0.35$ in.
$A_{2,\text{max}} = 0.042$ in.
$u_{2,\text{max}} = 0.74$ in.
$R_{1,\text{max}} = 130$ lb

6.9 Repeat Prob. 6.8 for the three-story building frame of Prob. 3.5.

6.10 Determine the design story shears according to the SEAOC recommendation for the building frame of Prob. 3.5.

7
Blast-resistant Design

7.1 Introduction

Since the end of World War II a great deal of research has been conducted on the response of structures to the effects of nuclear weapons. This effort has not only resulted in the development of techniques by which structures may be designed to resist nuclear attack, but has also contributed very appreciably to the field of structural dynamics in general.

The methods of dynamic analysis which will be used in this chapter have been presented previously. The problem discussed herein differs from earlier examples, primarily in the nature of the loading. Unfortunately, the loading effects of nuclear explosions cannot be precisely specified. The data commonly used and presented below are empirical and based on a mixture of theoretical results for ideal conditions and actual field observations. The methods of analysis used in this chapter are generally approximate with regard to both loading and structural response. An approximate approach has been adopted, not only to simplify this introductory presentation, but also because precise methods are probably not justified in view of the uncertainties in the loading. Where approximations have been made, they are generally conservative from a design viewpoint; i.e., the loading may be overestimated, and the structural resistance underestimated. Thus the techniques presented are appropriate for defensive or design purposes, but may not be proper for offensive purposes, e.g., military target-analysis.

7.2 Loading Effects of Nuclear Explosions[37]

Data are given below for the characteristics of the nuclear-blast pressures, which provide the basis for the computation of forces on structural configurations. The data are restricted to the effects occurring at or near the ground surface and resulting from a surface burst of the weapon. Blast phenomena for bursts appreciably above or below ground surface are somewhat more complex.

When an explosion occurs, a circular shock front is propagated away from the point of burst. At any instant of time the distribution of overpressure (the excess above atmospheric pressure) along a radial line is as shown in Fig. 7.1a. The shock front travels with a velocity U and has a peak pressure p_{so} which decays behind the front as indicated. When the shock front strikes an object such as a building, there is a "diffraction" effect producing forces which result from the higher pressures due to reflection of the wave on the front face of the object and also from the time lag before the overpressure acts on the rear face. At the same time the air behind the shock front is moving outward at high velocity, and this "wind" produces drag forces on any objects encountered. Thus the total loading consists of three parts: (1) the initial diffraction effect, (2) the effects of the general overpressure p_s, and (3) the drag loading. At a fixed point on the ground the variation of overpressure and dynamic pres-

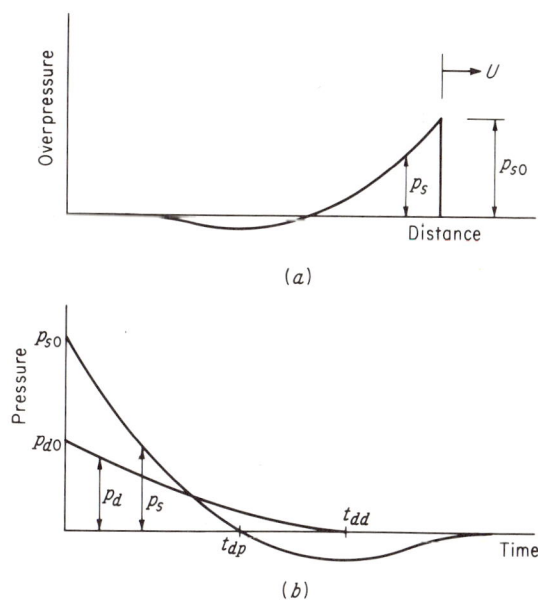

FIGURE 7.1 Pressure-pulse shapes.

278   Introduction to Structural Dynamics

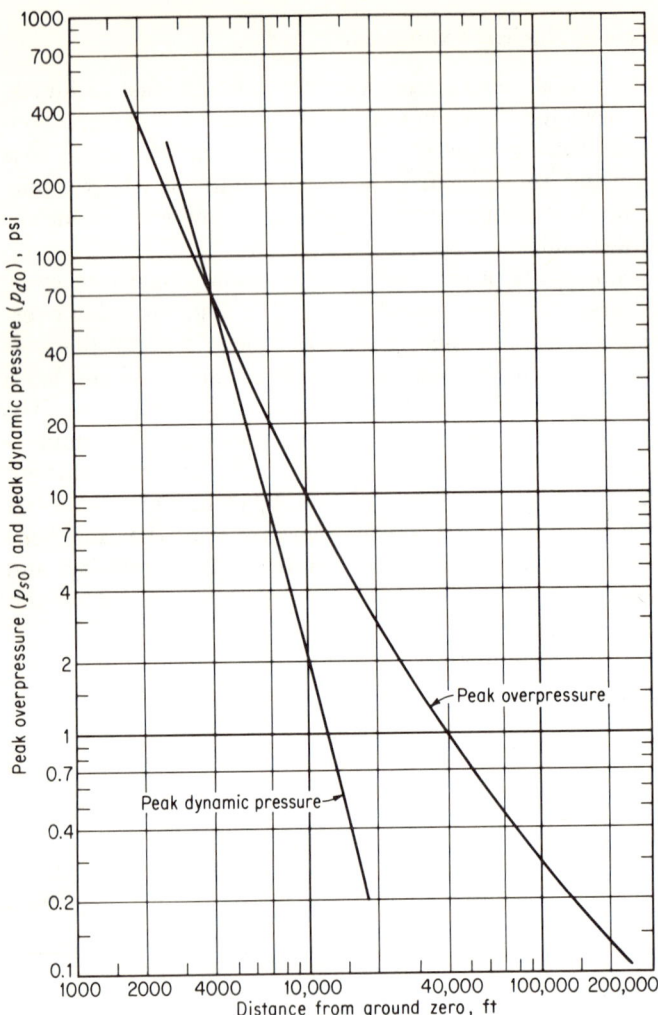

FIGURE 7.2 Overpressure and dynamic pressure versus range. 1-MT weapon. (*U.S. Department of Defense and Atomic Energy Commission.*[37])

sure with time is as indicated in Fig. 7.1b. The dynamic pressure p_d is merely $\frac{1}{2}\rho v^2$, where ρ is the air density and v is the velocity of the air particles. The drag pressure on an object in the path of the wind is the dynamic pressure times the appropriate drag coefficient C_d. The negative overpressure phase, or suction, indicated in Fig. 7.1, is relatively unimportant and may normally be ignored for structural-design purposes. Structures belowground are subjected to the effects of the overpressure

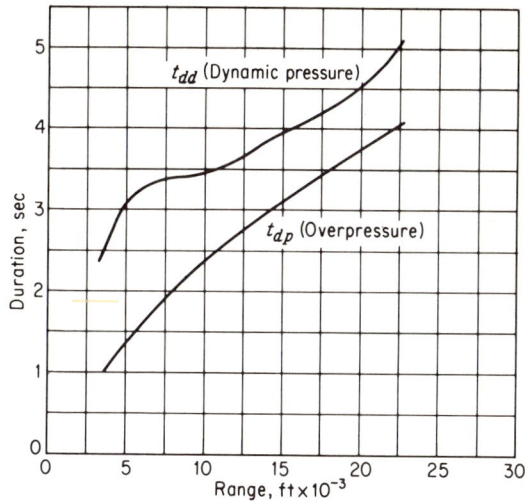

FIGURE 7.3 Overpressure and dynamic-pressure positive-phase durations versus range. 1-MT weapon. (*U.S. Department of Defense and Atomic Energy Commission.*[37])

and to ground-transmitted shock, but obviously not to the diffraction and drag effects.

The variations of the peak values of overpressure and dynamic pressure with distance (or range) from the point of burst (ground zero) are given in Fig. 7.2 for a 1-MT (megaton) weapon. For other values of yield (i.e., weapon size), the range for a given pressure may be determined by the following scaling law:

$$\frac{\mathcal{R}_1}{\mathcal{R}_2} = \left(\frac{Y_1}{Y_2}\right)^{1/3} \quad (7.1)$$

where $\mathcal{R}_1$ is the distance at which the pressure occurs with a yield of Y_1, and $\mathcal{R}_2$ the distance for a yield of Y_2. Thus the ranges for a yield of 0.001 MT (1 kiloton) are exactly one-tenth of those given in Fig. 7.2 for the same pressures.

The durations of the positive phases of overpressure and dynamic pressure (Fig. 7.1b) versus range are given in Fig. 7.3 for a 1-MT yield. These values may be scaled to other weapon yields by the following relationship:

$$\frac{t_{d1}}{t_{d2}} = \left(\frac{Y_1}{Y_2}\right)^{1/3} \quad (7.2)$$

where t_{d1} and t_{d2} are the durations for the same overpressure (or dynamic pressure) but different yields. Thus, to obtain the duration for a yield and range of Y_1 and $\mathcal{R}_1$, we should first compute $\mathcal{R}_2$ for $Y_2 = 1$ MT by Eq. (7.1), read t_{d2} on Fig. 7.3, and then compute t_{d1} by Eq. (7.2).

The velocity of the shock front (Fig. 7.1) depends only on the peak overpressure, and is given by

$$U = U_s \left(1 + \frac{6p_{so}}{7p_o}\right)^{\frac{1}{2}} \tag{7.3a}$$

where U_s is the velocity of sound, and p_o is atmospheric pressure. Under normal atmospheric conditions at sea level, this becomes

$$U = 1120 \left(1 + \frac{6p_{so}}{103}\right)^{\frac{1}{2}} \quad \text{fps} \tag{7.3b}$$

The rate of decay with time of the pressure at a point on the ground depends upon the intensity of the peak pressure and the positive-phase duration. This may be represented by the normalized curves of Fig. 7.4, where t is the time after arrival of the shock front. Note that the dynamic pressure decays more rapidly than does the overpressure.

When a shock front strikes a solid surface placed normal to the direction of shock travel, there is an instantaneous increase in pressure above that of the shock front itself. This is in part due to the formation of a reflected wave, which has the effect of doubling the overpressure, and, in addition, to the sudden onslaught of dynamic pressure. The total pressure, which is normally referred to as the reflected pressure, is given by

$$p_r = 2p_{so}\left(\frac{7p_o + 4p_{so}}{7p_o + p_{so}}\right)$$

or

$$p_r = 2p_{so}\left(\frac{103 + 4p_{so}}{103 + p_{so}}\right) \quad \text{psi} \tag{7.4}$$

the latter being applicable at sea level under normal atmospheric conditions. If the surface is inclined, i.e., the angle between the shock front and the surface is not zero, the reflected pressure is decreased. However, the decrease is not appreciable unless the angle mentioned is greater than about 35°, and Eq. (7.4) may be used for all smaller angles.

The reflection effect may be assumed to diminish linearly and to disappear at the clearing time t_c, which is approximately

$$t_c = \frac{3S_c}{U} \tag{7.5}$$

where U is given by Eq. (7.3), and S_c is either the height of the reflecting surface aboveground or one-half the width, whichever is smaller. Thus, for the rectangular building in Fig. 7.5a, S_c would be the smaller of H or $B/2$. After time t_c, the pressure on the surface is the overpressure plus

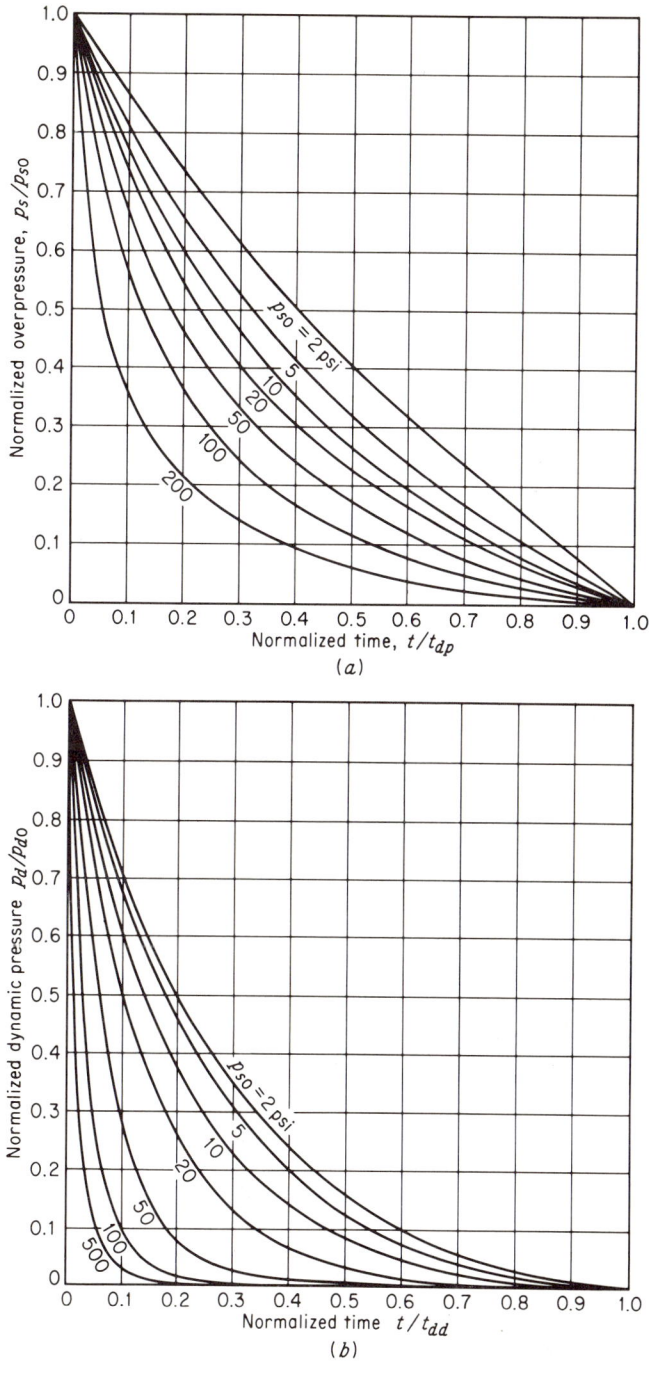

FIGURE 7.4 Overpressure and dynamic-pressure decay curves. (*U.S. Department of Defense and Atomic Energy Commission.*[37])

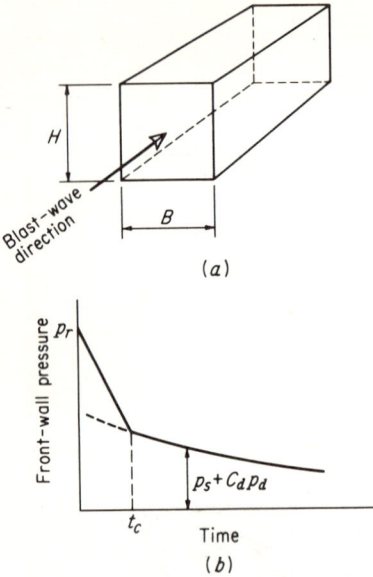

FIGURE 7.5 Pressure pulse on front face of rectangular building.

the drag pressure, both of which decay as described previously. Thus the complete pressure-time variation for a surface normal to the direction of shock propagation is as shown in Fig. 7.5b.

The sides of the rectangular building shown in Fig. 7.5a are subjected to the overpressure plus the drag pressure, which in this case would be negative. The rear face is subject to the same combination of loading, except that a certain time is required for the pressure to build up to the steady-state condition. This time (after the shock front reaches the rear face) may be approximated by $4S_c/U$, where S_c and U are as defined in connection with Eq. (7.5).

The total horizontal force on the building is merely the algebraic sum of the front- and rear-wall forces. It is apparent that the presence of openings in the walls would complicate the loading appreciably. However, if the function of the building is to protect the contents from blast effects, it would normally be windowless.

7.3 Aboveground Rectangular Structures

To illustrate the principles of blast-resistant design, we now consider in some detail the design of the one-story, windowless building shown in Fig. 7.6. The structure consists of a series of steel rigid frames supporting an outer shell of reinforced concrete slabs. It is assumed that the building is sufficiently long so that each interior frame may be analyzed independently of the rest of the structure.

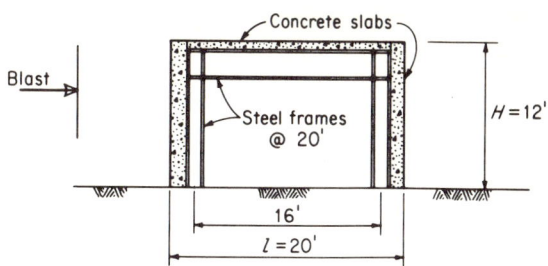

FIGURE 7.6 Example. Rectangular building.

Although the structure under consideration is about as simple as one could imagine, it is in reality a complex dynamic system. It consists of a group of elements, each with distributed mass, and hence infinite degrees of freedom, all interacting with one another in a complicated manner. Clearly, a rigorous solution is impractical. We shall apply the approximate methods developed in Chap. 5, considering each element of the total structure to be an independent one-degree system. This approach will be justified as the analysis develops. More precise methods than those used below, even if possible, are probably not worthwhile, in view of the inherent uncertainties in the blast loading.

The building is to be designed for a peak overpressure p_{so} of 20 psi and a weapon yield Y of 0.5 MT. With the exception of the roof girder, all elements will be permitted to undergo a plastic deformation corresponding to a ductility ratio μ of 5. This implies a moderate degree of damage, and the building could thereafter be restored to usefulness by relatively minor repairs. The girder will be designed for elastic response in order to ensure the integrity of the frame (Sec. 7.3c). The dynamic material strengths are given as follows:

$$\text{Concrete compressive strength } \sigma'_{dc} = 5200 \text{ psi}$$
$$\text{Reinforcement and structural steel yield strength } \sigma_{dy} = 60,000 \text{ psi}$$

The blast wave is considered to be traveling perpendicularly to the long axis of the building, since this is the most severe condition for the frame.

a. Loading

Using Fig. 7.2, it is found that the range for a 1-MT weapon and 20 psi overpressure is 7100 ft. By Eq. (7.1), the range for the weapon under consideration is

$$\mathcal{R}(0.5 \text{ MT}) = 7100(0.5/1)^{1/3} = 5650 \text{ ft}$$

Also from Fig. 7.2, the peak dynamic pressure is

$$p_{do} = 8.1 \text{ psi}$$

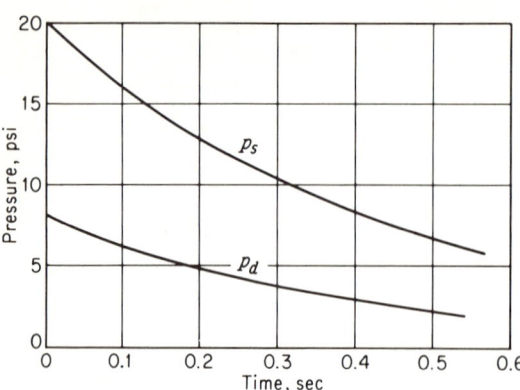

FIGURE 7.7 Pressure-time curves. $W = 0.5$ MT, $p_{so} = 20$ psi.

The durations of the overpressure and dynamic pressure for a 1-MT weapon are provided by Fig. 7.3; $t_{dp} = 1.8$ sec, and $t_{dd} = 3.4$ sec. For the actual weapon yield, Eq. (7.2) gives

$$t_{dp}(0.5 \text{ MT}) = 1.8(0.5/1)^{1/3} = 1.4 \text{ sec}$$
$$t_{dd}(0.5 \text{ MT}) = 3.4(0.5/1)^{1/3} = 2.7 \text{ sec}$$

The foregoing information and the normalized decay curves of Fig. 7.4 (for $p_{so} = 20$ psi) permit us to plot the actual pressure-time relationships as shown in Fig. 7.7.

The velocity of the shock front is given by Eq. (7.3b):

$$U = 1120\left(1 + \frac{6 \times 20}{103}\right)^{1/2} = 1650 \text{ fps}$$

The reflected pressure is given by Eq. (7.4):

$$p_r = 2 \times 20\left(\frac{103 + 4 \times 20}{103 + 20}\right) = 59.5 \text{ psi}$$

The clearing time for this pressure, according to Eq. (7.5), is

$$t_c = \frac{3 \times 12}{1650} = 0.022 \text{ sec}$$

The foregoing results represent the basic loading data, and we now proceed to consider the individual elements.

b. Roof Slab

The roof will be designed as a two-way slab with sides of 20 and 16 ft. The latter dimension is based on the assumption that support is provided by a steel spandrel spanning between columns. It might be economical to support the roof slab with the wall, in which case the transverse span would be somewhat larger. It will be assumed that all four edges of the two-way slab are fully restrained. For the edges along the roof girders, this assumption is justified by the fact that adjacent monolithic roof panels are loaded simultaneously. Along the wall edges the condition of full restraint is approached if the two slabs are monolithic, since the wall will be considerably thicker than the roof slab.

The total roof pressure equals overpressure plus drag. The drag coefficient for the roof of this configuration is approximately -0.4. Therefore the initial peak pressure is

$$p_i = p_{so} - 0.4 p_{do} = 20 - 0.4 \times 8.1 = 16.8 \text{ psi}^*$$

By combining the two decay curves of Fig. 7.7, the time variation of total roof pressure is constructed as shown in Fig. 7.8. The rise time of the loading equals the transit time of the shock front across the slab span, or

$$t_r = \frac{l}{U} = \frac{16}{1650} = 0.01 \text{ sec}$$

which is so small that it may be ignored. Actually, the pressure across the roof slab is not uniformly distributed, but may be assumed to be so because of the very short transit time.

For design purposes the load-time curve will be assumed triangular, and as a first approximation, the initial slope of the actual pressure variation is used. As indicated in Fig. 7.8, this leads to an effective duration t_{de} of 0.48 sec. Furthermore, for a first-trial computation of required slab strength, this rather long duration will be assumed infinite. According to Eq. (5.14), the required strength based on this assumption is

$$\text{Required } R_m = F_1 \left(\frac{1}{1 - 1/2\mu} \right) = 16.8 \left(\frac{1}{1 - 1/2 \times 5} \right) = 18.7 \text{ psi}$$

which corresponds to a total slab resistance of

$$\text{Required } R_m = 18.7 \times 16 \times 20 \times 144 = 860{,}000 \text{ lb}$$

According to Table 5.5, the maximum resistance for an aspect ratio a/b of $16/20 = 0.8$ is

$$R_m = \frac{1}{a} [12(\mathfrak{M}_{Pfa} + \mathfrak{M}_{Psa}) + 10.3(\mathfrak{M}_{Pfb} + \mathfrak{M}_{Psb})]$$

---

* The roof pressure would be slightly greater if the blast were traveling in the long direction of the building since the suction due to drag would be smaller. However, to simplify the example, this condition is not considered here.

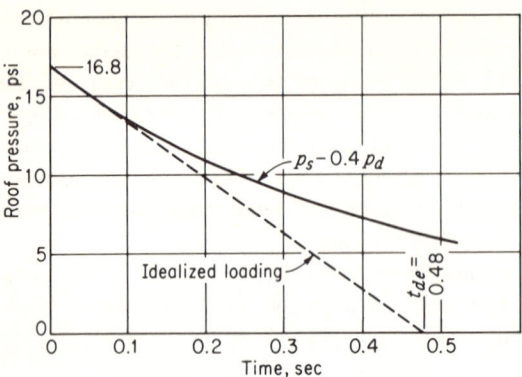

FIGURE 7.8 Example. Roof-slab loading.

which, if we make the bending resistance equal at all points and in both directions, becomes

$$R_m = \tfrac{1}{16}[12(2\mathfrak{M}_P \times 16) + 10.3(2\mathfrak{M}_P \times 20)] = 49.8\mathfrak{M}_P$$

where $\mathfrak{M}_P$ is the bending resistance per unit of width. Therefore

$$\text{Required } \mathfrak{M}_P = \frac{\text{Required } R_m}{49.8} = \frac{860{,}000}{49.8} = 17{,}300 \text{ lb-ft/ft}$$

For bending strength we use the expression

$$\mathfrak{M}_P = \rho_s b d^2 \sigma_{dy}\left(1 - \frac{\rho_s \sigma_{dy}}{1.7\sigma'_{dc}}\right)$$

which becomes

$$\mathfrak{M}_P = 560 d^2 \quad \text{lb-ft/ft}$$

if we arbitrarily let the steel ratio $\rho_s = 0.01$ and insert the material properties given. Thus the required effective thickness is obtained by

$$d = \left(\frac{17{,}300}{560}\right)^{1/2} = 5.6 \text{ in.}$$

which corresponds to a total slab thickness of about 7 in.

We must now refine the design by computing the response more exactly. In Sec. 5.6c it was determined that, for a two-way slab having an aspect ratio of 0.8, the effective stiffness is

$$k_E = \frac{430 E I_a}{a^2}$$

Using the approximate expression for moment of inertia (Sec. 5.6a),

$$I_a = \frac{bd^3}{2}(5.5\rho_s + 0.083) = \frac{(1)(5.6)^3}{2}(5.5 \times 0.01 + 0.083)$$
$$= 12.1 \text{ in.}^4/\text{in.}$$

we find the stiffness to be

$$k_E = \frac{430(4 \times 10^6)12.1}{(16 \times 12)^2} = 5.65 \times 10^5 \text{ lb/in.}$$

The total mass of the slab is

$$M_t = \frac{150 \times \frac{7}{12} \times 16 \times 20}{386} = 72.5 \text{ lb-sec}^2/\text{in.}$$

The load-mass factor K_{LM} for the slab is 0.54 in the plastic range and 0.69 in the elastic range (Table 5.5). It is estimated that a proper value for this case ($\mu = 5$) is 0.57. Therefore

$$T = 2\pi\sqrt{\frac{K_{LM}M_t}{k_E}} = 2\pi\sqrt{\frac{0.57 \times 72.5}{5.65 \times 10^5}} = 0.054 \text{ sec}$$

We enter Fig. 2.24, the response chart for triangular pulses, with the values

$$\frac{t_{de}}{T} = \frac{0.48}{0.054} = 8.9 \qquad \frac{R_m}{F_1} = \frac{18.7}{16.8} = 1.11$$

and read

$$\mu \approx 4.0 \qquad \frac{t_m}{t_d} = 0.12 \qquad t_m = 0.12 \times 0.48 = 0.058 \text{ sec}$$

The computed μ value is sufficiently close to that desired ($\mu = 5$) since only a very slight decrease in R_m could be permitted. Furthermore, at the time of maximum response, the idealized load (Fig. 7.8) has not departed significantly from the actual pressure-time curve. We therefore conclude that the triangle selected to represent the load is sufficiently accurate and that the 7-in. slab is satisfactory.

c. Roof Girder

The design of the roof girder is complicated by the fact that this element must perform two functions: (1) it must support the roof slab; and (2) it must perform its part of the frame action resisting the horizontal forces. If the girder is forced into the plastic range by the vertical load, its ability to restrain the columns may be impaired. It is therefore advisable to design the girder so as to remain elastic. Furthermore, since the horizontal response causes plastic hinges in the columns at the girder connections, it is doubtful that the columns can provide effective moment restraint for the girder acting under vertical load. The girder will therefore be designed as though simply supported. This approach, i.e., elastic design assuming simple supports, is conservative but prudent. A more accurate method of design would have to consider horizontal and vertical responses simultaneously, which is not only cumbersome, but also unreliable.

The vertical load on the girder is the sum of the dynamic reactions from the adjacent roof-slab panels. Using the expression given for the short-edge reaction in Table 5.5 and neglecting the decay in applied load, we compute the slab reaction as

$$V_A = 0.07F + 0.13R_m$$
$$= 0.07(770{,}000) + 0.13(860{,}000)$$
$$= 166{,}000 \text{ lb}$$

for one panel, or twice this amount for the total girder load. This value is attained when the slab reaches its elastic limit. If we assume a suddenly applied constant load (Sec. 2.2c), the time at which the maximum girder load is attained may be computed by

$$y_{el} = \frac{R_m}{k_E} = \frac{F_1}{k_E}(1 - \cos \omega t_{el})$$

where all quantities are for the slab, and t_{el} is the time desired. Using the values computed for the slab in Sec. 7.3b,

$$\frac{R_m}{F_1} = 1 - \cos \omega t_{el}$$

or
$$1.11 = 1 - \cos 121 t_{el}$$
from which
$$t_{el} = 0.014 \text{ sec}$$

Thus we may assume the total girder load to vary with time as shown in Fig. 7.9a, where the rise is approximated by a straight line and the decay in overpressure is ignored. The load decreases suddenly at $t = 0.058$ sec, the time of maximum slab response. It may be assumed that the spanwise distribution of this load is triangular, as indicated in Fig. 7.9b, since this is the type of edge reaction developed along the short edge of a two-way slab in the plastic range.

The elastic (DLF)$_{max}$ for the load function shown in Fig. 7.9a is given by Fig. 2.9 if, as will probably be the case, the maximum response occurs before 0.058 sec. Since it is expected that the girder natural period will be much larger than the rise time, we take (DLF)$_{max} = 2$ as a first trial.

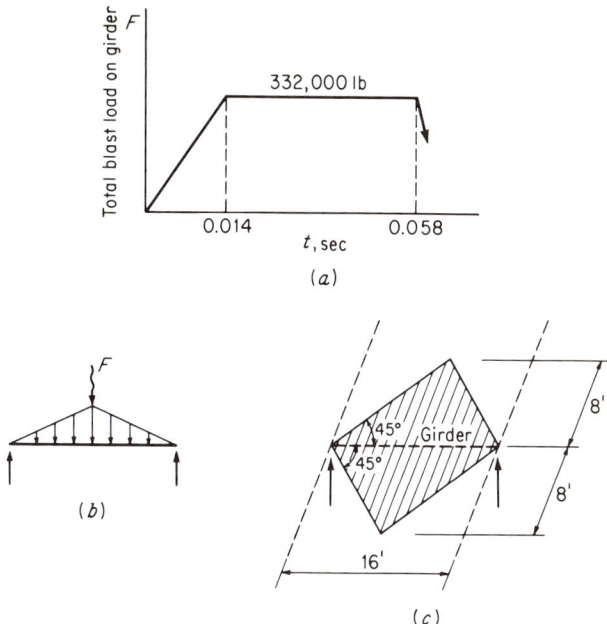

FIGURE 7.9 Example. Roof-girder loading.

Thus the estimated required strength including dead load is

Dynamic load	$332{,}000 \times 2 =$	664,000 lb
Slab weight $\frac{7}{12} \times 150 \times 16 \times 8 =$		11,200
Girder weight	$=$	2,400
Total		677,600 lb

The slab weight included above is that of the tributary area included within the 45° yield lines extending from the slab corners (Fig. 7.9c). Based on the triangular dead- and live-load distributions,

$$\text{Required } \mathfrak{M}_P = \frac{FL}{6} = \frac{677{,}600 \times 16 \times 12}{6} = 21.7 \times 10^6 \text{ in.-lb}$$

For the given dynamic yield strength of 60,000 psi, the most economical steel section providing this plastic bending strength is a 30WF116, for which $I = 4919$ in.^{4}.

In order to make a more accurate analysis of the steel member selected above, we must first compute the natural period. Following the procedure of Sec. 5.3, the load and mass factors are determined on the basis of the static-load shape. For a simply supported beam with triangular load, this shape is expressed by

$$\phi(x) = \frac{x}{8L^5}(5L^2 - 4x^2)^2$$

The distributions of mass and load for this case are

$$m = \frac{2x}{L}m_o \qquad p = \frac{2x}{L}p_o \qquad x < \frac{L}{2}$$

where m_o and p_o are the intensities per unit length at midspan. The total values of mass and load are

$$M_t = \tfrac{1}{2}m_o L \qquad F_t = \tfrac{1}{2}p_o L$$

and according to Sec. 5.2, the mass and load factors are expressed by

$$K_M = \frac{2\int_o^{L/2} m[\phi(x)]^2\, dx}{M_t} = 0.70$$

$$K_L = \frac{2\int_o^{L/2} p[\phi(x)]\, dx}{F_t} = 0.81$$

The numerical values given are obtained merely by evaluating the integrals shown. The stiffness for the girder with triangular load distribution is

$$k = \frac{60EI}{L^3} = \frac{60(30 \times 10^6)4919}{(16 \times 12)^3} = 1.25 \times 10^6 \text{ lb/in.}$$

Finally, the natural period is

$$T = 2\pi\sqrt{\frac{K_M M_t}{K_L k}} = 2\pi\sqrt{\frac{0.70 \times 13{,}600/386}{0.81 \times 1.25 \times 10^6}} = 0.031 \text{ sec}$$

To determine the maximum response, we enter Fig. 2.9 with $t_r/T = 0.014/0.031 = 0.45$ and read $(DLF)_{max} = 1.70$. This is somewhat less than the value of 2.0 originally assumed, and the girder is somewhat overdesigned. However, a second cycle need not be demonstrated here.

We should now reconsider some of the basic assumptions made in the slab and girder design. The time of maximum girder response may be determined by reading on Fig. 2.9 $t_m/t_r = 1.63$, from which $t_m = 1.63 \times 0.014 = 0.023$ sec. Thus the use of Fig. 2.9 is valid, since $t_m < 0.058$, the time of maximum slab response at which the girder load decreases (Fig. 7.9a). Furthermore, the decay in slab reaction due to the decrease of overpressure is not appreciable in the time range of interest.

Also in question is the assumption that the slab and girder may be treated as independent one-degree systems. This may be investigated by comparing the deflection of the two elements. The dynamic girder deflection is given by

$$y_{max} = (\text{DLF})_{max} \frac{F_1}{k} = 1.63 \times \frac{332{,}000}{1.25 \times 10^6} = 0.43 \text{ in.}$$

and the maximum slab deflection by

$$y_{max} = \frac{R_m}{k_E} \mu = \frac{860{,}000}{5.65 \times 10^5} \times 4.0 = 6.1 \text{ in.}$$

It is apparent that the girder deflection is too small to have an appreciable effect on the slab response, and hence on the applied girder load itself. Thus it is permissible to treat the two elements separately.

d. Wall Slab

The wall of the building will be considered to be a one-way, simply supported slab. It is presumably supported at the bottom by a wall footing, which is assumed to lack the rigidity necessary to provide rotational restraint, and at the top by the roof slab, which has considerably less thickness, and hence insufficient stiffness and strength to restrain the wall appreciably. The slab could also be supported by the vertical steel columns, thus becoming a two-way slab, but this arrangement has little advantage. Both exterior walls would of course be designed for face-on blast exposure since the explosion could occur on either side.

In Sec. 7.3a, the reflected pressure was computed to be 59.5 psi and the clearing time for this pressure 0.022 sec. The drag coefficient for the front face is about 0.9, and hence the total pressure after reflection is $p_s + 0.9p_d$. The total pressure-time curve is therefore as shown in Fig. 7.10. Since the natural period of the wall, and hence the time of response, will be short, the loading may be considered as a first approximation to be the single triangle defined by the initial peak of 59.5 psi and the duration of 0.039 sec (Fig. 7.10). This will be correct if the time of maximum response is less than 0.022 sec.

At this point an estimate must be made of the slab natural period. Suppose a value of 0.03 sec is assumed so that $t_d/T = 1.3$, where t_d is 0.039 sec. Then, by Fig. 2.24, the response chart for triangular load-time functions, we obtain, for $\mu = 5$,

$$\text{Required } \frac{R_m}{F_1} = 0.75$$
$$\text{Required } R_m = 0.75 \times 59.5 = 44.6 \text{ psi}$$

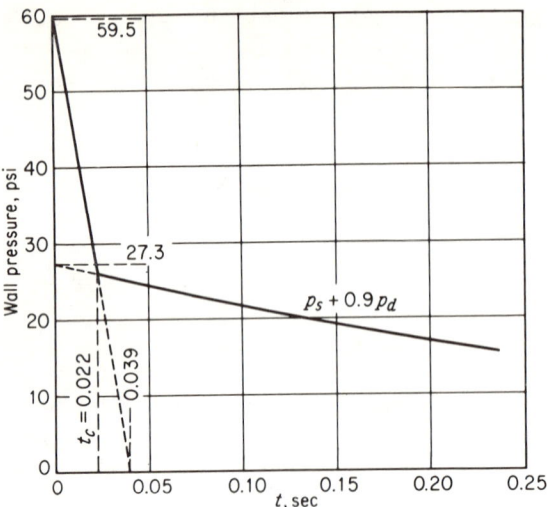

FIGURE 7.10 Example. Pressure-time curve for wall slab.

Considering a 1-in. strip of the simply supported slab, we find

$$\text{Required } \mathfrak{M}_p = \frac{R_m L^2}{8} = \frac{44.6 \times (12 \times 12)^2}{8} = 116{,}000 \text{ lb-in./in.}$$

As in the roof slab design, the required slab depth is given by (if we take $\rho_s = 0.01$)

$$\text{Required } d = \left(\frac{\text{Required } \mathfrak{M}_P}{560}\right)^{1/2} = 14.4 \text{ in.}$$

which corresponds to a total slab thickness of about 16 in. Making use of Table 5.1 and the empirical expression for moment of inertia, the natural period is computed as follows:

$$I_a = \frac{bd^3}{2}(5.5\rho_s + 0.083) = \frac{(1)(14.4)^3}{2}(5.5 \times 0.01 + 0.083)$$
$$= 206 \text{ in.}^4/\text{in.}$$

$$k = \frac{384 E I_a}{5 L^3} = \frac{384(4 \times 10^6) 206}{5(12 \times 12)^3} = 21{,}200 \text{ lb/in. per inch of width,}$$

or 254,000 lb/in. per foot of width

$$M_t = {}^{16}\!/_{12} \times 150 \times 12 \times {}^{1}\!/_{386} = 6.21 \text{ lb-sec}^2/\text{in. per foot of width}$$

$$K_{LM} \cong 0.68$$

$$T = 2\pi \sqrt{\frac{K_{LM} M_t}{k}} = 2\pi \sqrt{\frac{0.68 \times 6.21}{254{,}000}} = 0.026 \text{ sec}$$

We now reenter Fig. 2.24 with

$$\frac{t_d}{T} = \frac{0.039}{0.026} = 1.5 \quad \text{and} \quad \frac{R_m}{F_1} = 0.75$$

to obtain $\mu = 5.6$ and $t_m/t_d = 0.70$. From the latter we compute $t_m = 0.70 \times 0.039 = 0.027$, which is greater than 0.022, and hence the idealized loading which departed from the actual value at the latter time is not valid (Fig. 7.10)

The slab should be strengthened to reduce μ below the design value of 5. This should be done before revising the idealized load because t_m will be affected by a change in strength. For example, if the effective slab thickness were increased to 15.25 in. (total thickness = 16.5 in.) and the procedure exactly as given above repeated, we should find $R_m = 50$ psi, $\mu = 4.2$, and $t_m = 0.023$ sec. The latter time is only slightly greater than 0.022 sec, and the idealized loading is considered to be satisfactory. If t_m were still appreciably larger than 0.022, a possible procedure would be to adopt two triangles for the idealized load function and to compute μ by the use of Eq. (5.13) in Sec. 5.5a.

In the above analysis the effect of the horizontal frame motion on the wall-slab response has been ignored. The justification for this simplified approach will be considered after the frame has been designed.

e. Rigid Frame

The total horizontal force on the steel frame is the algebraic sum of the front- and rear-wall reactions. That for the front wall is given by the expression for dynamic reaction in Table 5.1,

$$V = 0.38R_m + 0.12F$$

where F is the external wall pressure (Fig. 7.10) multiplied by the tributary wall area (20×12 ft), and R_m is 50 psi (see above), or 1.73×10^6 lb for one frame. Since it is probably short compared with the natural period of the frame, the time required to develop this reaction may be ignored. After the maximum wall-slab response ($t_m = 0.023$ sec), it may be assumed that the wall reaction is merely one-half the applied load F. The residual elastic wall vibration after 0.023 sec actually results in a variation about this mean ($\frac{1}{2}F$), but this has little effect on total frame response because it occurs with relative rapidity. Based on these idealizations, the force applied to the frame by the front wall is as shown in Fig. 7.11a.

The total pressure on the rear wall is

$$p = p_s - 0.5p_d$$

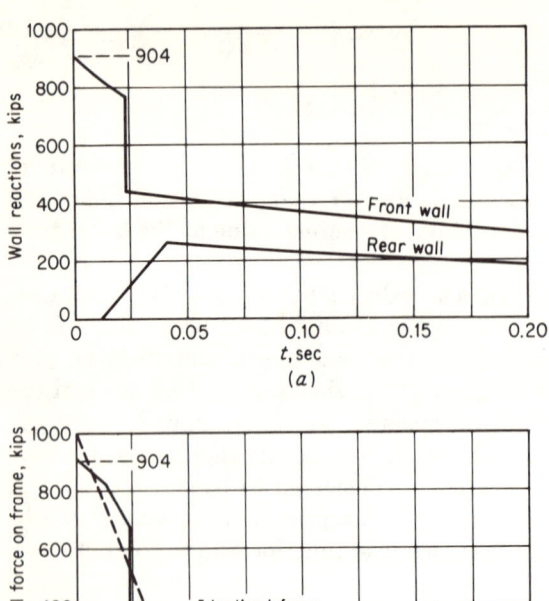

FIGURE 7.11 Example. Horizontal loading on building frame.

where -0.5 is the drag coefficient. This begins to develop when the shock front reaches the rear of the building ($t = l/U = 20/1650 = 0.012$ sec) and has a rise time equal to $4S_c/U = 4 \times 12/1650 = 0.029$ sec. Under this loading the wall would remain elastic, and if the vibration is ignored, the reaction on the frame is

$$\tfrac{1}{2}(p \times 144)(12 \times 20) \text{ lb}$$

which is also plotted in Fig. 7.11a.

The net horizontal force on one frame is merely the difference between the two wall reactions. This is plotted in Fig. 7.11b. As a first approximation this load function is idealized by the dashed line, which is defined by $F_1 = 1000$ kips and $t_d = 0.05$ sec. Therefore, assuming that the frame period will be about 0.08 sec, we estimate, with the aid of Fig. 2.24, that the required strength should be approximately one-half of the initial load peak if μ is to be 5.

On the above basis the required resistance of the frame to horizontal load is approximately

$$R_m = \tfrac{1}{2}F_1 = 500 \text{ kips}$$

If the unbraced height of the columns is 9 ft, the required bending strength is

$$\mathfrak{M}_p = \frac{R_m h}{4} = \frac{500 \times 9}{4} = 1125 \text{ kip-ft}$$

Using the given yield strength of 60,000 psi, it is found that an 18WF105 steel section ($I = 1852$ in.^{4}) provides the required plastic bending moment. Note that, in the above calculation, the effect of axial column stress has been ignored. This is permissible because the amount of direct stress is not sufficient to reduce appreciably the column bending strength. However, a more refined analysis would take this effect into consideration.

In order to check the above preliminary column design, we first compute the natural period based upon effective values of stiffness and mass (see Sec. 5.3c). The stiffness is merely the inverse of the horizontal deflection due to a unit horizontal load at the top of the frame. With the moments of inertia of the girder and columns previously determined, conventional elastic analysis leads to the expression

$$k = \frac{18.4 E I_c}{h^3}$$

where I_c is the column moment of inertia, and h is the effective column length (9 ft). Thus

$$k = \frac{18.4(30 \times 10^3)1852}{(9 \times 12)^3} = 810 \text{ kips/in.}$$

The effective mass [Eq. (5.6)] is that at roof level plus one-third of the wall masses.

Roof slab	$\tfrac{7}{12} \times 150 \times 20 \times 17.25 =$	30,200 lb
Roof girder	$116 \times 17.25 =$	2,000
Walls $\tfrac{1}{3} \times 2 \times 16.5/12 \times 150 \times 12 \times 20 =$		33,000
Columns	$\tfrac{1}{3} \times 2 \times 105 \times 9.0 =$	600
Total weight		65,800 lb

$$M_t = 65{,}800/386 = 170 \text{ lb-sec}^2/\text{in.}$$

The natural period is therefore

$$T = 2\pi \sqrt{\frac{M_t}{k}} = 2\pi \sqrt{\frac{170}{810{,}000}} = 0.091 \text{ sec}$$

296   *Introduction to Structural Dynamics*

We now reenter Fig. 2.24 with the parameters

$$\frac{t_d}{T} = \frac{0.050}{0.091} = 0.55 \quad \text{and} \quad \frac{R_m}{F_1} = 0.5$$

and read $\quad \mu = 4.3 \quad$ and $\quad \dfrac{t_m}{t_d} = 1.2$

Therefore $t_m = 1.2 \times 0.050 = 0.060$ sec. Referring back to Fig. 7.11b, we may now make a better load approximation. The proper criterion is that the areas under the actual and idealized load functions should be approximately equal up to time t_m. On this basis it appears that a better approximation for the idealized load would be $F_1 = 950$ kips and $t_d = 0.058$ sec. For this loading the revised parameters are

$$\frac{t_d}{T} = \frac{0.058}{0.091} = 0.64 \quad \text{and} \quad \frac{R_m}{F_1} = {}^{500}\!/\!_{950} = 0.53$$

and we obtain from Fig. 2.24

$$\mu = 4.9$$
$$\frac{t_m}{t_d} = 1.1$$
$$t_m = 1.1 \times 0.058 = 0.064 \text{ sec}$$

The load approximation now appears reasonable, and we conclude that the ductility ratio is indeed about 5, the desired value. Therefore the 18WF105 column section is satisfactory.

The method of frame analysis given above is approximate in several respects, and it may be desirable to make a final design based on more exact procedures.[10] This would have to be executed by numerical analysis, because there are several time-varying effects and the procedure might differ from the above analysis in the following respects: (1) the actual load function rather than the triangular-load idealization would be used; (2) the effect of direct stress on the column bending strength would be included; and (3) the effect of the vertical load acting on the horizontal deflection might be considered. The last would take into account the eccentricity of the vertical load, the effect of which is to reduce the frame resistance by Fy/h, where F is the total vertical load and y is the horizontal deflection. In both procedures 2 and 3, the vertical column loads should be based on the dynamic reactions of the roof girder. Thus, even in the plastic range, the frame resistance would be recomputed at each time station of the numerical analysis.

It will be recalled that the wall slab was analyzed as though on rigid supports; i.e., the frame motion was ignored. This leads to only slight error because, as is now known, the frame responds more slowly and the inertia forces on the wall due to frame motion are small compared with those due to distortion of the wall panel itself. As discussed in Sec. 5.7, the alternative is to consider the two elements as coupled systems.

If the transverse rigid frame of the current example were to be designed in reinforced concrete rather than steel, the method of dynamic analysis would not differ from that given above. However, an important difference in column design results from the fact that the effect of axial compression on bending strength is significant in the case of reinforced concrete and should be included. The vertical roof load may cause an appreciable increase in column bending strength, and to ignore this fact is unduly conservative. If, as above, standard response charts are being used, some average value of column compression must be assumed to remain constant during the response. If this is considered too crude, the frame must be analyzed by numerical step-by-step methods.

For higher design overpressures, rigid frames of either steel or reinforced concrete are not economical, and the designer may resort to the use of reinforced concrete shear walls in transverse planes. In addition to the fact that the space within the building is obstructed, shear walls have the disadvantage (from the viewpoint of required strength) of low ductility capacity and great stiffness. As a result, the lateral natural period of the building is small and the design μ value must be set at a relatively low level. Both of these facts tend to increase the required strength. This is of course compensated for by the great inherent shear capacity of reinforced walls.[41]

7.4 Aboveground Arches and Domes

The behavior of arches and domes when subjected to the effects of nuclear blast is obviously a complex phenomenon. The loading, both with respect to spatial distribution and time variation, is difficult to determine because of the curvature of the exposed surfaces. In addition, the stress condition is more complicated than for other types of structural elements, such as beams or frames. Fairly exact methods of analysis are possible, but these are not practical for design purposes. We shall therefore limit ourselves to approximate procedures. The scope of the following discussion is restricted to circular arches and spherical domes, both of reinforced concrete.

The loading may be considered to consist of three parts: (1) the general overpressure, which by itself produces uniform compression in the arch or dome, (2) the reflected pressure, which is largest at the windward base of the structure where the surface is most nearly vertical, and (3) the drag pressures, which are generally positive on the windward side and negative on the opposite side. The latter two components are unsymmetrical and

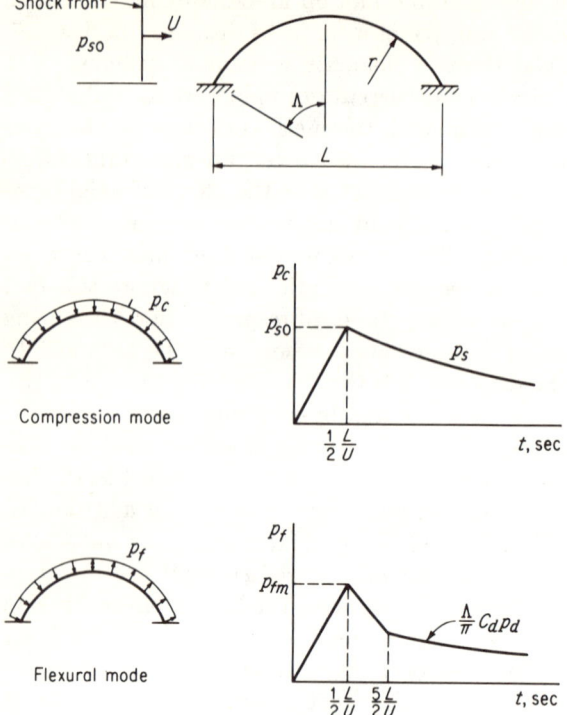

FIGURE 7.12 Idealized modal loads for arches and domes.

produce bending in the arch and nonuniform membrane stresses in the dome.

It has been suggested[38] that the total load could be represented by two components: (1) a "compression" mode, consisting of a uniform radial pressure, and (2) a "flexural" mode, consisting of an antisymmetrical but uniformly distributed pressure. These are shown in Fig. 7.12. The load-time function and magnitude for the compression mode are taken to be the same as the overpressure, except that there is a finite rise time which is conservatively estimated to be one-half the transit time of the shock front, i.e., one-half the time required for the front to cross the structure. The flexural mode is given a peak-pressure intensity of

$$p_{fm} = \tfrac{1}{2} p_r \qquad \text{domes}$$

or
$$p_{fm} = \left(0.5 + \frac{\Lambda}{\pi}\right) p_{so} \qquad \text{arches} \tag{7.6}$$

where Λ is one-half the central angle of the arch or dome, and p_r is the maximum reflected pressure on the dome. The latter occurs at the base,

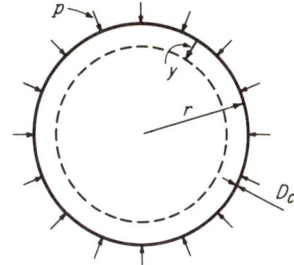

FIGURE 7.13 Circular ring under radial pressure.

and is given by Eq. (7.4), provided that $\Lambda > 55°$. These peak values are attained after about one-half the transit time (i.e., the shock front has reached the center of the structure). The flexural mode pressure then decreases until at 2.5 transit times and thereafter it is equal to $(\Lambda/\pi)p_d C_d$, where p_d is the dynamic pressure and C_d is the drag coefficient. The latter coefficient is approximately 0.4 for $p_{so} < 50$ psi and 1.0 for $p_{so} > 75$ psi. This change in C_d results from the fact that the drag on curved objects displays a Reynolds-number effect. Between the two pressure levels given, C_d may be assumed to vary linearly.

The load components given above are obviously rather crude approximations, but are believed to be conservative and sufficiently accurate for design purposes. It should be apparent that one reason for selecting these distributions (symmetrical and antisymmetrical) was that internal stresses may be readily computed for these cases. This method of load definition should not be used for small weapons, say, $Y < 0.1$ MT.

Having defined the load in the manner outlined above, the structure may be assumed to have two degrees of freedom, one mode corresponding to each of the two modal loads. The dynamic analysis may then be executed, using any of the methods previously developed.

a. Reinforced Concrete Barrel Arches

The compression mode of an arch corresponds to a condition of uniform axial strain and is therefore analogous to a circular ring under uniform radial pressure as in Fig. 7.13. The natural period may be derived as follows:

$$\text{Ring strain} = \frac{pr}{D_c E_c}$$

where r and D_c are the radius and thickness of the ring, and E_c is the modulus of elasticity of concrete. Reinforcing steel has, in most cases, a negligible effect on the strain. The radial deflection is

$$y = \frac{pr^2}{D_c E_c}$$

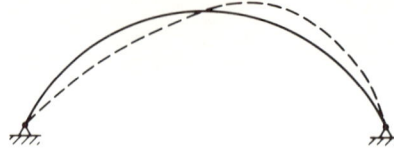

FIGURE 7.14 Flexural mode distortion of arch.

and the stiffness, defined as total radial load per unit width divided by radial deflection, is

$$k = \frac{2\pi r p}{p r^2 / D_c E_c} = \frac{2\pi D_c E_c}{r}$$

The total mass per unit width is given by

$$M = \frac{2\pi r D_c \rho}{g}$$

where ρ is the concrete density. The natural period of the compression mode is therefore

$$T_c = 2\pi \sqrt{\frac{M}{k}} = 2\pi \sqrt{\frac{\rho r^2}{g E_c}}$$

and if we take $\rho = 144$ lb/ft^3 and $E_c = 4 \times 10^6$ psi, this becomes

$$T_c = \frac{r}{1810} \quad \text{sec} \tag{7.7}$$

where r is the radius in feet.

As seen in Fig. 7.12, the rise time for the compression mode load is $L/2U$. For the overpressures of interest in connection with aboveground arches (say, $p_{so} < 100$ psi) and the internal angles usually encountered, this rise time is of about the same magnitude as the natural period T_c. Such being the case, the dynamic increase in load effect is generally small (Fig. 2.9), or in other words, the loading is only slightly more severe than a static load of the same magnitude. Furthermore, in the case of an aboveground arch, the mode of failure is one of excessive distortion due to flexure, and the major effect of the compressive force is to modify the bending resistance of the reinforced concrete cross section. This effect is significant throughout the time of flexural response, and therefore we seek, not the maximum compressive stress, but an average over this time, which is essentially the mean, or static, value. For both of these reasons, the compression in the arch may be approximated as a constant with a value equal to that corresponding to p_{so}. In other words,

$$P_c = p_{so} r \tag{7.8}$$

where P_c is the force per unit width of arch.

In the flexural mode the distortion is as shown in Fig. 7.14. Note that this is the fundamental flexural mode, which is the only mode of signifi-

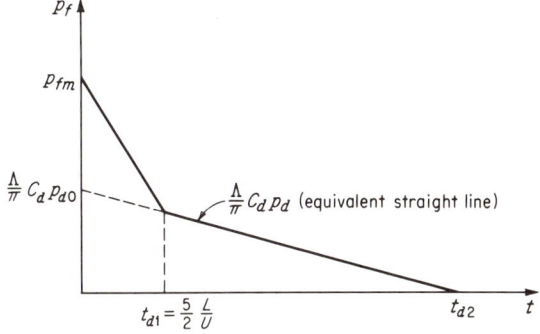

FIGURE 7.15 Two-triangle representation of flexural mode loading.

cance in the response due to the assumed flexural mode loading (Fig. 7.12). For this mode the natural period may be computed as that for a beam having a span equal to Λr, or half the arch arc, modified by a factor to take into account the lateral deflection of the crown. The beam natural period for the case of hinged arch supports may be defined by the empirical expression

$$T = \frac{(\Lambda r)^2}{300{,}000 d(\rho_s + 0.015)^{1/2}}$$

where d is the effective depth, and ρ_s the steel ratio in the tension face of the cross section. This expression is consistent with the approximate values of effective moment of inertia for reinforced concrete sections used previously (Sec. 5.6a). The modifying factor for translation of the crown is approximately

$$\frac{(\pi/\Lambda)^2 + 1.5}{(\pi/\Lambda)^2 - 1}$$

and therefore the approximate natural period of the flexural mode is

$$T_f = \frac{(\pi/\Lambda)^2 + 1.5}{(\pi/\Lambda)^2 - 1} \frac{(\Lambda r)^2}{300{,}000 d(\rho_s + 0.015)^{1/2}} \quad \text{sec} \quad (7.9)$$

where both r and d are in inches.

In the flexural-mode load-time function (Fig. 7.12) the rise time $L/2U$ is generally small compared with T_f and may be ignored. Furthermore, we may replace the decaying drag effect by an equivalent straight line (selected so as to be compatible with the time of maximum response as in Sec. 7.3b), so that the total load may be represented by two triangles, as in Fig. 7.15. Having accomplished this simplification, inelastic response may be determined by the use of Eq. (5.13), which combines the effects of two triangular functions.

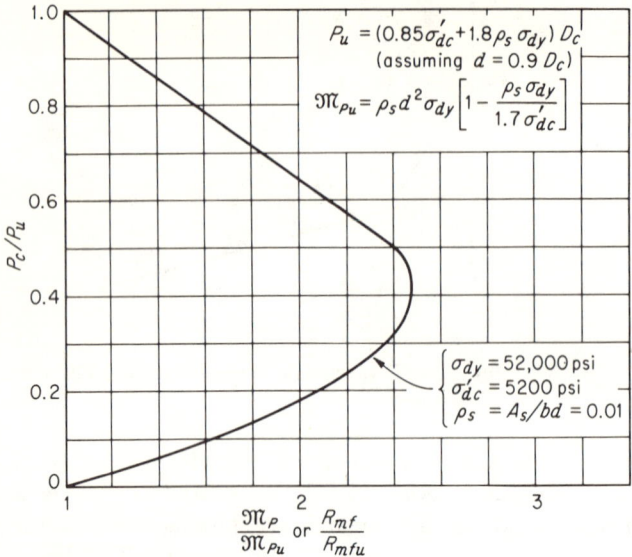

FIGURE 7.16 Interaction curve for rectangular reinforced concrete section in direct stress and bending.

The flexural resistance of the arch may, like the natural period, be based upon a beam of span Λr, modified to take into account translation of the arch crown. If the arch has hinged bases, the resistance per unit arc length is given by

$$R_{mf} = \frac{8 \mathfrak{M}_P}{(\Lambda r)^2} \frac{(\pi/\Lambda)^2 - 1}{(\pi/\Lambda)^2} \qquad (7.10)$$

where $\mathfrak{M}_P$ is the ultimate bending strength of the concrete cross section. The effect of axial compression on bending strength is significant and, in most cases, results in an increase of flexural resistance. The relationship between ultimate bending strength and compression is well known,[39] and an interaction curve for one set of parameters is shown in Fig. 7.16. For this figure it has been assumed that the curve for the given dynamic material properties is the same as that for the corresponding static properties ($\sigma_c' = 4{,}000$ psi, $\sigma_y = 40{,}000$ psi).

The approximate methods of analysis outlined above are now applied to the following example. We desire to determine the maximum flexural response in terms of the ductility ratio μ of the reinforced concrete arch shown in Fig. 7.17, which is subjected to a peak overpressure of 50 psi from a 1-MT weapon. Design would consist of a series of such analyses for various section properties converging on those which result in a desired μ.

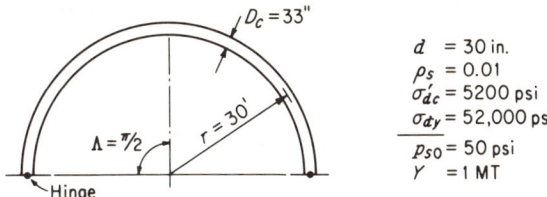

FIGURE 7.17 Example. Reinforced concrete barrel arch.

By Eq. (7.8), the arch compression per inch of width is

$$P_c = p_{so}r = 50 \times 30 \times 12 = 18{,}000 \text{ lb/in.}$$

The ultimate compressive strength for the given parameters is

$$P_u = (0.85\sigma'_{dc} + 1.8\rho_s\sigma_{dy})D_c$$
$$= (0.85 \times 5200 + 1.8 \times 0.01 \times 52{,}000)33$$
$$= 176{,}000 \text{ lb/in.}$$

Therefore $\dfrac{P_c}{P_u} = 0.102$

Using the interaction curve in Fig. 7.16, we obtain

$$\frac{R_{mf}}{R_{mfu}} = 1.64$$

where R_{mfu} is the flexural resistance for $P_c = 0$.

By Eq. (7.6), the peak value of flexural load is

$$p_{fm} = \left(0.5 + \frac{\Lambda}{\pi}\right)p_{so} = 50 \text{ psi}$$

The shock-front velocity is given by Eq. (7.3b):

$$U = 1120\left(1 + \frac{6p_{so}}{103}\right)^{\frac{1}{2}} = 2220 \text{ fps}$$

and therefore (Fig. 7.15)

$$t_{d1} = \frac{5}{2}\frac{L}{U} = \frac{5}{2} \times \frac{60}{2220} = 0.0675 \text{ sec}$$

Referring to Fig. 7.2, we find that the given overpressure occurs at a range of 4800 ft and that the corresponding peak dynamic pressure (p_{do}) is 40 psi. From Fig. 7.3, we determine the duration of the dynamic pressure t_{dd} to be 3.15 sec. Thus the initial value of the dynamic-pressure

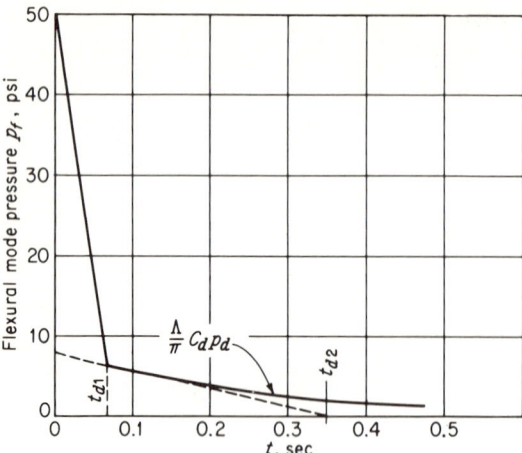

FIGURE 7.18 Example. Flexural mode loading for arch.

portion of the flexural mode loading (Fig. 7.15) is

$$\frac{\Lambda}{\pi} p_{do} C_d = 8.0 \text{ psi}$$

taking C_d at 0.4. Based on the above data and the appropriate decay curve of Fig. 7.4b, the complete load-time function for the flexural mode is as shown in Fig. 7.18.

The bending resistance of the cross section without axial compression by the equation given in Fig. 7.16, with $d = 30$ in., is 440,000 lb-in. per inch of arch width. By Eq. (7.10), the corresponding flexural resistance is

$$R_{mfu} = \frac{8 \times 440,000}{[(\pi/2) \times 360]^2} \frac{(2)^2 - 1}{(2)^2} = 8.3 \text{ psi}$$

It was found previously that, because of P_c, this is increased by 1.64, and therefore the actual resistance is

$$R_{mf} = 8.3 \times 1.64 = 13.6 \text{ psi}$$

As formulated here, the resistance is given in terms of external pressure.

Evaluating Eq. (7.9), we obtain the natural period of the flexural mode,

$$T_f = \frac{(2)^2 + 1.5}{(2)^2 - 1} \times \frac{[(\pi/2) \times 360]^2}{300,000 \times 30(0.01 + 0.015)^{1/2}} = 0.41 \text{ sec}$$

At this point we must estimate the effective duration of the dynamic-pressure load component $(\Lambda/\pi)p_d C_d$. It is expected that the maximum response will occur in the vicinity of $t_m = \frac{1}{2}T_f$, and on this basis it is estimated that a proper value of effective duration, t_{d2}, would be 0.35 sec (Fig. 7.18).

We now determine the maximum response by the approximate equation (5.13):

$$\left(\frac{F_1}{R_m}\right)_1 C_1(\mu) + \left(\frac{F_1}{R_m}\right)_2 C_2(\mu) = 1$$

where the first term relates to the triangle defined by $p_{fm} - (\Lambda/\pi)p_{do}C_d$ and t_{d1}, and the second term to that defined by $(\Lambda/\pi)p_{do}C_d$ and t_{d2}. For the current problem we have the following data:

$$\left(\frac{F_1}{R_m}\right)_1 = \frac{42}{13.6} = 3.1 \qquad \frac{t_{d_1}}{T_f} = \frac{0.0675}{0.41} = 0.165$$

$$\left(\frac{F_1}{R_m}\right)_2 = \frac{8}{13.6} = 0.59 \qquad \frac{t_{d_2}}{T_f} = \frac{0.35}{0.41} = 0.85$$

We now solve Eq. (5.13) for μ, using a trial-and-error procedure. For an assumed μ, $C_1(\mu)$ and $C_2(\mu)$ are read from Fig. 2.24a, each being the value of R_m/F_1 corresponding to that μ and the appropriate t_d/T_f. The calculations are given in the following table:

Assumed μ	$C_1(\mu)$	$C_2(\mu)$	$(F_1/R_m)_1 C_1(\mu)$	$(F_1/R_m)_2 C_2(\mu)$	Σ
5	0.17	0.61	0.53	0.36	0.89
4	0.18	0.67	0.56	0.39	0.95
3	0.21	0.75	0.65	0.44	1.09

It is apparent that Eq. (5.13) is satisfied by a μ value slightly less than 4. In other words, the maximum deflection of the arch in the flexural mode is about four times that corresponding to the formation of plastic hinges at or near the quarter points of the arch.

For very flat arches it may be necessary to consider buckling due to the compression mode loading. However, in most cases, flexure is the important consideration, and the compressive force is not large enough to cause an unstable condition.

If the arch were a rib supporting a cylindrical surface, the above procedure would be modified in an obvious manner to account for the increase in load relative to the rib width.

The foregoing procedure is approximate in several respects. If the analyst desires a more precise solution, or if the response is completely elastic, a numerical method of analysis should be used. This can be accomplished using only the system parameters computed above and the nondimensional equation of motion given in Sec. 2.8a.

$$\frac{1}{4\pi^2}\ddot{\eta} + \eta = \frac{F_1}{R_{mf}}f(\xi)$$

where $\eta = y/y_{el}$
$\xi = t/T_f$
$F_1 f(\xi)$ = load-time function for flexural mode

The preceding equation applies in the elastic range only, and beyond the elastic limit it should be changed to

$$\frac{1}{4\pi^2}\ddot{\eta} + 1 = \frac{F_1}{R_{mf}}f(\xi)$$

The maximum value of η so obtained is, by definition, μ. If, in addition, it is desired to take into account the variation in P_c and the consequent effect on R_{mf}, the above equations may be rewritten as

$$\frac{1}{4\pi^2}\ddot{\epsilon} + \epsilon = f(\xi) \qquad \text{elastic range}$$

and

$$\frac{1}{4\pi^2}\ddot{\epsilon} + \frac{R_{mf}}{F_1} = f(\xi) \qquad \text{plastic range}$$

where ϵ is y/y_{st}, and y_{st} is the flexural mode deflection due to the static application of F_1. In each step of the numerical analysis R_{mf} in the last equation may be determined for the value of $P_c = p_s r$ occurring at that time.

b. Reinforced Concrete Spherical Domes

Spherical domes may be treated in a manner similar to that given for arches. In this case, however, the so-called flexural mode loading does not produce bending, since it is resisted by membrane stresses.

The natural period in the compression mode may be derived following the procedure given above for arches. Noting that the radial deflection for a spherical shell under uniform pressure is

$$y = \frac{pr^2}{2D_c E_c}(1 - \nu)$$

where ν is Poisson's ratio, we find that the natural period is approximately expressed by

$$T_c = \frac{r}{2500} \qquad \text{sec}$$

where r is the radius in feet. The natural period in the flexural mode may be taken approximately as the same as that for the compression mode.

The natural periods for a dome are extremely short, and for the pressure levels ($p_{so} < 100$ psi) and weapon yields ($Y > 0.1$ MT) considered here, the rise times for both modal loads (Fig. 7.12) are never appreciably less than the periods. Therefore the dynamic increase in the load effect is generally small. Furthermore, the decay of the peak loads is relatively slow, which indicates that the required strength is not very sensitive to the design-deflection criterion; i.e., a resistance slightly greater than the peak load would result in elastic behavior, while a slightly smaller resistance would result in very large deflection. For this reason it is sufficiently accurate for design purposes to consider both load components to be statically applied, regardless of the amount of shell deformation to be permitted. Thus the ultimate membrane strength is made equal to the internal forces corresponding to the maximum applied pressures.

The maximum membrane force due to the compression mode load is

$$P_c = \tfrac{1}{2} p_{so} r \qquad \text{lb per unit width}$$

where P_c is uniform throughout the shell and occurs in all directions. For the antisymmetrical flexural mode, the maximum local compression which occurs on the windward side is approximately given by

$$P_f = \tfrac{3}{2} p_{fm} r = \tfrac{3}{4} p_r r \qquad \text{lb per unit width}$$

the last substitution being made according to Eq. (7.6). The ultimate strength is equal to

$$P_u = (0.85 \sigma'_{dc} + 1.8 \rho_s \sigma_{dy}) D_c \qquad \text{lb/in.}$$

where ρ_s is the steel ratio for each face and in each of two perpendicular directions. The modal membrane stresses are additive, and hence the design equation is

$$P_c + P_f = P_u$$

or
$$\tfrac{1}{2} p_{so} r + \tfrac{3}{4} p_r r = (0.85 \sigma'_{dc} + 1.8 \rho_s \sigma_{dy}) D_c \qquad (7.11)$$

Note that, even though the loads are considered to be applied statically, the dynamic material strengths may be used.

To illustrate the above, we shall determine the required thickness of a dome for the following parameters:

$$p_{so} = 50 \text{ psi} \qquad r = 30 \text{ ft} \qquad \rho_s = 0.005$$
$$\sigma'_{dc} = 5200 \text{ psi} \qquad \sigma_{dy} = 52{,}000 \text{ psi}$$

By Eq. (7.4), the reflected pressure is 198 psi, and using Eq. (7.11), we obtain

$$\tfrac{1}{2}(50)(360) + \tfrac{3}{4}(198)(360) = (0.85 \times 5200 + 1.8 \times 0.005 \times 52{,}000) D_c$$

or
$$D_c = 12.8 \text{ in.}$$

which is the required total thickness for the shell.

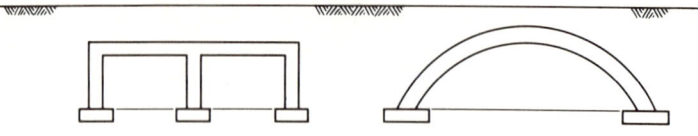

FIGURE 7.19 Shallow-buried structures.

In addition to the membrane stresses considered above, there is some bending resulting from boundary restraints or load irregularities. However, these would normally not cause failure, and excessive cracking may be prevented by maintaining at least a nominal amount of reinforcement in each face.

7.5 Belowground Structures

Aboveground protective construction is economical only up to overpressures ranging between 50 and 100 psi, the exact limit depending upon the type and function of the structure. For higher design overpressures, such structures are placed belowground. In this way it is possible to provide protection against very high overpressures, e.g., 1000 psi or even greater.

Most of the problems peculiar to the design of underground structures are in the realm of soil mechanics or wave propagation, and not in the area of structural dynamics as treated in this text. Our discussion will therefore be limited to some general observations regarding the dynamic behavior of underground structures.

If the structure is *shallow-buried*, i.e., if the earth just covers the top of the structure as in Fig. 7.19, the primary effect of the burial is to eliminate the reflection and drag components of loading. In the case of a boxlike structure, the design of the roof is not appreciably different from that for the aboveground situation. There is, of course, no suction due to drag, and the earth cover increases both the dead load and the natural period of the roof slab or beam. None of these factors is normally of great importance. However, the vertical walls of the rectangular structure are subjected to much smaller loads than in the aboveground case. First, the very important reflected pressure is eliminated, and second, the pressure normal to the wall is only about one-fourth to one-half of the overpressure (p_{so}), depending upon the type of soil. Finally, there is no appreciable horizontal loading on the transverse frames of the structure because, if the backfill is properly placed and compacted, very little distortion of this type is possible.

Arches and domes which are shallow-buried are particularly resistant to blast effects. The behavior in the compression mode is not appreciably different from that in the aboveground case, except that the natural period is lengthened by the weight of soil. However, the compression mode is not severe, because arch or dome structures are ideally suited to resist this type of loading. The important point is that the flexural mode loading to which they are not so resistant is greatly reduced by the burial. The only flexural load present is that due to transit of the shock over the structure. The time function for this load may be considered to be an isosceles triangle with a peak value of $\frac{1}{2}p_{so}$ and a duration equal to the transit time. This is much less severe than that used for aboveground structures.

If the structure is *deeply buried*, i.e., if the earth cover is equal to or greater than about half the width, there are three major effects in addition to those mentioned for the shallow-buried case: (1) the overpressure is attenuated with depth; i.e., the pressure at the structure is less than the surface overpressure p_{so}; (2) the duration of the positive phase is increased; and (3) the soil acting as an arch above the structure takes an appreciable part of the vertical load. The last is extremely important and can perhaps be best visualized in the case of an arch structure, where, if properly compacted, the soil above the structure acts as an arch itself, thus relieving the structure of load. For deeply buried arches and domes, the flexural mode may be ignored, since the surrounding soil prevents significant distortion of this sort. The reduction in compression mode loading, together with the elimination of the flexural mode, makes deeply buried arches and domes extremely blast-resistant.

The highest level of protection is provided by rock tunnels deep below the surface. Such protection is limited only by the strength of the rock and its ability to prevent closure of the cavity. At somewhat lower pressure levels, spalling of the tunnel walls may occur, but damage can be minimized by rock bolting and by placing a liner against the tunnel wall. Structures are sometimes built within cavities, and these need be designed only to withstand the ground motion resulting from the blast, as discussed in the next section.

7.6 Ground Motions

A nuclear explosion causes sizable motions of the ground, which may be important in the design of hardened facilities. If the structure itself has been designed to withstand overpressure (such as the examples in previous sections), it is unlikely that ground shock would be an important consideration, and is often ignored. However, for structural elements within but not part of a protective enclosure, ground shock may be the only blast effect. These elements would include isolated floors or pieces of equipment supported on the ground within the enclosure, as well as structures within tunnels belowground. Facilities involving electronic equipment are often more vulnerable to ground shock than to other blast effects.

The ground motion may be considered to consist of two parts: (1) the air-induced shock resulting from the application of overpressure to the ground surface, and (2) the shock transmitted directly through the ground from a burst at or below the surface. Only the first part is considered herein, since this is usually the more severe.

The actual motion can at best be only approximately predicted. It depends on the properties of the soil, not only at the point of interest, but at points far removed, particularly in the region below the point under consideration. Since the earth is not a homogeneous medium, the phenomena become quite complex. The expressions given below, which have been taken from Ref. 38, are estimates of the peak air-induced effects believed to be reasonable for typical conditions.

The maximum vertical displacement at the ground surface is divided into two parts: (1) the elastic, or transient, displacement (y_{soe}), and (2) the plastic, or permanent, displacement (y_{sop}). These may be taken as follows:

$$y_{soe} = 10\left(\frac{p_{so}}{100}\right)^{0.4}\left(\frac{1000}{c_s}\right)Y^{1/3} \quad \text{in.} \tag{7.12}$$

$$y_{sop} = \frac{p_{so} - 40}{30}\left(\frac{1000}{c_s}\right)^2 \quad \text{in.} \tag{7.13}$$

where p_{so} = peak overpressure at surface, psi
c_s = seismic velocity of soil, fps
Y = weapon yield, MT

The seismic velocity varies with type of soil from about 1000 fps for a soft material to about 2000 for a sandy silt and to 12,000 or higher for rock. The maximum horizontal displacements may be taken as one-third of the vertical displacements given above.

The maximum vertical velocity at the surface is approximated by

$$\dot{y}_{so} = 50\left(\frac{p_{so}}{100}\right)\left(\frac{1000}{c_s}\right) \quad \text{in./sec} \tag{7.14}$$

and the peak horizontal velocity may be taken as two-thirds of this value. Both the maximum vertical and horizontal accelerations are given by

$$\ddot{y}_{so} = 150\left(\frac{p_{so}}{100}\right)\left(\frac{1000}{c_s}\right) \quad \text{in units of } g\text{'s} \tag{7.15}$$

However, it is recommended that c_s in this equation be taken as 2000 fps for all soils having greater seismic velocities.

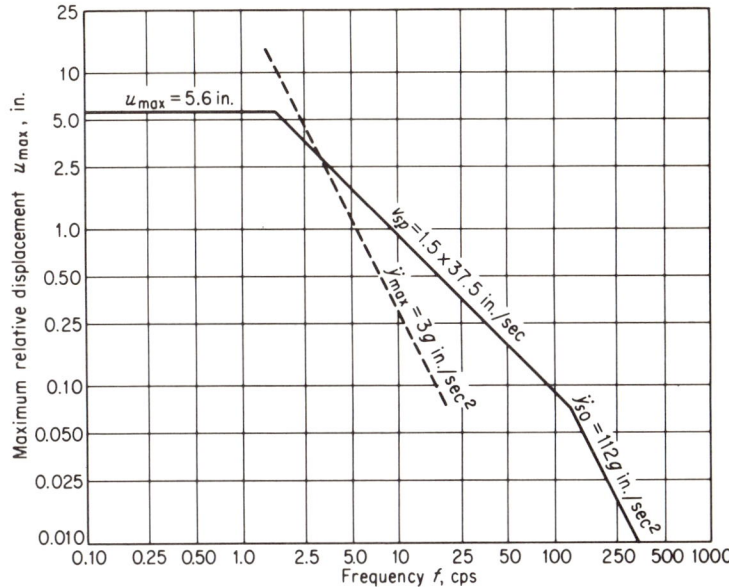

FIGURE 7.20 Idealized response spectrum for vertical ground motion. $W = 0.5$ MT, $p_{so} = 150$ psi, $c_s = 2000$ fps.

At depths below the surface, the air-induced ground motion is of course less severe than the values given above. At moderate depths (say, less than 100 ft), the displacement and velocity are not appreciably different, but the peak acceleration is greatly reduced. Reference 38 provides procedures for estimating these below-surface values.

Equations (7.12) to (7.15) provide peak values of displacement, velocity, and acceleration. It is extremely difficult to predict the actual time function of the motion. For this reason the response-spectrum approach as discussed in Sec. 6.4 in connection with earthquake design is most useful. It has been suggested[38] that an approximate spectrum for design purposes may be constructed on a log-log plot (Fig. 6.9) as follows: (1) a straight line of constant displacement equal to the total maximum ground displacement, (2) a straight line of constant spectral velocity equal to 1.5 times the maximum ground velocity, and (3) a straight line of constant acceleration equal to the maximum ground acceleration. A response spectrum for vertical motion constructed on this basis is shown in Fig. 7.20. This is for the surface burst of a 0.5-MT weapon at a range such that $p_{so} = 150$ psi. The seismic velocity has been taken as 2000 fps, and the peak values for the ground motion as given by the expressions above are

$$y_{so} = 5.6 \text{ in.} \qquad \dot{y}_{so} = 37.5 \text{ in./sec} \qquad \ddot{y}_{so} = 112g$$

312 Introduction to Structural Dynamics

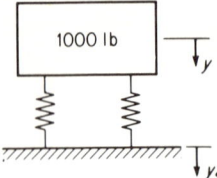

FIGURE 7.21 Shock-mounting example.

The three lines specified above are therefore given, respectively, by the equations

$$u_{max} = 5.6 \text{ in.} \qquad u_{max} = \frac{37.5 \times 1.5}{2\pi f} \qquad u_{max} = \frac{112g}{(2\pi f)^2}$$

where u_{max} is the maximum displacement relative to the ground. For any given frequency of the responding system f, Fig. 7.20 provides the maximum relative displacement. The absolute acceleration of the mass of the responding system is given by

$$\ddot{y}_{max} = (2\pi f)^2 u_{max}$$

To illustrate application of the above, we consider the problem of shock-mounting a piece of sensitive equipment. Usually the equipment may be considered to be a rigid mass which can be protected from the ground shock by supporting it on springs as shown in Fig. 7.21. Suppose that the weight is 1000 lb and that the equipment would be damaged if the vertical acceleration, up or down, not including the acceleration of gravity, exceeded $3g$. Our problem is to design the supporting springs. Only vertical motion will be considered. Protection must also be provided against horizontal motion, but the two motions are generally uncoupled. The criterion of $\ddot{y}_{max} = 3g$ can be expressed by the equation

$$u_{max} = \frac{3g}{(2\pi f)^2} = \frac{29.4}{f^2} \text{ in.}$$

which, when constructed on the log-log plot, is the straight dashed line in Fig. 7.20. To the right of the intersection of this line with the response spectrum, the acceleration of the mass would be greater than $3g$. Since the intersection occurs at $f = 3.3$ cps, we conclude that the design criterion will be satisfied if the frequency is less than this amount. Since the frequency of the mass on springs is given by

$$f = \frac{1}{2\pi} \sqrt{\frac{k}{M}}$$

the requirement for the design is that the combined spring constant for all supporting springs must be less than $(2\pi f)^2 M$, which, for this example, means that

$$k < (2\pi)^2(3.3)^2(1000/386) = 1110 \text{ lb/in.}$$

If this upper limit were actually selected, the motion of the mass relative to the support as given by Fig. 7.20 would be

$$u_{\max} = 2.7 \text{ in.}$$

and sufficient clearance to permit this vertical motion must be provided. A smaller value of k would result in reduced acceleration of the mass but larger relative displacement.

If the equipment were mounted on the enclosing structure rather than on the ground, the support motion, for which the shock mount would be designed, is of course the motion of the structure at the point of support. The latter would be obtained by analysis of the response of the structure to air blast or, in some cases, to ground shock. Having this support motion, the equipment and its mounting would be analyzed using the procedures of Sec. 2.6 or 6.2.

Problems

7.1 A one-way reinforced concrete slab is subjected to side-on overpressure only. The slab is fixed against rotation at both supports and has the following properties:
   $d = 6$ in. (total depth $= 7.5$ in.)
   $\rho_s = 0.01$ (for both positive and negative moment)
   $\sigma'_{dc} = 5200$ psi; $\sigma_{dy} = 52,000$ psi
   $E = 4 \times 10^6$ psi; span $= 12$ ft
If the peak overpressure p_{so} is 10 psi and produced by a 0.2-MT weapon, what is the maximum midspan deflection? Consider only flexural behavior.
Answer
   $\mu \approx 1.6$
   $y_m = 0.32$ in.

7.2 What would be the required thickness of the slab in Prob. 7.1 for a peak overpressure of 50 psi and $\mu = 3$, all other data remaining the same?

7.3 A steel door is built up of 16WF96 sections, spaced at 20 in., with a light, nonstructural covering. The beams are simply supported on a span of 15 ft, and the dynamic yield strength is 60,000 psi. If the door is face on to the blast of a 5-MT weapon, and the peak overpressure is 50 psi, what would be the maximum midspan deflection? The clearing distance S_c is 20 ft, and the weight of the door is 200 lb/ft^2 of surface, in addition to the weight of the steel sections given.

7.4 The building shown in Fig. 7.6 and designed in Sec. 7.3 is subjected to the blast wave from a 10-MT weapon having a peak overpressure of 10 psi. Estimate the maximum horizontal deflection of the frame, neglecting the vibration of the walls, i.e., assuming the blast pressure on the wall surfaces to be transmitted to the frame without modification.

7.5 A reinforced concrete barrel arch has the same dimensions and properties as those given for the slab in Prob. 7.1 (except span). It has a radius of 10 ft, an internal half angle (Λ) of 60°, and hinged supports. If subjected to a blast with $p_{so} = 20$ psi and $Y = 0.5$ MT, what would be the resulting value of μ in the flexural mode?

7.6 Referring to Prob. 7.3, what would be the maximum elastic bending stress in the door due to the horizontal ground motion resulting from the weapon specified?

7.7 A machine weighs 5 tons and is supported by a heavy mat, which may be assumed to have the same motion as the undisturbed ground. Springs are to be placed between the mat and the machine. The design criteria are that the vertical acceleration of the machine should not exceed $4g$ and that the displacement relative to the mat should not exceed 5 in. The seismic velocity of the soil is 3000 fps, and the design weapon is 20 MT at a range of 6000 ft. What values of spring stiffness would satisfy these requirements?

Answer
  $k = 5100$ to $12,500$ lb/in.

8

Beams Subjected to Moving Loads

8.1 Introduction

A particular class of problem which has long been of interest to engineers involves the determination of the dynamic response of a beam or girder resulting from the passage of a force or mass across the span. Examples include the analysis of crane beams and of highway and railway bridges under the effect of moving vehicles. Although solutions to some of these problems have been available for some time, it is only in recent years that numerical results in quantity have been attainable by the use of electronic computation.

8.2 Constant Force with Constant Velocity

We consider first the relatively simple case of a constant force F moving across the span of a beam at constant velocity v as indicated in Fig. 8.1. In Sec. 4.3, it was found that the modal equation of motion (neglecting damping) for a beam with a single concentrated load is

$$\ddot{A}_n + \omega_n^2 A_n = \frac{F\phi_n(c_F)}{\int_o^l m[\phi_n(x)]^2 \, dx} \tag{8.1}$$

where ϕ_n is the modal-shape function for the nth mode, and c_F is the distance from the end of the span to the force. In the present case c_F is

316   *Introduction to Structural Dynamics*

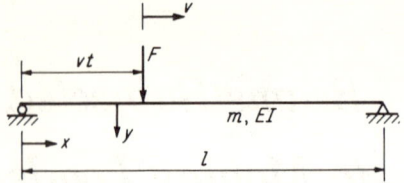

FIGURE 8.1 Constant force crossing beam with constant velocity.

a function of time and is equal to vt, where t is measured from the instant at which the force entered the span. If the beam is simply supported and prismatic,

$$\phi_n(x) = \sin \frac{n\pi x}{l}$$

and Eq. (8.1) becomes, after substitution of $c_F = vt$,

$$\ddot{A}_n + \omega_n^2 A_n = \frac{2F}{ml} \sin \frac{n\pi vt}{l} \quad (8.2)$$

The modal solution is expressed by

$$A_n = A_{nst}(\text{DLF})_n \quad (8.3a)$$

where

$$A_{nst} = \frac{2F}{ml\omega_n^2} \quad (8.3b)$$

and $(\text{DLF})_n$ is determined from the time function

$$f_n(t) = \sin \frac{n\pi vt}{l} = \sin \Omega_n t \quad (8.3c)$$

It should now be apparent that the modal solution is the same as that for a one-degree system subjected to a sinusoidal force, as discussed in Sec. 2.5. Therefore, by Eq. (2.34b), the dynamic load factor when the load is on the span is

$$(\text{DLF})_n = \frac{1}{1 - \Omega_n^2/\omega_n^2} \left(\sin \Omega_n t - \frac{\Omega_n}{\omega_n} \sin \omega_n t \right)$$

where

$$\Omega_n = \frac{n\pi v}{l}$$

Inserting this expression for $(\text{DLF})_n$ and Eq. (8.3b) into Eq. (8.3a) and combining modes according to the expression

$$y = \sum_{n=1}^{N} A_n \phi_n(x)$$

we obtain the total solution for deflection:

$$y = \frac{2F}{ml} \sum_{n=1}^{N} \frac{1}{\omega_n^2 - \Omega_n^2} \left(\sin \Omega_n t - \frac{\Omega_n}{\omega_n} \sin \omega_n t \right) \sin \frac{n\pi x}{l} \quad (8.4)$$

If we assume viscous damping in each mode, the solution becomes

$$y = \frac{2F}{ml} \sum_{n=1}^{N} \frac{\sin(n\pi x/l)}{(\omega_n^2 - \Omega_n^2)^2 + 4(\beta_n \Omega_n)^2} \left\{ (\omega_n^2 - \Omega_n^2) \sin \Omega_n t \right.$$

$$- 2\beta_n \Omega_n \cos \Omega_n t + e^{-\beta_n t} \left[2\beta_n \Omega_n \cos \omega_n t \right.$$

$$\left. \left. + \frac{\Omega_n}{\omega_n} (2\beta_n^2 - \Omega_n^2 - \omega_n^2) \sin \omega_n t \right] \right\} \quad (8.5)$$

where β_n/ω_n is the fraction of critical damping in the nth mode. For most beams of interest here, damping would be small and can often be neglected, especially if one is interested only in the first few cycles of response in any mode.

Since the case under consideration is similar to a sinusoidal force applied to a one-degree system, it might be concluded that resonance is an important possibility. This is not so for two reasons: first, the loading exists only for a limited number of cycles (e.g., the duration of the first-mode loading is only one-half cycle), and second, extremely high load velocities are required for resonance. To illustrate the second point, consider a beam with a 50-ft span and a typical fundamental period of 0.25 sec. For resonance in the first mode, Ω_1 must equal ω_1, and therefore

$$\frac{\pi v}{l} = \omega_1 = \frac{2\pi}{T_1}$$

or
$$v = \frac{2l}{T_1} = \frac{2 \times 50}{0.25} = 400 \text{ fps, or } 272 \text{ mph}$$

This velocity would be highly improbable in most applications of the theory. Even larger velocities would be required for resonance in higher modes.

Equations (8.4) and (8.5) apply only while the force is on the span. The free vibration occurring thereafter may be determined simply by computing the conditions at the instant the force leaves the span and using these as initial conditions for the ensuing analysis.

As an example of the application of the above theory, we shall determine the midspan deflection of a simple beam traversed by a constant force, ignoring damping and including only the fundamental mode (higher modes are of negligible importance). The parameters of the system are given as

$$m = 0.1 \text{ lb-sec}^2/\text{in.}^2 \qquad EI = 2 \times 10^{10} \text{ lb-in.}^2$$
$$l = 40 \text{ ft} \qquad v = 50 \text{ fps}$$

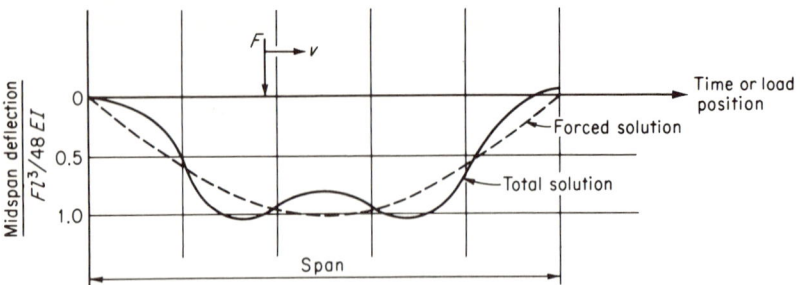

FIGURE 8.2 Response of simple beam to constant force crossing span with constant velocity.

from which we compute

$$\omega_1 = \frac{\pi^2}{l^2}\sqrt{\frac{EI}{m}} \quad \text{Eq. (4.7)}$$

$$= \frac{\pi^2}{(480)^2}\sqrt{\frac{2 \times 10^{10}}{0.1}} = 19.2 \text{ rad/sec}$$

$$\Omega_1 = \frac{\pi v}{l} = \frac{\pi(50)}{40} = 3.93 \text{ rad/sec}$$

Substituting in Eq. (8.4), we obtain the midspan deflection at any time.

$$y_{x=l/2} = \frac{F}{8470}(\sin 3.93t - 0.204 \sin 19.2t) \quad 0 \le t \le 0.8 \quad (8.6)$$

The first term in the parentheses is the forced, and the second is the free, vibration.

The deflection given by Eq. (8.6) is plotted in Fig. 8.2 as a fraction of maximum midspan static deflection. The abscissa may be considered to be either time or the position of the load on the span. Plotted separately is the forced part of the solution, which is very nearly equal to the static "crawl" deflection, i.e., the plot of deflection versus load position for the case in which the force moves very slowly. The residual vibration after the force leaves the span would merely be a continuation of the free part of Eq. (8.6). Since the midspan bending moment is very nearly proportional to midspan deflection, the ordinate of Fig. 8.2 may also be regarded as the ratio of dynamic to maximum static moments.

8.3 Pulsating Force with Constant Velocity

As an extension of the case given in the preceding section, we now consider a force which moves across the beam span and, in addition, has a harmonic variation in magnitude. This situation is illustrated in Fig. 8.3. Histor-

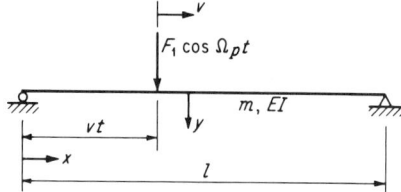

FIGURE 8.3 Pulsating force crossing beam with constant velocity.

ically, this problem has been of interest in connection with the vibration of railway bridges due to the passage of steam locomotives.[42] The unbalanced weight on the driving wheels produces, in addition to the gravity force, a harmonic alternating force.

Considering the applied force to be $F_1 \cos \Omega_p t$ (which implies maximum downward force at the time of entry to the span), we simply substitute this expression for F in Eq. (8.2) to obtain the modal equation of motion.

$$\ddot{A}_n + \omega_n^2 A_n = \frac{2F_1}{ml} \cos \Omega_p t \sin \Omega_n t$$

$$= \frac{F_1}{ml} [\sin (\Omega_p + \Omega_n)t - \sin (\Omega_p - \Omega_n)t] \qquad (8.7)$$

where Ω_n is as defined in Eq. (8.3c). It is apparent that the solution may be obtained by superimposing those corresponding to the two sine terms on the right side of Eq. (8.7). Each of these is given by Eq. (8.4) if we substitute for Ω_n the expression $(\Omega_p + \Omega_n)$ in one case and $(\Omega_p - \Omega_n)$ in the other and note that, for the latter solution, the sign must be reversed. In this way we obtain the solution for deflection due to the pulsating force.

$$y = \frac{F_1}{ml} \sum^n \left\{ \frac{1}{\omega_n^2 - (\Omega_p + \Omega_n)^2} \left[\sin (\Omega_p + \Omega_n)t - \frac{\Omega_p + \Omega_n}{\omega_n} \sin \omega_n t \right] \right.$$
$$\left. - \frac{1}{\omega_n^2 - (\Omega_p - \Omega_n)^2} \left[\sin (\Omega_p - \Omega_n)t - \frac{\Omega_p - \Omega_n}{\omega_n} \sin \omega_n t \right] \right\} \sin \frac{n\pi x}{l} \qquad (8.8)$$

If, as in many cases, there is a constant moving gravity force in addition to the pulsating force, the total solution may be obtained by adding Eqs. (8.4) and (8.8).

If the pulsating force had been $F_1 \sin \Omega_p t$, a solution could have been obtained in a similar manner. This is given by

$$y = \frac{F_1}{ml} \sum^n \left[\frac{\cos (\Omega_p - \Omega_n)t - \cos \omega_n t}{\omega_n^2 - (\Omega_p - \Omega_n)^2} \right.$$
$$\left. - \frac{\cos (\Omega_p + \Omega_n)t - \cos \omega_n t}{\omega_n^2 - (\Omega_p + \Omega_n)^2} \right] \sin \frac{n\pi x}{l} \qquad (8.9)$$

When a beam is subjected to a traveling pulsating force, resonance is a distinct possibility and may be of importance. Neglecting damping and considering only resonance with the fundamental mode, we see that the maximum response occurs when $\Omega_p = \omega_1$ and approximately at the time when the load leaves the span, i.e., when $t = l/v$ and the maximum number of load cycles has occurred. At this time

$$\sin(\Omega_p + \Omega_n)t = \sin\left(\omega_1 \frac{l}{v} + \pi\right) = -\sin \omega_1 \frac{l}{v}$$

$$\sin(\Omega_p - \Omega_n)t = \sin\left(\omega_1 \frac{l}{v} - \pi\right) = -\sin \omega_1 \frac{l}{v}$$

and from Eq. (8.8), we obtain by substitution

$$y = \frac{F_1}{ml}\left[\frac{2 + \pi v/\omega_1 l}{2\omega_1(\pi v/l) + (\pi v/l)^2} + \frac{2 - \pi v/\omega_1 l}{2\omega_1(\pi v/l) - (\pi v/l)^2}\right] \sin \frac{\pi x}{l} \sin \omega_1 \frac{l}{v}$$

which may be simplified to

$$y = \frac{2F_1}{\omega_1 m \pi v} \sin \frac{\pi x}{l} \sin \omega_1 \frac{l}{v} \qquad (8.10a)$$

To obtain the maximum midspan deflection, we let $x = l/2$ and take $\sin(\omega_1 l/v) = 1$. Although the latter substitution is inconsistent, it provides a close estimate of the first peak deflection after the load has left the span. Therefore the maximum midspan deflection in the resonant condition is given by

$$(y_{x=l/2})_{\max} = \frac{2F_1}{m \pi v \omega_1} \qquad (8.10b)$$

which may also be written as

$$(y_{x=l/2})_{\max} = y_{st} \frac{\omega_1 l}{\pi v} = y_{st} \frac{\omega_1}{\Omega_1} = y_{st} \frac{2T_c}{T} \qquad (8.10c)$$

where y_{st} is the static deflection due to F_1 applied at midspan, and T_c is the crossing time of the force.

If considerable damping is present, the free part of the vibration might be essentially eliminated by the time the force reaches the end of the span. If the free terms (that is, $\sin \omega_n t$) are removed from Eq. (8.8), it is found that the maximum resonant deflection is very nearly one-half of that given by Eq. (8.10c).

For the example beam of Sec. 8.2, it was computed that $\omega_1 = 19.2$ rad/sec and $\Omega_1 = 3.93$ rad/sec. Thus, if the force had been pulsating in resonance ($\Omega_p = \omega_1$), the maximum deflection without damping would have been $\omega_1/\Omega_1 = 19.2/3.92 = 4.9$ times the static deflection due to the same force.

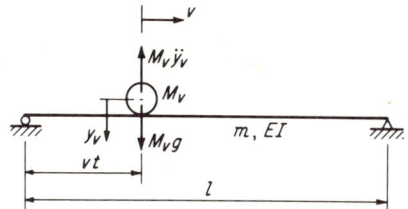

FIGURE 8.4 Beam traversed by rolling mass at constant velocity.

8.4 Beam Traversed by a Rolling Mass

Up to this point we have discussed only beams subjected to forces which were either constant or varied in some specified manner while crossing the span. In most practical cases the force is actually due to gravity acting on a moving mass, and the presence of that mass affects the solution. The analyses given previously are approximately correct for a moving mass if that mass is small compared with the beam itself.

To set up a more precise solution we refer to Fig. 8.4, where M_v is the mass of a vehicle or other object crossing the span. The force applied to the beam at any instant is the gravity force minus the inertia force due to acceleration of the mass. Therefore

$$\text{Force} = M_v g - M_v \ddot{y}_v$$

where $\ddot{y}_v$ is the acceleration of the beam at the mass location. It is assumed that contact is always maintained, i.e., that $\ddot{y}_v$ is also the acceleration of the mass. The expression for force given above may be inserted into Eq. (8.2) to provide the modal equation of motion,

$$\ddot{A}_n + \omega_n^2 A_n = \frac{2M_v}{ml}(g - \ddot{y}_v)\sin\frac{n\pi vt}{l} \tag{8.11}$$

We note that the mass acceleration must include the effects of all N beam modes, and therefore

$$\ddot{y}_v = \sum_{n=1}^{N} \ddot{A}_n \sin\frac{n\pi vt}{l}$$

Substituting the latter in Eq. (8.11) and rearranging, we obtain

$$\ddot{A}_n + \left(\frac{2M_v}{ml}\sin\frac{n\pi vt}{l}\right)\left(\sum_{n=1}^{N} \ddot{A}_n \sin\frac{n\pi vt}{l}\right) + \omega_n^2 A_n = \frac{2M_v g}{ml}\sin\frac{n\pi vt}{l} \tag{8.12}$$

If we consider only the fundamental mode and let $A_1 = y_c$, the midspan deflection, this becomes

$$\ddot{y}_c\left(1 + \frac{2M_v}{ml}\sin^2\frac{\pi vt}{l}\right) + \omega_1^2 y_c = \frac{2M_v g}{ml}\sin\frac{\pi vt}{l} \tag{8.13a}$$

322 Introduction to Structural Dynamics

or

$$\ddot{y}_c \left(\frac{ml}{2} + M_v \sin^2 \frac{\pi vt}{l} \right) + \frac{ml\omega_1{}^2}{2} y_c = M_v g \sin \frac{\pi vt}{l} \qquad (8.13b)$$

In Eq. (8.13b), the second term in the parentheses may be considered to be an "effective" value of the moving mass which varies with time (or mass position), and $M_v g$ may be considered to be a constant moving force.

A convenient rigorous solution for the problem formulated above has not been found. However, Eqs. (8.12) and (8.13) are in suitable form for numerical analysis, and solutions to the problem may be easily obtained by electronic computation.

8.5 Beam Vibration Due to Passage of Sprung Masses

We now turn to the more complex case of a mass supported by a spring, both of which cross the span at constant velocity. This is of practical interest in connection with highway-bridge vibration, as discussed in the next section. Such a system is shown in Fig. 8.5, where, for generality, two masses are included, a sprung mass M_{vs} supported by a spring of stiffness k_v and an unsprung mass M_{vu} which is assumed to be always in contact with the beam. The force applied to the beam may be expressed by

$$\text{Force} = M_{vu}(g - \ddot{y}_v) + [k_v(z - y_v) + M_{vs}g]$$

where the first term is the same as given previously for unsprung masses, and the term in brackets is the force in the spring. Note that z is the absolute deflection of M_{vs}, and the term $M_{vs}g$ is included because z is measured from the neutral spring position.

Proceeding as before, we substitute the above expression for force into Eq. (8.2) to obtain the modal equation of motion for the beam. If, at the same time, we insert

$$y_v = \sum_{n=1}^{N} A_n \sin \frac{n\pi vt}{l}$$

$$\ddot{y}_v = \sum_{n=1}^{N} \ddot{A}_n \sin \frac{n\pi vt}{l}$$

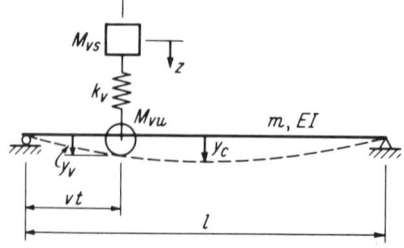

FIGURE 8.5 Beam with sprung and unsprung masses moving at constant velocity.

multiply both sides by $ml/2$, and rearrange, the result is

$$\frac{ml}{2}\ddot{A}_n + \left(M_{vu}\sin\frac{n\pi vt}{l}\right)\left(\sum_{n=1}^{N}\ddot{A}_n \sin\frac{n\pi vt}{l}\right) + \frac{ml\omega_n^2}{2}A_n$$

$$= \left[W_{vt} + k_v\left(z - \sum_{n=1}^{N} A_n \sin\frac{n\pi vt}{l}\right)\right]\sin\frac{n\pi vt}{l} \quad (8.14)$$

where W_{vt} is the total weight Mg of both masses. Equation (8.14) indicates a set of equations, one for each normal mode of the beam. However, the complete system has one additional mode, since z represents an additional degree of freedom. One more equation is required, and this is the dynamic-equilibrium equation for the sprung mass, which may be written

$$M_{vs}\ddot{z} + k_v\left(z - \sum_{n=1}^{N} A_n \sin\frac{n\pi vt}{l}\right) = 0 \quad (8.15)$$

Equations (8.14) and (8.15) provide a set of $N + 1$ equations, where N is the number of beam modes considered, which may be solved by numerical analysis for the motions of the beam and the sprung mass.

If we include only one beam mode, let $A_1 = y_c$, the midspan beam deflection, and introduce viscous-damping terms, the foregoing equations may be written as

$$\left(\tfrac{1}{2}ml + M_{vu}\sin^2\frac{\pi vt}{l}\right)\ddot{y}_c + \frac{ml\omega_1^2}{2}y_c + c_B\dot{y}_c$$

$$= \left[W_{vt} + k_v\left(z - y_c\sin\frac{\pi vt}{l}\right)\right]\sin\frac{\pi vt}{l} \quad (8.16)$$

$$M_{vs}\ddot{z} + k_v\left(z - y_c\sin\frac{\pi vt}{l}\right) + c_v\left(\dot{z} - \dot{y}_c\sin\frac{\pi vt}{l}\right) = 0 \quad (8.17)$$

where c_B and c_v are the damping coefficients for the bridge and sprung-mass system, respectively. These equations represent an equivalent two-degree system, which, for clarification, might be represented by the two coupled one-degree systems shown in Fig. 8.6. Solution may be accomplished by a straightforward numerical procedure as for any two-degree system.

8.6 Bridge Vibration Due to Moving Vehicles

The vibration of bridges due to moving traffic is important for two reasons. First, the stresses are increased above those due to static-load application. This is normally accounted for by the "impact" factor in design. The second reason is that excessive vibration may be noticeable to persons on the bridge. Although not related to safety, this may have the psycholog-

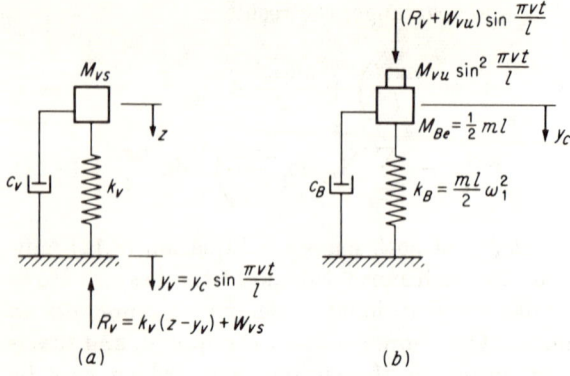

FIGURE 8.6 System of Fig. 8.5 represented by separate but coupled one-degree systems. (*a*) Idealized vehicle dynamic system [Eq. (8.17)]; (*b*) idealized beam dynamic system [Eq. (8.16)].

ical effect of impairing public confidence in the structure. There have been cases in which the latter phenomenon has been of considerable importance to highway officials.

Equations (8.16) and (8.17) may be used to investigate the vibration of actual simple-span highway bridges resulting from the passage of a single heavy vehicle. This implies the following assumptions:

1. The actual bridge, which consists of a floor system and several stringers or girders, may be represented by a single beam of equivalent rigidity.
2. Only the fundamental mode of the bridge beam need be considered.
3. The vehicle, although having two or more axles and a corresponding number of springs and flexible tires, may be considered to be a one-degree system.
4. The entire vehicle weight is applied to the bridge at the center of vehicle mass, rather than at the actual wheels.

Assumption 1 produces little error if the bridge is relatively narrow (e.g., two lanes) and if the vehicle is positioned on the center line. Assumption 2 is permissible for most purposes since the higher modes contribute little to the deflection or bending moment at midspan. With regard to assumption 3, the vehicle actually has many degrees of freedom associated with the individual springs and tires. However, the important vehicle motion with respect to bridge vibration appears to be the mode in which all these flexible elements act in phase. The acceptability of assumption 4 obviously depends upon the ratio of bridge span to vehicle-axle spacing. The error is not serious if this ratio is greater than about 5.

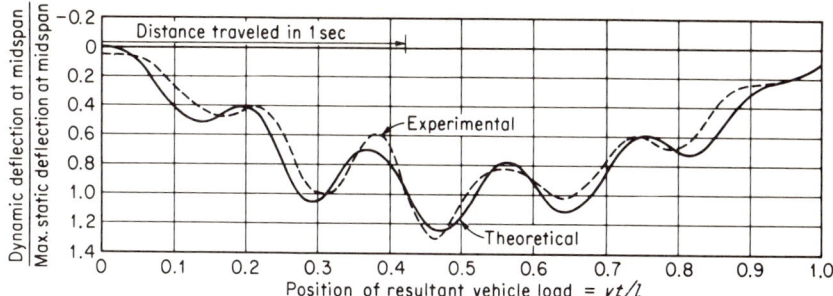

FIGURE 8.7 Midspan bridge deflections. Comparison of theory and actual field test. (*Biggs, Suer, and Louw.*[44])

Comparisons have been made between the theory, as represented by Eqs. (8.16) and (8.17), and experimental results obtained from laboratory models and from actual bridge structures in the field.[44] One such comparison is shown in Fig. 8.7, where the ratio of dynamic to maximum static midspan deflection is plotted against time (or vehicle position). The field test and the theoretical analysis were for a two-axle heavy truck with a velocity of 37 fps and an eight-stringer steel bridge of 88-ft span. In the theoretical analysis, the spring constant for the vehicle and the damping coefficients for both vehicle and bridge were given values which had been obtained experimentally. In addition, the vertical displacement and velocity of the vehicle mass at the point of entry into the span had been measured during the field test, and these values were inserted as initial conditions in the analysis. As may be observed in Fig. 8.7, the agreement between experiment and theory is very good.

Other studies have been made which include the effect of multiple axles and transverse bridge flexibility and which extend the theory to continuous spans.[45,46] An extensive series of field tests were made in connection with the AASHO Road Test at Ottawa, Ill.[47]

As a result of the investigations referred to above, it may be concluded that the primary causes of large bridge vibration are the initial "bounce" of the vehicle on its own springs as it enters the span (caused by roughness on the approach) and surface irregularities on the bridge itself. The former may be accounted for in the analysis by assigning initial values of z and $\dot{z}$ at $t = 0$. The effect of surface roughness on the bridge may be included in Eqs. (8.16) and (8.17) by adding to the term $[z - y_c \sin(\pi v t/l)]$ a quantity giving the deviation of the bridge profile from a straight line in terms of the position vt.

It is of interest to consider the effects of variations in the parameters of the system on the maximum dynamic bridge deflection. The results presented below are for a vehicle having an initial amplitude of "bounce" defined by the parameter

$$\alpha = \frac{z_{mo}\omega_v^2}{g}$$

where z_{mo} is the initial amplitude, and ω_v is the vehicle natural frequency given by

$$\omega_v = \sqrt{\frac{k_v}{M_v}}$$

Since M_{vu} is normally small compared with M_{vs}, the distinction has been dropped, and it is assumed that the total vehicle mass M_v is spring-supported. In arriving at these results the vehicle velocity has been varied (with an upper limit of about 50 mph for most span lengths), and in addition the phase of the initial bounce has also been varied. The dynamic deflections given below represent the most severe combination of these two parameters and therefore are an upper bound. Damping in both bridge and vehicle has been ignored because the former is of little importance and the latter could easily be counteracted by surface roughness on the bridge deck. Bridge surface roughness has not been included directly because, if within reasonable limits, it has the same general effect as the initial bounce.

The most important parameters are the magnitude of initial vehicle oscillation and the ratio of bridge to vehicle natural frequencies. The effect of the latter is shown in Fig. 8.8a, where the ratio of maximum dynamic to maximum static deflection is plotted against the frequency ratio for a given value of α and of the ratio of vehicle to bridge mass. It may be observed that a substantial peak occurs when the frequency ratio is unity. This is similar to the resonant condition for a system subjected to sinusoidal load. It is not true resonance, however, since the peak value is limited by the fact that there is a limited amount of energy in the bouncing vehicle as it enters the span.

The effect of the magnitude of initial vehicle oscillation is shown in Fig. 8.8b, where the ratio of maximum deflection to that for $\alpha = 0.3$ is plotted versus the parameter α. It may be seen that the effect is almost linear. The third parameter which is necessary in the analysis, $2M_v/ml$, has only a secondary effect on the dynamic deflection and may often be ignored.

To demonstrate use of Fig. 8.8, suppose that the ratio of bridge to vehicle natural frequency is 4 and that the initial vehicle oscillation is such that $\alpha = 0.2$. We read from plot a, 1.41, and from plot b, 0.93. Therefore the expected maximum dynamic deflection at midspan is $1.41 \times 0.93 = 1.31$ times the static deflection due to the vehicle. The same factor when applied to the static bending moment provides a good estimate of the maximum dynamic moment.

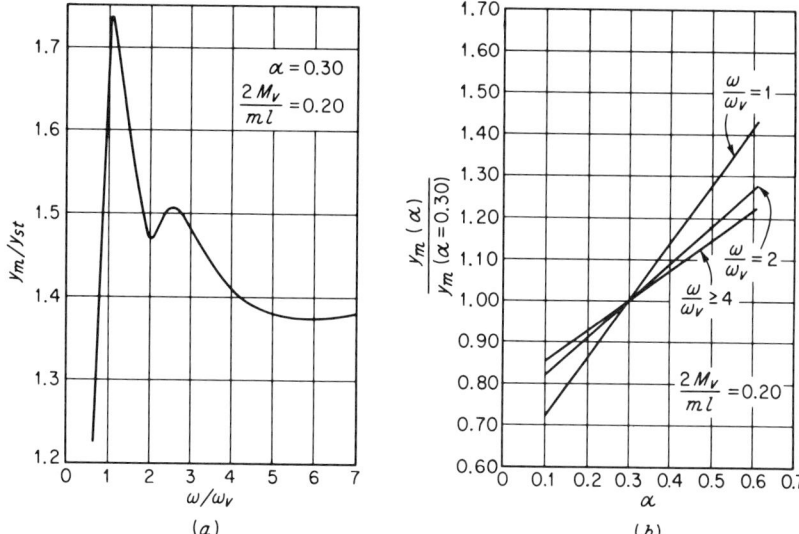

FIGURE 8.8 Maximum dynamic bridge deflection. Effect of parameter variations. (*Biggs, Suer, and Louw.*[44])

In current bridge-design practice (1964), the dynamic effects are accounted for by increasing the static live load by an empirical "impact" factor which depends only upon span. This approach may be justified because there are several major obstacles to the application of the theory outlined above. First, a proper design value of α has not been established. Second, the dynamic properties of future vehicles are not known. Third, the theory has not been extended to include more than one vehicle simultaneously on the span, and even if it were, a proper design assumption in this regard has not been established. All these are statistical problems, and until the application of statistics and probability theory to structural design has been more completely developed, the empirical approach will have to suffice.[48] Meanwhile, the theory presented above provides a valuable insight into the general behavior of bridges under moving traffic.

Problems

8.1 A constant force crosses a simple beam with a constant velocity of 40 fps. The beam has a span of 40 ft and a natural frequency of 30 rad/sec, and the force, if applied statically at midspan, would cause a deflection of 0.25 in. Neglecting damping and considering only the first mode, determine the maximum midspan deflection and the amplitude of the residual vibration after the force has left the span.
Answer
0.330 in., 0.084 in.

8.2 Repeat Prob. 8.1 for the case in which there is an *additional* force given by $P \cos 4\pi t$, where P is one-quarter of the constant force and always located at the same point.

8.3 An unsprung mass rolls across a simple beam with a constant velocity of 50 fps. The beam has a span of 40 ft and a natural frequency of 20 rad/sec. The weight of the rolling mass is one-half the total weight of the beam. Determine the maximum midspan deflection, considering only the first mode and neglecting damping.

Answer
  1.19 in.

8.4 Repeat Prob. 8.3 for the case where the moving mass is supported on a spring and the natural frequency of this spring-mass system is 9 rad/sec. When this mass enters the span, the spring is elongated by an amount $z = 1.2$ in. ($\dot{z} = 0$).

8.5 Estimate the upper bound (for all practical velocities of the mass and phasing of the initial condition) for the deflection computed in Prob. 8.4. The variation in the parameter $2M_v/ml$ may be ignored.

Answer
  1.37 in.

Appendix Matrix Formulation of Modal Analysis

This appendix contains an alternative derivation of the modal equations of motion making use of matrix notation. The development is specifically for lumped-parameter systems and is exactly parallel to the derivation given in Chap. 3. It is assumed that the reader is familiar with matrix algebra.[12]

The following discussion contains only brief descriptions and explanations of the steps in the derivations. The complete treatment of the subject in relation to the physical phenomena is given in Chap. 3. The author has deliberately chosen not to use matrix notation in the body of this text. The condensation of a set of equations into one is of course a convenience, but it tends to obscure the true meaning of the equations and to make derivations seem to be mere mathematical manipulation. The computational methods given in the main text for obtaining numerical results are of course not affected by the manner in which the problem is formulated.

Having mastered the basic concepts, the student will find it convenient to use matrix shorthand when working with the equations for a multidegree system. For this purpose the basic equations are restated below in matrix notation.

a. Normal Modes

The equations of motion for a multidegree lumped-mass system may be written as

$$[M]_D\{\ddot{y}\} + [K]\{y\} = \{F(t)\} \tag{A.1}$$

where $[M]_D$ = a diagonal matrix containing masses of the system
$\{\ddot{y}\}, \{y\}$ = column matrices of accelerations and displacements, respectively
$[K]$ = square stiffness matrix
$\{F(t)\}$ = a column matrix of applied dynamic forces

It is important to note that, for linear structural problems, $[K]$ is symmetric. Equation (A.1) is the matrix form of Eq. (3.3).

If the system is vibrating in a normal mode, we may make the substitutions

$$\{y\} = \{a_n\} \sin \omega_n t \qquad \{\ddot{y}\} = -\omega_n^2\{a_n\} \sin \omega_n t \qquad \{F(t)\} = 0$$

to obtain

$$-\omega_n^2[M]_D\{a_n\} + [K]\{a_n\} = 0$$

or

$$([K] - \omega_n^2[M]_D)\{a_n\} = 0 \tag{A.2}$$

where $\{a_n\}$ is the column matrix, or vector, of the modal displacements for the nth mode. Noting that $\{a_n\}$ cannot be zero and using Cramer's rule, we write

$$|[K] - \omega_n^2[M]_D| = 0 \tag{A.3}$$

Thus we have a characteristic-value problem, and the roots of Eq. (A.3) are the characteristic numbers, or eigenvalues, which are equal to the squares of the natural frequencies of the modes. Note that Eqs. (A.2) and (A.3) correspond to Eqs. (3.8) and (3.9) in Chap. 3. For each root there is a characteristic vector solution $\{a_n\}$, having an

330 Introduction to Structural Dynamics

arbitrary magnitude and representing the characteristic shape of that mode. Numerical methods for the solution of Eq. (A.3) are discussed in Secs. 3.2 to 3.5.

b. Orthogonality

For any two roots corresponding to the nth and mth modes, we may write Eq. (A.2) as

$$\omega_n{}^2[M]_D\{a_n\} = [K]\{a_n\} \tag{A.4}$$
$$\omega_m{}^2[M]_D\{a_m\} = [K]\{a_m\} \tag{A.5}$$

If we postmultiply the transpose of (A.4) by $\{a_m\}$, we obtain

$$(\omega_n{}^2[M]_D\{a_n\})^T\{a_m\} = ([K]\{a_n\})^T\{a_m\}$$

or
$$\omega_n{}^2\{a_n\}^T[M]_D^T\{a_m\} = \{a_n\}^T[K]^T\{a_m\} \tag{A.6}$$

Premultiplying (A.5) by $\{a_n\}^T$, we write

$$\omega_m{}^2\{a_n\}^T[M]_D\{a_m\} = \{a_n\}^T[K]\{a_m\} \tag{A.7}$$

It is known that $[M]_D = [M]_D^T$, as for any diagonal matrix, and also that $[K] = [K]^T$, since $[K]$ is symmetric. It is apparent that the right sides of Eqs. (A.6) and (A.7) are equal, and therefore subtracting (A.7) from (A.6) yields

$$(\omega_n{}^2 - \omega_m{}^2)\{a_n\}^T[M]_D\{a_m\} = 0 \tag{A.8}$$

Since $\omega_n \neq \omega_m$,

$$\{a_n\}^T[M]_D\{a_m\} = 0 \tag{A.9}$$

which is the *orthogonality condition* and the same as Eq. (3.13a).

c. Modal Equations

Since the modal displacements may be given any amplitude, it is now convenient to replace $\{a_n\}$ by $\{\phi_n'\}$ such that

$$\{\phi_n'\}^T[M]_D\{\phi_n'\} = 1 \tag{A.10}$$

The modal displacements $\{\phi_n'\}$ are evaluated so as to satisfy Eq. (A.10) and at the same time keep the elements in the same proportion as those in $\{a_n\}$. The characteristic vector is then said to be *normalized*. Note that Eq. (A.8) has not been violated since, if $n = m$, $\omega_n{}^2 - \omega_m{}^2 = 0$, and the remaining terms may be given any desired value.

Equation (A.2) may now be written for the nth mode as

$$[K]\{\phi_n'\} = \omega_n{}^2[M]_D\{\phi_n'\}$$

Now we let $[\Phi']$ be a square matrix containing all normalized characteristic vectors such that the nth column is the set of characteristic displacements for the nth mode. The last matrix equation may then be written so as to include all modes.

$$[K][\Phi'] = [M]_D[\Phi'][\omega_n{}^2]_D \tag{A.11}$$

where $[\omega_n{}^2]_D$ is a diagonal matrix of all characteristic numbers. We now premultiply both sides of (A.11) by $[\Phi']^T$ to obtain

$$[\Phi']^T[K][\Phi'] = [\Phi']^T[M]_D[\Phi'][\omega_n{}^2]_D \tag{A.12}$$

It may be shown that

$$[\Phi']^T[M]_D[\Phi'] = [I]_D \tag{A.13}$$

where $[I]_D$ is the unit diagonal matrix. Equation (A.13) can be easily verified by expansion and follows from the orthogonality condition and the fact that $[\Phi']$ has been normalized. It therefore follows that Eq. (A.12) may be written as

$$[\Phi']^T[K][\Phi'] = [\omega_n^2]_D \tag{A.14}$$

Returning now to the equation of motion (A.1), we let

$$\begin{aligned}\{y\} &= [\Phi']\{A_n\}\\ \{\ddot{y}\} &= [\Phi']\{\ddot{A}_n\}\end{aligned} \tag{A.15}$$

and

where A_n is the modal amplitude. This merely states that the true modal displacements equal the characteristic displacements times the modal amplitude determined by the response calculations and, further, that the total displacements are linear combinations of the modal values. If we now premultiply Eq. (A.1) by $[\Phi']^T$ and substitute Eqs. (A.15), we obtain

$$[\Phi']^T[M]_D[\Phi']\{\ddot{A}_n\} + [\Phi']^T[K][\Phi']\{A_n\} = [\Phi']^T\{F(t)\}$$

Finally, substituting for the left sides of Eqs. (A.13) and (A.14) in the last provides

$$\{\ddot{A}_n\} + [\omega_n^2]_D\{A_n\} = [\Phi']^T\{F(t)\} \tag{A.16}$$

which represents the modal equations of motion. These are of course uncoupled, and one of the equations represented by this matrix equation may be written as

$$\ddot{A}_n + \omega_n^2 A_n = \sum_{r=1}^{j} \phi'_{rn} F_r(t) \tag{A.17}$$

where $r = 1 \rightarrow j$ identifies the masses of the system. The final displacement obtained by superimposing the modes is

$$\begin{aligned}\{y_r(t)\} &= [\Phi']\{A_n(t)\} & (A.18)\\ &= [\Phi']\{A_{nst}(\text{DLF})_n\} & (A.19)\end{aligned}$$

where $A_n(t)$ is the solution of Eq. (A.17).

Equation (A.17) is exactly the same as Eq. (3.46), except that, in the latter, the right side is in terms of ϕ rather than ϕ' and is divided by $\sum_{r=1}^{j} M_r \phi_{rn}^2$. This is true because ϕ' is the normalized shape and Eq. (A.10) indicates that $\sum_{r=1}^{j} M_r \phi_{rn}'^2 = 1$. Thus, in reality, the two equations are identical. Equation (3.46) is perhaps more convenient to use since it is easier to evaluate the right-side denominator in that equation than to normalize the shapes according to Eq. (A.10).

Equation (A.19) is of course equivalent to Eq. (3.49). $A_n(t)$ or A_{nst} is different by the two procedures, since ϕ and ϕ' are different, but the final result, $y(t)$, is identical.

d. Damped Systems

If damping had been included in the foregoing development, the equations of motion (A.1) would have been

$$[M]_D\{\ddot{y}\} + [K]\{y\} + [c]\{\dot{y}\} = \{F(t)\} \tag{A.20}$$

where $[c]$ is the matrix of damping coefficients. This equation is the same as Eq. (3.53). If we add to Eqs. (A.15)

$$\{\ddot{y}\} = [\Phi']\{\ddot{A}_n\}$$

and follow the same procedure thereafter, the modal equations of motion (A.16) become

$$\{\ddot{A}_n\} + [\omega_n{}^2]_D\{A_n\} + [\Phi']^T[c][\Phi']\{\dot{A}_n\} = [\Phi']^T\{F(t)\} \tag{A.21}$$

In any one modal equation of motion the coefficient of $\dot{A}_n$ is equal to $2\omega_n C_n$, where C_n is the ratio of actual to critical damping in the nth mode. Therefore the complete set of damping coefficients is defined by

$$[\Phi']^T[c][\Phi'] = [2\omega_n C_n]_D \tag{A.22}$$

Equation (A.22) is equivalent to Eq. (3.54) and, when solved, provides all damping coefficients in terms of the damping ratios C_n. Procedures for accomplishing this are discussed in Sec. 3.10.

e. Support Motion

As discussed in Secs. 2.6 and 6.2, solutions for support motion may be obtained if $\{F(t)\}$ is replaced by $-\ddot{y}_s(t)\{M\}$, where $\ddot{y}_s(t)$ is the prescribed support acceleration. Thus the modal equation may be written as

$$\{\ddot{A}_n\} + [\omega_n{}^2]_D\{A_n\} + [2\omega_n C_n]_D\{\dot{A}_n\} = -\ddot{y}_s(t)[\Phi']^T\{M\} \tag{A.23}$$

This represents a set of equations any one of which has the form

$$\ddot{A}_n + \omega_n{}^2 A_n + 2\omega_n C_n \dot{A}_n = -\ddot{y}_s(t) \sum_{r=1}^{j} \phi'_{rn} M_r \tag{A.24}$$

which is the same as Eq. (6.5), except that, in the latter, the characteristic shape has not been normalized. Note that, in Eqs. (A.23) and (A.24), A_n is the relative motion with respect to the support.

The participation factors for the modes (Sec. 6.2) are given by

$$\{\Gamma'_n\} = [\Phi']^T\{M\} \tag{A.25}$$

which corresponds to Eq. (6.6). Finally, the solution for relative displacement is given by

$$\{u_r(t)\} = [\Phi']\{\Gamma'_n u_n{}^0(t)\} \tag{A.26}$$

where u_r is relative displacement, and $u_n{}^0$ is the response of a one-degree system having a frequency of ω_n. The last equation is identical with Eq. (6.10), although Φ' and Γ' are numerically different from the corresponding terms in Eq. (6.10).

References

1. Hildebrand, F. B.: "Introduction to Numerical Analysis," McGraw-Hill Book Company, New York, 1956.
2. Salvadori, M. G., and M. L. Baron: "Numerical Methods in Engineering," Prentice-Hall, Inc., Englewood Cliffs, N.J., 1952.
3. Crandall, S. H.: "Engineering Analysis," McGraw-Hill Book Company, New York, 1956.
4. Newmark, N. M.: Computation of Dynamic Structural Response in the Range Approaching Failure, *Proc. Symp. on Earthquake and Blast Effects*, Earthquake Engineering Research Institute, Los Angeles, Calif., 1952.
5. Newmark, N. M.: A Method of Computation for Structural Dynamics, *Trans. ASCE*, vol. 127, pt. 1, pp. 1406–1435, 1962.
6. Timoshenko, S.: "Vibration Problems in Engineering," D. Van Nostrand Company, Inc., Princeton, N.J., 1955.
7. Den Hartog, J. P.: "Mechanical Vibrations," McGraw-Hill Book Company, New York, 1956.
8. Jacobsen, L. S., and Robert S. Ayre: "Engineering Vibrations," McGraw-Hill Book Company, New York, 1958.
9. Norris, C. H., et al.: "Structural Design for Dynamic Loads," McGraw-Hill Book Company, New York, 1959.
10. U.S. Army Corps of Engineers: "Design of Structures to Resist the Effects of Atomic Weapons," Manual EM 1110-345-415, 1957.
11. Rogers, G. L.: "Dynamics of Framed Structures," John Wiley & Sons, Inc., New York, 1959.
12. Hildebrand, F. B.: "Methods of Applied Mathematics," Prentice-Hall, Inc., Englewood Cliffs, N.J., 1952.
13. Norris, C. H., and J. B. Wilbur: "Elementary Structural Analysis," McGraw-Hill Book Company, New York, 1960.
14. Synge, J. L., and B. A. Griffith: "Principles of Mechanics," McGraw-Hill Book Company, New York, 1959.
15. Bisplinghoff, R. L., H. Ashley, and R. L. Halfman: "Aeroelasticity," Addison-Wesley Publishing Company, Inc., Reading, Mass., 1955.
16. Salvadori, M. G.: Earthquake Stresses in Shear Buildings, *Trans. ASCE*, vol. 119, p. 171, 1954.
17. Cohen, E., L. S. Levy, and L. E. Smollen: Impulsive Motion of Elasto-plastic Shear Buildings, *Trans. ASCE*, vol. 122, p. 293, 1957.
18. Berg, G. V., and D. A. DaDeppo: Dynamic Analysis of Elasto-plastic Structure, *Proc. ASCE*, vol. 86, no. EM 2, p. 35, April, 1960.
19. Rayleigh, Lord: "Theory of Sound," Dover Publications, New York, 1945 (reprint).
20. Veletsos, A. S., and N. M. Newmark: Natural Frequencies of Continuous Flexural Members, *Trans. ASCE*, vol. 122, p. 249, 1957.
21. Darnley, E. R.: The Transverse Vibration of Beams and the Whirling of Shafts Supported at Intermediate Points, *Phil. Mag.*, vol. 41, p. 81, 1921.
22. Looney, C. T. G.: Behavior of Structures Subjected to a Forced Vibration, *Proc. ASCE*, vol. 80, separate no. 451, 1954.

23. Bleich, H. H.: Frequency Analysis of Beam and Girder Floors, *Trans. ASCE*, vol. 115, p. 1023, 1950.
24. Timoshenko, S., and S. Woinowsky-Krieger: "Theory of Plates and Shells," McGraw-Hill Book Company, New York, 1959.
25. Baron, M. L., H. H. Bleich, and P. Weidlinger: Dynamic Elastic-plastic Analysis of Structures, *Trans. ASCE*, vol. 127, pt. 1, p. 604, 1962.
26. Melin, J. W., and S. Sutcliffe: "Development of Procedures for Rapid Computation of Dynamic Structural Response," Final Report, Contract AF 33(600)-24994, University of Illinois for U.S. Air Force, Urbana, Ill., January, 1959.
27. Anderson, A. W., et al.: Lateral Forces of Earthquake and Wind, *Trans. ASCE*, vol. 117, p. 716, 1952.
28. Biot, M. A.: Analytical and Experimental Methods in Engineering Seismology, *Trans. ASCE*, vol. 108, p. 365, 1943.
29. Blume, J. A., N. M. Newmark, and L. H. Corning: "Design of Multistory Reinforced Concrete Buildings for Earthquake Motions," Portland Cement Association, Chicago, 1961.
30. Hudson, D. E.: Response Spectrum Techniques in Engineering Seismology, *Proc. World Conf. on Earthquake Eng.*, Earthquake Engineering Research Institute, Berkeley, Calif., 1956.
31. Clough, R. W.: Dynamic Effects of Earthquakes, *Trans. ASCE*, vol. 126, p. 847, 1961.
32. Housner, G. W.: Behavior of Structures during Earthquakes, *Proc. ASCE*, vol. 85, no. EM 4, p. 109, October, 1959.
33. Rosenblueth, E.: Some Applications of Probability Theory in Aseismic Design, *Proc. World Conf. on Earthquake Eng.*, Earthquake Engineering Research Institute, Berkeley, Calif., 1956.
34. Rosenblueth, E., and J. I. Bustamente: Distribution of Structural Response to Earthquakes, *Proc. ASCE*, vol. 88, no. EM 3, p. 75, June, 1962.
35. Veletsos, A. S., and N. M. Newmark: Effect of Inelastic Behavior on the Response of Simple Systems to Earthquake Motions, *Proc. Second World Conf. on Earthquake Eng.*, Tokyo, vol. 2, p. 895, 1960.
36. Goodman, L. E., E. Rosenblueth, and N. M. Newmark: Aseismic Design of Firmly Founded Elastic Structures, *Trans. ASCE*, vol. 120, p. 782, 1955.
37. "The Effects of Nuclear Weapons," U.S. Department of Defense and Atomic Energy Commission, 1962.
38. "Design of Structures to Resist Nuclear Weapons Effects," ASCE Manual 42, New York, 1961.
39. Ferguson, P. M.: "Reinforced Concrete Fundamentals," John Wiley & Sons, Inc., New York, 1959.
40. Newmark, N. M.: An Engineering Approach to Blast Resistant Design, *Trans. ASCE*, vol. 121, p. 45, 1956.
41. Benjamin, J. R., and H. A. Williams: Behavior of Reinforced Concrete Shear Walls, *Trans. ASCE*, vol. 124, p. 669, 1959.
42. Inglis, C. E.: "A Mathematical Treatise on Vibration in Railway Bridges," Cambridge University Press, London, 1934.
43. Hillerborg, A.: "Dynamic Influences of Smoothly Running Loads on Simply Supported Girders," Royal Institute of Technology, Institute of Structural Engineering and Bridge Building, Stockholm, 1951.
44. Biggs, J. M., H. S. Suer, and J. M. Louw: Vibration of Simple-span Highway Bridges, *Trans. ASCE*, vol. 124, p. 291, 1959.

45. Walker, W. H., and A. S. Veletsos: Response of Simple Span Highway Bridges to Moving Vehicles, *Univ. Illinois, Civil Eng. Studies, Structural Res. Ser.*, no. 272, 1963.
46. Huang, T., and A. S. Veletsos: Dynamic Response of Three-span Continuous Highway Bridges, *Univ. Illinois Civil Eng. Studies, Structural Res. Ser.*, no. 190, 1960.
47. Fenves, S. J., A. S. Veletsos, and C. P. Siess: Dynamic Studies of the AASHO Road Test Bridges, *Highway Res. Board Spec. Rept.* 73, p. 83, 1962.
48. Deflection Limitations of Bridges, Progress Report of the ASCE Committee on Deflection Limitations of Bridges, *Proc. ASCE*, vol. 84, no. ST3, paper 1633, May, 1958.

Index

A

Alternating force, step force, 64
  (*See also* Pulsating force)
Arches, design for blast effects, 297–306, 308–309
Atomic weapons (*see* Nuclear weapons effects)

B

Beams, 2, 88, 150–188
  approximate analysis, 206–212
  beam-girder system, 183–188, 237–242
  boundary conditions, 152–158, 208
  characteristic shapes, 152
  continuous, 174–183
  dynamic stresses, 165–169
  elasto-plastic analysis, 72–76, 192–195, 201
    effective spring constant, 204–205
  flexibility coefficients, 104
  forced vibration, 158–173, 180–183
  with moving force, 315–318
  under moving mass, 321–323
  nonprismatic, 170–173
  reactions, dynamic, 217–219, 227, 230, 236–237
  reinforced concrete, 217, 224–228
    effective stiffness, 226
  relative importance of modes, 162–165, 168, 183, 324
  roof girder, design of, 287–291
  shear, 218, 227–228, 230
  steel, 50, 72–76, 80, 229–230, 289, 295
  support motion, analysis for, 262–263
Bilinear resistance function, 21, 69, 203, 231
  (*See also* Elasto-plastic systems)

Blast effects (*see* Nuclear weapons effects)
Bridges, 319, 322–327
Building frames (*see* Frames, structural)
Buildings subjected to blast effects, 282–297
Buried structures, 308–309

C

Characteristic amplitude, 93
Characteristic functions, 161
Characteristic shapes, 88, 93, 127, 152, 250
  continuous beams, 174–182
  normalized, 119, 330
  by Rayleigh method, 105–111, 170–172
  single-span beams, 152–158
  by Stodola-Vianello method, 97–105, 127, 131
Characteristic values, 91, 329
Computers, electronic, 1–2, 195, 254
Coordinates, generalized, 112
Coupling, 87, 233, 237
Critical damping (*see* Damping, critical)

D

D'Alembert's principle, 4, 95
Damping, 17, 51–58, 331–332
  beam with moving force, 317
  coefficient, 17, 18
  Coulomb, 56–58
  critical, 17, 52, 64, 332
  effect of, 19, 52, 55, 320, 326
  logarithmic decrement, 54
  in multidegree systems, 140–147, 332

337

Damping, percent of critical, 18, 52, 141, 146–147
  system, with sinusoidal force, 62–64
  with support motion, 68
  viscous, 17, 51
Degrees of freedom, 3, 11, 85, 150
Design criteria, 224, 283, 288
Distributed mass systems, 150, 199, 216
Domes, design for blast effects, 297–299, 306–309
Ductility ratio, 224
Ductility of structures, 21, 203–204
  blast-resistant design, 283
  earthquake design, 270
Dynamic equilibrium, 4, 12, 218, 250
  of beam element, 151, 194
  rotational equilibrium, 138, 152
Dynamic load factor, 39, 246, 251, 258
  applied to modal analysis, 120, 316

E

Earthquake analysis, 245–273
  response spectrum, 264–265
  significance of plastic behavior, 252–255, 266–269
Earthquake design codes, 269–273
Earthquake ground motion, 263–265
Eigenvalues, 91, 329
Elastic limit, 22, 23, 140, 222
Elasto-plastic systems, 20, 69–81, 201
  analyzed for support motion, 253–255
  beam-girder, 237–242
  beams, 192–195
  charted maximum responses, 72–78
  design of, 222–224, 227–228
  multidegree, 137–140
  permanent distortion of, 26, 255
Energy methods, inelastic response, 222–224
  Lagrange equation, 111–116, 118
  Rayleigh method, 105–111
Equivalent mass, 2–3, 119, 202–203
  for frames, 216, 235

F

Finite-difference methods, 192–195
  (*See also* Numerical integration)
Flexibility coefficients, 102–105
Floor systems, 183–188, 237–242
Forced vibration, 37, 40, 58, 158, 318
  continuous beams, 180–183
  damped, 54
Foundations, 136–138
Frames, structural, 10, 16, 26
  analyzed for support motion, 250–257
  approximate analysis, errors in, 296–297
  designed for blast effects, 293–297
  earthquake analysis, 266–269
  earthquake design, 272–273
  elasto-plastic analysis, 138–140, 253–255, 293–297
  flexible foundation, effect of, 136–138
  girder flexibility, effect of, 87, 130–134
  idealization of, 215–217
  modal analysis of, 125–136
  reinforced concrete, 297
  stiffness and flexibility coefficients, 104
  vertical loading, 233–237
Free vibration, 35, 58, 318
  damped, 51
Frequency, natural (*see* Natural frequency)
Frequency equation, 91, 153, 155, 157
  continuous beams, 174–176, 178
  two-degree system, 92
Friction, 56
Fundamental mode, 93

G

Ground motions (*see* Earthquake ground motion; Nuclear weapons effects)

H

Harmonic motion, 36, 90, 152

I

Idealized systems, 2–3, 199–205
  beams, 206–212
  frames, 10, 16, 215–217
  slabs, 206–215
Impact, 323, 327
Impulse, 40, 223
Inelastic behavior (see Elasto-plastic systems)
Inertia forces, 4, 12
  in beams, 218, 238
  in multidegree systems, 95, 103
  work by, 105–106, 112
Initial conditions, 35, 36, 38, 51

K

Kinetic energy, 106, 113
  in beams, 158, 185
  due to impulse, 223
  in modal analysis, 116
  in plates, 189

L

Lagrange equation, 111–116, 118
  application, 114–116
    to beam-girder system, 185–188
    to beams, 158–160
    to slabs, 189–190
  derivation, 111–113
Linear systems theory, 40–41
Load function, 3
  combined, 220–221
  idealized, 205–206
    blast pressures, 285, 291, 294
      arches and domes, 298–299, 301
      for roof girder, 288
Logarithmic decrement, 54

M

Machinery, rotating, 58
  effect on building frame, 134
Magnification factor, 64
  (See also Dynamic load factor)
Matrix methods, 329–332
Modal analysis, 116–125
  beams, 158–170
    moving loads, 315–322
  frames, 125–138, 250–253
  matrix formulation, 330–332
  slabs, 188–192
  support motion, 247–253, 266–269
Modal displacement, 118
Mode (see Normal modes)

N

Natural frequency, 7, 36–37, 208
  arches, 300–301, 304
  beams, continuous, 176–180
    nonprismatic, 171
    single-span, 152–158, 167
  damping, effect of, 52
  direct determination, 89–93
  domes, 306
  by Rayleigh method, 105–111
  slabs, 190, 208, 232, 292
  by Stodola-Vianello method, 97–105, 127, 131
  structural frames, 127–128, 132, 236
Natural period, 7, 37, 219
  (See also Natural frequency)
Nonlinear systems (see Elasto-plastic systems)
Normal modes, 88, 144, 151
  equivalent one-degree system, 116, 119–121
  fundamental, 93, 102, 105
  matrix formulation, 329
  orthogonality of, 95–97
  participation factor, 108–109
    analysis for support motion, 248, 251, 332

Normalized shape, 119, 330
Nuclear weapons effects, 277–282
  on arches and domes, 298–299, 303
  on belowground structures, 308
  on buildings, 282–285, 291, 293
  dynamic pressure, 278–279, 281, 283
  ground shock, 309–313
  overpressure, 278–279, 281
  pressure duration, 279–280, 284
  reflected pressure, 280, 284
  shock front velocity, 280, 284
Numerical analysis, 1, 26, 138–140
  for arches, 305–306
  two-degree systems, 13–16, 240–241
Numerical integration, 4–9, 14, 24, 254
  beam-girder system, 240–242
  damped systems, 19, 20
  finite-difference methods, 30, 31
  linear-acceleration method, 27–29
  multidegree system, 142–143
  Newmark β method, 30
  time interval, 6, 8, 13, 29

O

One-degree system, 3, 85
  elastic responses, 42–51
  elasto-plastic, 20–26, 69–81
  nondimensional equations of motion, 78–89, 306
Orthogonality, 95–97, 101, 159
  matrix formulation, 330
  second orthogonality condition, 96–97, 117

P

Period, natural (*see* Natural period)
Permanent set, 26
Plates (*see* Slabs)
Pulsating force, beam-girder system, 183–188
  multistory frame, 134–136
  one-degree system, 58–65
  with damping, 62
  traversing beam, 318–320

R

Rayleigh method, 105–111
  applied to nonprismatic beam, 170–172
  for higher modes, 108
Rebound, 22, 240
Recurrence formulas, 6, 27, 28, 30, 31
Reinforced concrete beams (*see* Beams, reinforced concrete)
Residual vibration, 25, 71, 318
Resistance function, 21–23, 203–204
Resonance, 58, 61, 64
  limited cycle, 61, 62
  due to moving loads, 317, 319–320, 326
Rigid frames (*see* Frames, structural)
Rise time, 49, 80

S

Schmidt orthogonalization procedure, 108–111
Seismic coefficient, 270
Shear (*see* Beams, shear)
Shear buildings, 125, 138, 266
Shear walls, 297
Slabs, 188–192
  approximate analysis, 206–215
  design of, 230–233, 284–287, 291–293
  two-way and flat, 212–215
Spectral velocity, 247
Spectrum, for earthquake analysis, 264–265, 271
  for ground shock due to nuclear explosions, 311
  response, 257–263
Spectrum analysis, 262–263, 265–269, 312–313
Spring constant, 2, 10, 16
  effective value for fixed beams and slabs, 204–205, 230–231, 287
  (*See also* Stiffness coefficients)
Steady-state response, 60, 63, 134, 260
Steel beams (*see* Beams, steel)
Steel frames (*see* Frames, structural)

Stiffness coefficients, 89
  compared with flexibility coefficients, 101–105
  frames, 127, 130, 250
Stodola-Vianello procedure, 97–101, 127, 131
  compared with Rayleigh method, 107, 111
  stiffness and flexibility methods, 101–105
Strain energy, 106, 112, 185, 189
Strength of materials, dynamic, 225
Stresses, dynamic, in beams, 51, 165–169, 229–230
  due to support motion, 163
  in frames, 128–130
  in multidegree systems, 124–125
  in slabs, 191–192
Support flexibility, 183
Support motion, 65–69, 238, 312
  analysis of multidegree systems for, 246–256, 332
  response spectra, 257–263

T

Three-moment equation, 176
Transformation factors, 202–217
Transient response, 60
Trusses, 217
Two-degree systems, 11–17, 85
  analyzed for support motion, 250–255
  beam-girder combination, 183–188, 237–242
  bridge-vehicle combination, 323
  characteristic shapes, 93, 94, 114–117, 144
  damping, 144–145
  frequency equation, 92

V

Vehicles, bridge vibration due to, 322–327